Advances in Industrial Control

Other titles published in this series:

Digital Controller Implementation and Fragility
Robert S.H. Istepanian and James F. Whidborne (Eds.)

Optimisation of Industrial Processes at Supervisory Level
Doris Sáez, Aldo Cipriano and Andrzej W. Ordys

Robust Control of Diesel Ship Propulsion
Nikolaos Xiros

Hydraulic Servo-systems
Mohieddine Jelali and Andreas Kroll

Model-based Fault Diagnosis in Dynamic Systems Using Identification Techniques
Silvio Simani, Cesare Fantuzzi and Ron J. Patton

Strategies for Feedback Linearisation
Freddy Garces, Victor M. Becerra, Chandrasekhar Kambhampati and Kevin Warwick

Robust Autonomous Guidance
Alberto Isidori, Lorenzo Marconi and Andrea Serrani

Dynamic Modelling of Gas Turbines
Gennady G. Kulikov and Haydn A. Thompson (Eds.)

Control of Fuel Cell Power Systems
Jay T. Pukrushpan, Anna G. Stefanopoulou and Huei Peng

Fuzzy Logic, Identification and Predictive Control
Jairo Espinosa, Joos Vandewalle and Vincent Wertz

Optimal Real-time Control of Sewer Networks
Magdalene Marinaki and Markos Papageorgiou

Process Modelling for Control
Benoît Codrons

Computational Intelligence in Time Series Forecasting
Ajoy K. Palit and Dobrivoje Popovic

Modelling and Control of Mini-Flying Machines
Pedro Castillo, Rogelio Lozano and Alejandro Dzul

Ship Motion Control
Tristan Perez

Hard Disk Drive Servo Systems (2nd Ed.)
Ben M. Chen, Tong H. Lee, Kemao Peng and Venkatakrishnan Venkataramanan

Measurement, Control, and Communication Using IEEE 1588
John C. Eidson

Piezoelectric Transducers for Vibration Control and Damping
S.O. Reza Moheimani and Andrew J. Fleming

Manufacturing Systems Control Design
Stjepan Bogdan, Frank L. Lewis, Zdenko Kovačić and José Mireles Jr.

Windup in Control
Peter Hippe

Nonlinear H_2/H_∞ Constrained Feedback Control
Murad Abu-Khalaf, Jie Huang and Frank L. Lewis

Practical Grey-box Process Identification
Torsten Bohlin

Control of Traffic Systems in Buildings
Sandor Markon, Hajime Kita, Hiroshi Kise and Thomas Bartz-Beielstein

Wind Turbine Control Systems
Fernando D. Bianchi, Hernán De Battista and Ricardo J. Mantz

Advanced Fuzzy Logic Technologies in Industrial Applications
Ying Bai, Hanqi Zhuang and Dali Wang (Eds.)

Practical PID Control
Antonio Visioli

(continued after Index)

ZhiWu Li • MengChu Zhou

Deadlock Resolution in Automated Manufacturing Systems

A Novel Petri Net Approach

ZhiWu Li, PhD
School of Electro-Mechanical Engineering
Xidian University
2 South TaiBai Road
710071 Xi'an
China

MengChu Zhou, PhD
Department of Electrical and Computer Engineering
New Jersey Institute of Technology
323 MLK Blvd.
Newark
NJ 07102-1982
USA

ISBN 978-1-84996-830-0 e-ISBN 978-1-84882-244-3

DOI 10.1007/978-1-84882-244-3

Advances in Industrial Control series ISSN 1430-9491

A catalogue record for this book is available from the British Library

Cover design: eStudio Calamar S.L., Girona, Spain

Printed on acid-free paper

9 8 7 6 5 4 3 2 1

springer.com

Advances in Industrial Control

Professor (Emeritus) O.P. Malik
Department of Electrical and Computer Engineering
University of Calgary
2500, University Drive, NW
Calgary, Alberta
T2N 1N4
Canada

Professor K.-F. Man
Electronic Engineering Department
City University of Hong Kong
Tat Chee Avenue
Kowloon
Hong Kong

Professor G. Olsson
Department of Industrial Electrical Engineering and Automation
Lund Institute of Technology
Box 118
S-221 00 Lund
Sweden

Professor A. Ray
Department of Mechanical Engineering
Pennsylvania State University
0329 Reber Building
University Park
PA 16802
USA

Professor D.E. Seborg
Chemical Engineering
3335 Engineering II
University of California Santa Barbara
Santa Barbara
CA 93106
USA

Doctor K.K. Tan
Department of Electrical and Computer Engineering
National University of Singapore
4 Engineering Drive 3
Singapore 117576

Professor I. Yamamoto
Department of Mechanical Systems and Environmental Engineering
The University of Kitakyushu
Faculty of Environmental Engineering
1-1, Hibikino,Wakamatsu-ku, Kitakyushu, Fukuoka, 808-0135
Japan

in memory of my mother,
YuQing Zhang
(ZWL)

for my family, Fang Chen, Albert and Benjamin
(MCZ)

Series Editors' Foreword

The series *Advances in Industrial Control* aims to report and encourage technology transfer in control engineering. The rapid development of control technology has an impact on all areas of the control discipline. New theory, new controllers, actuators, sensors, new industrial processes, computer methods, new applications, new philosophies…, new challenges. Much of this development work resides in industrial reports, feasibility study papers and the reports of advanced collaborative projects. The series offers an opportunity for researchers to present an extended exposition of such new work in all aspects of industrial control for wider and rapid dissemination.

Much of the technological infrastructure of modern society is comprised of large networked dynamical systems. These systems include transportation (road, rail, air), energy networks (gas, electricity, oil), resource networks (water supply, wastewater disposal) and information networks (the Internet, information systems for transportation). Integrated with these are the primary industries that take raw material inputs and produce refined outputs like steel, paper, petroleum products, power and so on. The outputs of the primary industries supply secondary industries that manufacture both complex and simple products ranging from aircraft and automobiles, to computers, consumer white goods, food and pharmaceutical products. These are all process and manufacturing areas where control engineering plays an essential role.

The control communities approach to the modelling, analysis and design problems of industrial processes and networks has been threefold. Firstly, methods for continuous-time systems have been developed progressively since the 1940s and are now very well established but even these methods are still evolving especially in the nonlinear systems area. Secondly, there has been the rise of methods for discrete-event dynamical systems; this has often used ideas first devised by the computer science community. Finally, since about the 1980s, the idea of a hybrid system approach has gained credence and this paradigm is still under development.

In the *Advances in Industrial Control* monograph series and the *Advanced Textbooks in Control and Signal Processing* series, we have sought to feature titles that cover all aspects of this growth of the control field. For example, the modelling and analysis of discrete-event processes involves the challenging issues of resource allocation, logical decision-making, timed-event activity, and constraint handling. Within the *Advanced Textbooks in Control and Signal Processing* series, many of the latest developments in this field have been captured in *Modeling and Control of Discrete-event Dynamical Systems* by Branislav Hrúz end MengChu Zhou (ISBN 978-1-84628-872-2, 2007). This book is particularly relevant here because it contains an excellent introduction to the modelling and analysis tools of Petri nets, which have been used by various authors to solve some of the discrete, logical and continuous modelling, analysis and design problems of advanced industrial processes.

One entry to the *Advances in Industrial Control* series has been *Modelling and Analysis of Hybrid Supervisory Systems* by Emilia Villani, Paulo E. Miyagi, and Robert Valette (ISBN 978-1-84628-650-6, 2007). This reported a new technique that built on the capabilities of Petri nets and captured the realistic behaviour of large and small complex mixed dynamical discrete and continuous industrial systems. The method was demonstrated on three complex industrial examples, a heating, ventilation, and air conditioning system, an aircraft landing system and a cane-sugar production plant.

Continuing with the emphasis on solving real industrial control and supervisory control problems, this entry to *Advances in Industrial Control* by ZhiWu Li and MengChu Zhou tackles the deadlock problem using the formalism of Petri nets. It is an exhaustive text for this area of research and proposes new solutions for a long-standing problem. The reader who already has some familiarity with Petri nets and the associated analysis techniques will benefit directly; however, the inclusion of an introductory chapter on Petri nets makes this book self-contained. It can be usefully supplemented by reading the Petri net chapters in the above-mentioned Hrúz and Zhou textbook. The penultimate chapter of *Deadlock Resolution in Automated Manufacturing Systems* also compares a range of deadlock prevention policies and many readers interested in automated manufacturing will find this a useful source of ideas and further reading. The Editors of the *Advances in Industrial Control* series welcome this book as a valuable addition to the growing literature on these important, complex and large-scale industrial problems.

Industrial Control Centre
Glasgow
Scotland, UK
2008

M.J. Grimble
M.A. Johnson

Preface

The rapid evolution of computing, communication, control, and sensor technologies has brought about the proliferation of new man-made dynamic systems, mostly technological and often highly complex. Examples around us are air traffic control systems; automated manufacturing systems; computer and communication networks; embedded and networked systems; and software systems. The activity in these systems is governed by operational rules designed by humans and their dynamics is often driven by asynchronous occurrences of discrete events. This class of dynamic systems is therefore called discrete-event (dynamic) systems.

Based on finite-state automata and formal languages, the seminal work by Ramadge and Wonham in the early 1980s aims at providing a comprehensive and structural treatment of the modeling and control of discrete-event systems (DESs). The results in this area are gradually shaped and lead to supervisory control theory (SCT). SCT considers a DES as a generator of a formal language. Its behavior can be controlled by a supervisor that prevents event occurrences in order to satisfy a given specification.

Due to its generality, SCT is a paradigm that bridges the two worlds of control theory and computer science. In the latter, there exists a well-established Petri net community. As a natural and alternative modeling formalism, Petri nets are widely used for DES modeling and control. Their structural properties have been successfully exploited for the design of supervisors for supervisory control problems. Significant progress in this direction was made over the last two decades. The results obtained so far deal mainly with the safety of a plant, i.e., avoidance of dangerous or forbidden conditions given in a control specification. Liveness in Petri nets is an important behavioral property that leads to the safety of the supervised plant. It implies the freedom of deadlock–a highly undesired situation that an automated system must completely avoid. This property is equivalent to the non-blockingness in SCT. SCT is independent of the specific representation. That is to say, it is independent of a specific implementation technology.

A variety of theoretical results and computational algorithms have been developed in the literature to assess the liveness of certain classes of Petri nets. Most of these results are based on the fact that the liveness of a Petri net is closely related to the satisfiability of certain predicates on siphons. As a set of place elements, a siphon is a structural object in Petri nets. This relation between liveness and siphons becomes strong and apparent when we investigate the practical DES including a variety of resource allocation systems in a contemporary technological domain. Consequently, the siphon-based characterization of liveness and liveness-enforcing supervision for DESs modeled with Petri nets is usually considered to be one of the most interesting developments in the last decade from both theoretical and practical points of view.

However, the power of siphon-based liveness-enforcing approaches is degraded and deteriorated as the number of siphons grows quickly beyond practical limits and in the worst case grows exponentially fast with respect to the Petri net size. They suffer from the computational complexity problem since it is known that in general the complete siphon enumeration in a Petri net is NP-complete. Furthermore, they usually lead to a much more structurally complex liveness-enforcing Petri net supervisor than the plant net model that is originally built. This book tries to show how an elementary siphon-based methodology tackles these problems.

The book is intended for researchers, graduate students, and engineers who are interested in the control problems arising from manufacturing, transportation, workflow systems, communication, computer networks, complex software, and chemical industry. It is also appropriate for the students in automatic control, computer science, and applied mathematics and can be used as a supplementary textbook in the courses on Petri net theory and applications as well as the supervisory control of DESs.

Nevertheless, we try to maintain as a goal the presentation of a detailed discussion of the fundamental aspects of the related theory used throughout this book and hope to give readers a sufficiently solid foundation for their own advanced work and further study of the literature on this subject, it is highly desired that readers are familiar with the basics of linear algebra, set theory, and (integer) linear programming. In this sense, this book is self-contained. However, it is not intended to be an introductory textbook on Petri net theory. Being already familiar with net theory is hardly necessary to open this book but surely helpful if readers know its preliminaries.

Following the introduction in Chap. 1, the basics of Petri nets are presented in Chap. 2, which is used throughout this book. Explanatory examples are given to illustrate the concepts so that readers can understand the book without the prior knowledge of general Petri net theory. The concept of elementary and dependent siphons in a net is proposed in Chap. 3. As a natural extension to the concept of elementary siphons, Chap. 4 presents a novel monitor implementation to enforce generalized mutual exclusion constraints (GMECs). Chap. 5 presents a number of deadlock prevention policies that are developed on the basis of elementary siphons. The role of elementary siphons is fully shown in Chap. 6 by investigating the existence of a maximally permissive (optimal) liveness-enforcing monitor-based Petri net supervisor for a flexible manufacturing system. A survey and comparison of a variety of deadlock prevention policies in the literature are presented in Chap. 7. The comparison is conducted from the following points of view: computational complexity, structural complexity, and behavior permissiveness. The last chapter concludes this book by summarizing the results in the literature and presenting some interesting and open problems as well as some guidelines to tackle them.

Attached to the end of every chapter is a reference bibliography, and a glossary and a complete index in the final part, which should facilitate readers in using this book.

Readers of this book can learn the basics of Petri nets, siphon-based characterization of liveness, the theory of elementary siphons, and deadlock resolution methods and strategies for automated manufacturing systems. They can also learn a number of deadlock prevention policies developed on the basis of elementary siphons. They can finally master the concept of elementary siphons and related methods in designing structurally simple liveness-enforcing monitor-based Petri net supervisors.

Xidian University, China *ZhiWu Li*
New Jersey Institute of Technology, USA *MengChu Zhou*
August 2008

Acknowledgments

We are very grateful to Professor W. M. Wonham, Department of Electrical and Computer Engineering, University of Toronto, Professor M. D. Jeng, Department of Electrical Engineering, National Taiwan Ocean University, Professor X. L. Xie, INRIA, France, Professor Y. S. Huang, Department of Aeronautical Engineering, Chung Cheng Institute of Technology, National Defense University (Taiwan), Professor M. Uzam, Niğde Üniversitesi, Professor N. Q. Wu, Department of Mechatronics Engineering, Guangdong University of Technology, Professor Y. Chao, Department of Management and Information Science, National Cheng Chi University, L. Feng, KTH-Royal Institute of Technology, and F. Lewis, The University of Texas at Arlington, for their valuable comments and suggestions to our research.

We would like to express our sincere gratitude and appreciation to Professor M. Shpitalni for hosting the first author of this book as a visiting professor from February 2007 to February 2008 in the Laboratory for CAD & Life-cycle Engineering, Department of Mechanical Engineering, Technion-Israel Institute of Technology.

The first author would like to wholeheartedly thank his wife, TongLing Feng, and his son, BuZi Li, for their superhuman patience and sacrifice, and consistent encouragement. They have graciously endured many long nights and lonely weekends while he was immersed in his research.

A tribute is due to the unceasing efforts of the numerous investigators in this area, whose scientific contributions are directly responsible for the creation of this book. Among them are K. Barkaoui, Y. Chao, J. Ezpeleta, M. P. Fanti, A. Giua, Y. S. Huang, M. V. Iordache, M. D. Jeng, K. Lautenbach, F. Lewis, J. Park, S. A. Reveliotis, E. Roszkowska, F. G. Tricas, M. Uzam, N. Q. Wu, X. L. Xie, and K. Y. Xing.

Finally, the authors would like to thank the following students: A. R. Wang, H. S. Hu, M. Zhao, N. Wei, M. M. Yan, D. Liu, C. F. Zhong, M. Qin, G. Y. Liu, and Y. F. Hou. We appreciate their hard work in this area, particularly during the difficult times that we had in past years.

This work was in part supported by the National Nature Science Foundation of China under Grant No. 60228004, 60474018, and 60773001, the Scientific Research Foundation for the Returned Overseas Chinese Scholars, the Ministry of Education,

P. R. China, under Grant No. 2004-527, the Laboratory Foundation for the Returned Overseas Chinese Scholars, the Ministry of Education, P. R. China, under Grant No. 030401, Chang Jiang Scholars Program, the Ministry of Education, P. R. China, the National Research Foundation for the Doctoral Program of Higher Education, the Ministry of Education, P. R. China, under Grant No. 20070701013, Technion-Xidian Academic Exchange Program, "863" High-tech Research and Development Program of China, under Grant No. 2008AA04Z109, and the New Jersey Commission on Science and Technology.

Contents

Abbreviations

AMG	Augmented marked graph
cs-property	Controlled-siphon property
DES	Discrete-event system
ERCN	Extended resource control net
ES^3PR	Extended S^3PR
FBM	First-met bad marking
FMS	Flexible manufacturing system
GMEC	Generalized mutual exclusion constraint
LPP	Linear programming problem
LS^3PR	Linear S^3PR
MIP	Mixed integer programming
P-invariant	Place invariant
PNR	Process nets with resources
PPN	Production Petri net
PRT-circuit	Perfect resource transition circuit
RCN	Resource control net
SCT	Supervisory control theory
S^2LSPR	Systems of simple linear sequential processes with resources
S^2P	Simple sequential process
S^3PGR^2	System of simple sequential processes with general resource requirements
S^2PR	Simple sequential process with resources
S^3PR	System of simple sequential processes with resource
S^4R	System of sequential systems with shared resources
T-invariant	Transition invariant
WS^3PSR	Weighted system of simple sequential processes with several resources

Chapter 1
Introduction

Abstract This chapter first, from a historical viewpoint, shows why Petri nets are a widely used mathematical tool to investigate supervisory control of discrete-event systems, particularly for the deadlock analysis and control of automated manufacturing systems. The advantages and disadvantages of three major deadlock resolution strategies in the context of resource allocation systems, which are deadlock detection and recovery, deadlock avoidance, and deadlock prevention, are analyzed. A number of subclasses of Petri nets that can model various automated manufacturing systems are listed. Then, it reviews the existing deadlock prevention policies in the literature for automated manufacturing systems. The policies are qualitatively evaluated and compared briefly from computational complexity, supervisor complexity, and behavioral permissiveness. Finally, it outlines the book.

1.1 Background

A discrete-event system (DES) is a dynamical system that evolves according to asynchronous occurrences of discrete events. The examples of DES in the real world include a variety of man-made systems such as flexible manufacturing systems, complex computer programs, computer networks, communication systems, unmanned urban traffic systems, and workflow systems. A DES has a discrete set of states that may take symbolic values rather than real numbers. State transitions in these systems occur at asynchronous discrete instants of time in response to events, which may also take symbolic values. Usually, the relationships between state transitions and events cannot be described by differential or difference equations.

DES is a growing area that utilizes many interesting mathematical models and techniques. A DES is usually studied at two different levels: logical and performance levels. The models of the former are used to describe qualitative properties and control the sequences of events in a DES. The timing of event occurrences is ignored. At this level, typical problems are the avoidance of forbidden states or event sequences for the purpose of deadlock avoidance or liveness enforcement. The performance models deal with quantitative properties and aim to control the temporal behavior of a DES. In this case, the typical problems include the satisfaction of timing constraints, scheduling, and, particularly, optimization of some key performance criteria of a DES, for example, the production rate of a manufacturing system.

At the logical level, the most interesting and original approach to the control of a DES is supervisory control theory (SCT). The seminal theory by P. J. Ramadge and W. M. Wonham [65–67] considers a DES as a generator of a formal language. Its behavior can be controlled by a supervisor that prevents event occurrences in order to satisfy a control specification. SCT aims at providing a comprehensive and general framework that can deal with the control of DES represented by automata. It is concerned with a qualitative treatment with a control flavor of the discrete world.

In DES literature, a system to be controlled is usually called a plant. If it is modeled with a Petri net, the resultant Petri net is called a plant (Petri) net model. It is likely that the behavior of a plant may violate some constraints that must be enforced to the system. As a result, a plant often needs to be controlled by an external agent such that it behaves as one desires. A supervisor is referred to the external agent of a system to be controlled. Consequently, the plant model and its supervisor together are called a controlled system or controlled net if both take the form of Petri nets.

The framework proposed by Ramadge and Wonham is highly flexible with respect to the choice of models. The state space representation can be totally unstructured as in an automaton, or it can be structured as in a vector space, or it can be any combination thereof. Due to the fact that the state space of a Petri net is structured as a vector, Petri nets are widely used as a formalism in DES control theory. As stated in [9], the popularity of Petri nets as a formalism for the modeling and control of DES can be additionally attributed to the following reasons.

First, the well-established Petri net community that mainly consists of computer scientists has developed a large family of Petri net models across many disciplines. Different classes of Petri nets can represent different types of DES. Specifically, place/transition nets can be used to represent the logical level of a DES. Deterministic timed event graphs, a subclass of Petri nets, are equivalent to (max,+)-linear systems. More general timed deterministic and stochastic Petri nets can be used for performance evaluation. High-level nets can offer a compact model for complex systems. Hybrid nets can represent hybrid systems that involve both discrete and continuous processes. Generalized stochastic Petri nets can model general Markovian processes, which play a key role in stochastic optimization in DES. It is shown that the family of Petri nets developed in the literature can be used for simulation, control, verification, performance analysis, scheduling, and optimization.

Second, Petri nets can be used in all stages starting from modeling to control implementation. For example, Grafcet, a design paradigm of control programs for programmable logical controllers, is usually considered as a variation of Petri nets. Both plant and supervisor models can be represented with Petri nets. This feature can greatly facilitate modeling, open-loop system analysis and synthesis, control implementation, and closed-loop system analysis and evaluation.

Third, related computation can be made less extensive by fully utilizing the structural information of Petri nets. Also, Petri nets have a set of systematic mathematical analysis tools employing linear matrix algebra.

Last but not least, the research results using Petri nets as a formalism to deal with the modeling and control problems of DES over the past decades are very fruitful. Although the decision power of a Petri net is not unlimited, a good variety of DES control problems can be effectively and efficiently solved in a Petri net formalism [59]. For example, Petri nets have proved to be very successful in dealing with the forbidden-state problem, an important class of control specifications in supervisory control of DES. The achievements made by Petri net researchers in this area, however, in our own opinion, result partially from the supervisory control theory initialized by Ramadge and Wonham. In fact, many ideas in Petri net domain are borrowed from their theory, and most of research on Petri nets for DES has strongly been influenced by their supervisory control paradigm.

The facts mentioned above indicate that Petri nets are increasingly becoming an important and fully-fledged mathematical model to investigate the modeling and control of DES. In a Petri net formalism, liveness is an important property of system safety, which is equivalent to the non-blockingness in Ramadge and Wonham's supervisory control framework. Liveness implies the absence of global or local deadlock situations in a system.

A variety of theoretical results and computational algorithms have been developed in the literature to assess the liveness of certain classes of Petri nets. The liveness assessment can be performed by verifying the satisfiability of certain predicates on siphons, a well-known structural object in Petri nets. One of the most interesting past developments is the use of such structural objects to derive liveness-enforcing (Petri net) supervisors for DES.

However, the power of the siphon-based liveness-enforcing approaches is degraded and deteriorated by the fact that siphons' number in a Petri net grows quickly beyond practical limits and often grows exponentially with respect to the net size. They suffer from the computational complexity problem since it is known that in general the complete siphon enumeration in a Petri net is NP-complete. Furthermore, they usually lead to a much more structurally complex liveness-enforcing supervisor than the plant net model that is originally built. This book attempts to show (1) how Petri nets can be used to deal with deadlock control problems and (2) how the new concept of elementary siphons in a Petri net improves the existing deadlock control policies.

1.2 Literature Review

Deadlocks have been extensively investigated in computer operating systems [2, 13, 27–30, 32, 40, 60, 63, 74]. In general, they are an undesirable situation in a resource allocation system. Their occurrence implies the stoppage of the whole or partial system operation. In a production system, for example, deadlocks and related blocking phenomena often cause unnecessary costs such as long downtime and low utilization of some critical and expensive resources, and may lead to catastrophic results in highly automated systems, e.g., semiconductor manufacturing systems. Therefore, it is necessary to develop an effective control policy to make sure that deadlocks never occur in these systems. Over the last two decades, a great deal of research has been focused on solving deadlock problems in DES, resulting in a wide variety of approaches. This section is not intended to present a comprehensive overview of the deadlock control approaches in the literature. We instead concentrate on the most closely related approaches that are developed based on Petri nets.

The methods derived from a Petri net formalism for dealing with deadlocks either preclude the possibility of deadlock occurrence by breaking some necessary conditions for a deadlock to arise or detect and resolve a deadlock when it occurs. Generally, these deadlock resolution methods are classified into three strategies: deadlock detection and recovery [49, 85], deadlock avoidance [1, 5, 22, 34, 35, 80, 82–84], and deadlock prevention [19, 23, 24, 47, 52, 87].

- A deadlock detection and recovery approach permits the occurrence of deadlocks. When a deadlock occurs, it is detected and then the system is put back to a deadlock-free state, by simply reallocating the resources. The efficiency of this approach depends upon the response time of the implemented algorithms for deadlock detection and recovery. In general, these algorithms require a large amount of data and may become complex when several types of shared resources are considered [1].
- In deadlock avoidance, at each system state an on-line control policy is used to make a correct decision to proceed among the feasible evolutions. The main purpose of this approach is to keep the system away from deadlock states. Aggressive methods usually lead to higher resource utilization and throughput, but do not totally eliminate all deadlocks for some cases. In such cases if a deadlock arises, suitable recovery strategies are still required [49, 80, 85]. Conservative methods eliminate all unsafe states and deadlocks, and often some good states, thereby degrading the system performance. On the other hand, they are intended to be easy to implement.
- Deadlock prevention is considered to be a well-defined problem in DES literature. It is usually achieved by using an off-line computational mechanism to control the request for resources to ensure that deadlocks never occur. The goal of a deadlock prevention approach is to impose constraints on a system to prevent it from reaching deadlock states. In this case, the computation is carried out off-line in a static way and once the control policy is established, the system can no longer reach undesirable deadlock states. A major advantage of deadlock prevention algorithms is that they require no run-time cost since problems are solved in system design and planning stages. The major criticism is that they tend to be too conservative, thereby reducing the resource utilization and system productivity.

In the early work of Petri nets as a DES formalism, deadlock prevention is achieved by configuring proper initial markings under which a plant Petri net model is live. This idea can be originally traced back to the seminal works of Zhou and DiCesare in the 1990s [88–90]. In the last decade, a fair amount of work in this direction has been done by Jeng, Xie, Chu, Peng, Chung, and Barkaoui [6, 12, 41–46, 94]. The liveness of a Petri net model is tied to the absence of emptiable siphons. An emptiable siphon is a set of places whose marking becomes null during the net evolution and remains so in the subsequent markings. Most recent work in this direction utilizes this fact to analyze and control deadlocks in a DES.

One of the distinguishing features of Ramadge and Wonham's supervisory control framework is that there is a distinct boundary between a plant to be supervised and its supervisor such that the control implementation can be independent of the specific technology. Unfortunately, this boundary is not clearly shown in the work that was done in the early days of Petri nets as a DES formalism. In a deadlock resolution domain, the situation was changed after the seminal work of Ezpeleta et al. [19] and Lautenbach et al. [51], where liveness is enforced by adding monitors, also called control places, to prevent siphons from being emptied. This implies that both a plant and its supervisor are unified in a Petri net formalism. In addition, the significance of their work lies in the fact that a plant and its supervisor are successfully separated so that control implementation technology for the latter can be independently developed.

The success of separating a plant and its supervisor in a Petri net formalism becomes a spur that attracts much attention. Xing et al. [87] develop a deadlock prevention policy for a class of Petri nets, which is called Production Petri Nets, where the plant net model consists of resource places and production sequences. A deadlock structure is defined, which consists of a set of transitions. The set of resources used in the output places of the transition set is equal to the set of resources used in the input places of the transition set. The system is led to a deadlock state if the number of resources used by the deadlock structure equals the capacity of the resource. A control policy is accordingly developed by adding monitors, ensuring that for each involved resource, the deadlock structure always demands less resources than that the system has. Furthermore, the policy is minimally restrictive, i.e., it is optimal or maximally permissive.

As gradually recognized, the work by Ezpeleta et al. [19] suffers from a number of problems: application coverage, behavior permissiveness, computational complexity, and structural complexity. First of all, the policy in [19] can deal with only S^3PR, a class of Petri nets. It cannot model a manufacturing system with assembly and disassembly operations since an S^3PR is composed of state machines and resources and a state machine cannot represent assembly and disassembly operations. Second, the policy, in a general case, cannot lead to a maximally permissive supervisor. Third, the development of the policy depends on the complete siphon enumeration of a plant model. Such enumeration is expensive or impossible if the size of the plant is large since the number of siphons in a net grows exponentially fast with respect to the net size [18, 50]. The structural complexity problem of the supervisor results from the fact that for each strict minimal siphon in the plant net model, a monitor has to be added to prevent it from being emptied. The years following 1995 have seen a great deal of attention focused on these problems.

Many extensions to S^3PR nets have subsequently been proposed, which can be used to model more general automated flexible manufacturing systems (FMS).

- AMG (augmented marked graphs) [12]: An augmented marked graph is a Petri net mainly composed of two sets of places: operation places and resource places. The resultant net obtained by removing resource places and their related arcs is a marked graph.
- LS^3PR (linear system of simple sequential processes with resources) [20]: Strictly speaking, an LS^3PR is not an extended but a restrictive version of an S^3PR. Their difference is that a special constraint is imposed on the state machines in an LS^3PR. A state machine in it does not contain choices at internal states that are not the idle states. Note that idle states represent job requests.
- ES^3PR (extended S^3PR) [37]: Defined by Huang et al., an ES^3PR is an ordinary Petri net resulting from adding a set of resource places to a set of process nets that are state machines. An S^3PMR [38], from its definition, is equivalent to an ES^3PR in [37].
- ES^3PR (extended S^3PR) [77]: Composed of a set of state machines plus a set of resource places, this type of ES^3PR nets is more general than that defined in [37] since it may contain arcs from transitions to resource places with their weights perhaps being greater than one.
- WS^3PSR (weighted system of simple sequential processes with several resources) [76]: It is composed of state machines and resources. The usage of resources guarantees that they are neither destroyed nor created, i.e., conservativeness. In this sense, a WS^3PSR is a generalized Petri net.
- S^4R (system of sequential systems with shared resources) [1]: An S^4R is composed of a set of state machines plus a set of resource places. Compared with other classes of Petri nets that contain state machines, its usage of resources is almost arbitrary and requires only conservativeness.
- S^4PR [78]: An S^4PR is equivalent to an S^4R [1]. Both are developed independently.
- S^3PGR^2 (system of simple sequential processes with general resource requirements) [62]: An S^3PGR^2 is also equivalent to an S^4R.
- S^*PR [21]: This class of nets is a generalization of previously introduced classes that are composed of state machines. It properly includes S^4R.
- RCN (resource control nets)-merged nets [44]: An RCN-merged net includes S^3PR and some of augmented marked graphs.
- ERCN (extended resource control nets)-merged nets [86]: An ERCN-merged net includes RCN-merged nets and some of augmented marked graphs.
- ERCN*-merged nets [46]: An ERCN*-merged net includes ERCN-merged nets and some of augmented marked graphs.
- PNR (process nets with resources) [45]: A PNR is larger than the class of S^3PR, augmented marked graphs, and some of ERCN-merged nets.
- G-tasks [6]: A G-task is composed of acyclic state machines and a set of resource places. The resources can be arbitrarily used as long as their conservativeness is preserved.

- G-systems [94]: A G-system is the most general one among all the mentioned classes. It can properly contain each of the above classes. A G-system can model assembly (synchronization) and disassembly (splitting) operations in an FMS.

These classes can model various resource allocation systems. Their deadlock control policies are developed according to the relationship between liveness and siphons.

As known, the limited behavior permissiveness is a flaw in the notable deadlock prevention policy in [19]. Huang et al. claim that the deadlock prevention policy developed in [36] for S^3PR is in general more permissive than the one in [19]. This statement is not formally proved. Actually, in the opinion of the authors of this book, it may not be possible to develop a formal proof. The statistical investigation does support such a claim [36].

Huang's policy consists of two stages and performs the synthesis of a supervisor in an iterative way. The first stage, called siphon control, adds monitors to the plant model such that all the siphons in the plant are controlled. The siphon control stage is optimal or maximally permissive in the sense that no good states are removed due to the addition of monitors. In fact, the control of a siphon in this stage is implemented by enforcing a generalized mutual exclusion constraint (GMEC).

The second stage aims at making the newly generated siphons controlled, which result from the addition of the monitors in the first stage. To accelerate the convergence rate, the output arcs of the monitors added in the second stage point to only the source transitions of the plant model.

Sometimes termed optimality, maximal permissiveness is also an important parameter of a supervisor. In Ramadge and Wonham's approach, the existence and synthesis of an optimal non-blocking supervisor for a DES has been well addressed in a finite automaton and formal language paradigm. The existence of a synthesis approach for an optimal liveness-enforcing supervisor remains open until the work in [3,26,79]. By using the theory of regions [3] that can derive pure Petri nets from an automaton-based model, Uzam [79] develops an optimal liveness-enforcing supervisor synthesis method on the condition that such a supervisor exists. However, it is difficult to understand and use. Later, by using plain and popular linear algebraic notions, Ghaffari et al. [26] explore the conditions on the existence of an optimal supervisor that is maximally permissive, and develop a methodology to synthesize it.

These "explicit" approaches that need to generate the reachability graph of a Petri net require memory and time at least proportional to the number of reachable markings. Thus they are applicable to fairly small systems only. That is to say, a plant net model has to be small-sized. Also, its initial marking must be so small that its reachability set is limited to the computer's memory and processing capability.

As a result, the computational efficiency is the Achilles' heel of methods of this kind since the complete state enumeration is needed. This is not surprising since the theory of regions is a method to derive Petri nets from an existing automaton model. The work in [57] develops an optimal net supervisor design method that is based on the theory of regions. Its efficiency is improved by reducing the number of inequality systems that are used to separate events from some unsafe states.

Computational complexity has been a major problem when a deadlock prevention policy is developed [2, 61]. For a class of Petri nets, S^3PGR^2, Park and Reveliotis [62] propose a deadlock prevention policy that, originally developed under a finite-state automaton paradigm, is polynomial. Additional deadlock avoidance policies that are of polynomial-time complexity are presented in [21, 34]. They are not optimal in general.

Due to the inherent characteristics of Petri nets, the development of a polynomial-time algorithm to design a liveness-enforcing monitor-based supervisor is by no means an easy task. An efficient way of improving the computational efficiency of a siphon-based deadlock prevention policy is the introduction of the MIP-based deadlock detection method pioneered by Chu and Xie [12]. It was first used by Huang et al. in [36] to design a liveness-enforcing supervisor such that the complete siphon enumeration is successfully avoided. In this sense, this deadlock prevention policy enjoys high computational efficiency compared with the existing ones in the literature at that time. The MIP-based deadlock detection method is then used in [54] and [56].

A liveness-enforcing monitor-based supervisor derived from siphons reaches its high structural complexity when the number of siphons is large. This problem, having been recognized for a long time, has remained open for many years. By fully utilizing the structural information in a Petri net, the work by Li and Zhou proposes the concepts of elementary and dependent siphons in a Petri net [53, 55]. Siphons in a Petri net can be divided into elementary and dependent ones. The latter can be further distinguished by strongly and weakly dependent siphons with respect to elementary ones. It is shown that the number of the elementary siphons in a net is bounded by the smaller of place and transition counts. Moreover, a dependent siphon can be controlled by properly supervising the number of tokens that can stay at its elementary siphons.

The results concerning elementary siphons mentioned above can be naturally applied to most of the siphon-based deadlock prevention policies in the literature. For example, monitors can be added for elementary siphons only. The controllability of a dependent siphon can be ensured by properly supervising the initial number of tokens in the monitors that are added to its elementary siphons. That is to say, it is possible that we do not need to explicitly add a monitor for a dependent siphon any more. This is fully shown in [53] by an FMS example. In theory, the size of a supervisor that is computed by using elementary siphons is as a result less than that of the plant. Note that the method in [53] does not lower the computational complexity and improve the behavior permissiveness compared with the policy in [19]. On the positive side, it does lower the structural complexity of the supervisor notably.

It is worth noting that there is an established tool inside Petri net theory, which can be used to remove redundant monitors from a liveness-enforcing supervisor. It is called implicit places [14, 25, 68, 73]. Implicit places have the property that their addition to or removal from a net system does not change its behavior, i.e., an implicit place represents redundancy. In fact, the concept of implicit places has been proposed for many years before the existence of the structural complexity problem of a liveness-enforcing monitor-based supervisor. Unfortunately, no work in this direction is found in the literature except for [58].

For a dozen of years, we have witnessed that the results are much enriched in the area of liveness-enforcing supervisory control that is based on a Petri net formalism. On the other hand, many interesting problems remain unsolved, particularly the four above-mentioned hurdles, i.e., application scope, behavioral permissiveness, computational efficiency, and supervisor's structural complexity. This monograph represents the important research results that can be used to overcome these hurdles.

1.3 Outline of the Book

This monograph is intended to present a Petri net approach to deadlock resolution of automated manufacturing systems. It focuses on the role of elementary siphons of Petri nets in the development of a supervisor subject to liveness and other control requirements. It is outlined as follows.

Chapter 2 introduces the basics of Petri nets as well as the necessary notations used throughout this book. It also includes a brief comparison between Petri nets and automata.

Chapter 3 first defines the concepts of elementary and dependent siphons in Petri nets. Then, important results on elementary siphons such as their number in a net and the controllability of a dependent siphon are presented. The material in this chapter facilitates understanding of the development of deadlock prevention policies that are based on elementary siphons. Simple examples are given to illustrate these results.

Chapter 4 first presents a novel monitor implementation of a set of generalized mutual exclusion constraints that are divided into elementary and dependent ones, as motivated by the concept of elementary siphons. Conditions are then derived under which a dependent constraint is implicitly enforced. The constraint enforcement method is applied to a deadlock prevention policy developed in [62].

Chapter 5 introduces a well-established deadlock prevention policy via typical examples in the literature, and then shows the application of elementary siphons to the design of structurally simple liveness-enforcing monitor-based supervisors. The significance of elementary siphons is fully demonstrated. A few novel deadlock control strategies are accordingly presented.

For a class of Petri nets, Chap. 6 explores the existence and synthesis method of a liveness-enforcing monitor-based supervisor such that the controlled (net) system is maximally permissive on the assumption that all transitions are controllable and observable.

Chapter 7 presents and compares the existing deadlock prevention policies for flexible manufacturing systems via a case study. The comparison is conducted from the following points of view: computational complexity, structural complexity, and the behavior permissiveness.

Chapter 8 concludes this book by presenting and discussing a number of open and interesting problems in the field of DES control using a Petri net formalism and their relations with other DES formalisms.

1.4 Bibliographical Remarks

Before 1990, the work that used Petri nets as a formalism to deal with deadlock problems in DES was owing to E. Roszkowska [4,70,71]. However, Petri nets received more and more attention from academia and industry only after the publication of the research in [5,80,88].

There are several survey papers and books that investigate the supervisory control problems of DES using Petri nets: [31,39,59,69]. The paper [23] is a tutorial that surveys the deadlock control approaches in the literature. The edited volume [93] is the first comprehensive book that is dedicated to deadlock resolution methods in various computer-integrated systems. Other significant books published in the area of Petri nets and manufacturing automation include [8,15–17,64,81,90–92]. For the general problems of DES, the reader is referred to [7,10,11,33,48,72,75].

Problems

1.1. Some supervisory control problems are investigated and well addressed in the Ramadge–Wonham framework but this is not the case in a Petri net domain, e.g., the problems involving controllability and observability of events and decentralized control. Analyze and discuss the reasons from the development history of DES modeling and control theory. Reader can refer to [9].

References

1. Abdallah, I.B., ElMaraghy, H.A. (1998) Deadlock prevention and avoidance in FMS: A Petri net based approach. *International Journal of Advanced Manufacturing Technology*, vol.14, no.10, pp.704–715.
2. Araki, T., Sugiyama, Y., Kasami, T., Okui, J. (1977) Complexity of the deadlock avoidance problems. In *Proc. 2nd IBM Symposium on the Mathematical Foundations of Computer Science*, pp.229–252.
3. Badouel, E., Darondeau, P. (1998) Theory of regions. *Lectures on Petri Nets I: Basic Models, Lecture Notes in Computer Science*, vol.1491, W. Reisig and G. Rozenberg (Eds.), pp.529–586.
4. Banaszak, Z., Roszkowska, E. (1988) Deadlock avoidance in pipeline concurrent processes. *Podstawy Sterowania (Foundations of Control)*, vol.18, no.1, pp.3–17.
5. Banaszak, Z., Krogh, B.H. (1990) Deadlock avoidance in flexible manufacturing systems with concurrently competing process flows. *IEEE Transactions on Robotics and Automation*, vol.6, no.6, pp.724–734.

6. Barkaoui, K., Chaoui, A., Zouari, B. (1997) Supervisory control of discrete event systems based on structure theory of Petri nets. In *Proc. IEEE Int. Conf. on Systems, Man, and Cybernetics*, pp.3750–3755.
7. Ben-Naoum, L., Boel, R., Bongaerts, L., De Schutter, B., Peng, Y., Valckenaers, P., Vandewalle, J., Wertz, V. (1995) Methodologies for discrete event dynamic systems: A survey. *Journal A*, vol.36, no.4, pp.3–14.
8. Bogdan, S., Lewis, F.L., Kovacic, Z., Mireles, J. (2006) *Manufacturing Systems Control Design*. London: Springer.
9. Cao, X.R., Cohen, G., Giua, A., Wonham, W.M., Van Schuppen, J.H. (2002) Unity in diversity, diversity in unity: Retrospective and prospective views on control of discrete event systems. *Journal of Discrete Event Dynamic Systems: Theory and Applications*, vol.12, no.3, pp.253–264.
10. Cassandras, C.G., Lafortune, S. (1999) *Introduction to Discrete Event Systems*. Boston, MA: Kluwer.
11. Cassandras, C.G., Lafortune, S. (2008) *Introduction to Discrete Event Systems*. Springer.
12. Chu, F., Xie, X.L. (1997) Deadlock analysis of Petri nets using siphons and mathematical programming. *IEEE Transactions on Robotics and Automation*, vol.13, no.6, pp.793–804.
13. Coffman, E.G., Elphick, M.J., Shoshani, A. (1971) System deadlocks. *ACM Computing Surveys*, vol.3, no.2, pp.67–78.
14. Colom, J.M., Silva, M. (1989) Improving the linearly based characterization of P/T nets. In *Proc. 10th Int. Conf. on Applications and Theory of Petri Nets*, G. Rozenberg (Ed.), *Lecture Notes in Computer Science*, vol.483, pp.113–145.
15. David R., Alla, H. (1992) *Petri Nets and Grafcet*. Englewood Cliffs, NJ: Prentice-Hall.
16. Desrocher, A.A., Al-Jaar, R.Y. (1995) *Applications of Petri Nets in Manufacturing Systems: Modeling, Control, and Performance Analysis*. Piscataway, NJ: IEEE Press.
17. DiCesare, F., Harhalakis, G., Porth, J.M., Vernadat, F.B. (1993) *Practice of Petri Nets in Manufacturing*. Chapman and Hall.
18. Ezpeleta, J., Couvreur, J.M., Silva, M. (1993) A new technique for finding a generating family of siphons, traps, and st-components: Application to colored Petri nets. In *Advances in Petri Nets, Lecture Notes in Computer Science*, vol.674, G. Rozenberg (Ed.), pp.126–147.
19. Ezpeleta, J., Colom, J.M., Martinez, J. (1995) A Petri net based deadlock prevention policy for flexible manufacturing systems. *IEEE Transactions on Robotics and Automation*, vol.11, no.2, pp.173–184.
20. Ezpeleta, J., García-Vallés, F., Colom, J.M. (1998) A class of well structured Petri nets for flexible manufacturing systems. In *Proc. 19th Int. Conf. on Applications and Theory of Petri Nets, Lecture Notes in Computer Science*, vol.1420, J. Desel and M. Silva (Eds.), pp.64–83.
21. Ezpeleta, J., Tricas, F., García-Vallés, F., Colom, J.M. (2002) A banker's solution for deadlock avoidance in FMS with flexible routing and multiresource States. *IEEE Transactions on Robotics and Automaton*, vol.18. no.4, pp.621–625.
22. Ezpeleta, J., Recalde, L. (2004) A deadlock avoidance approach for non-sequential resource allocation systems. *IEEE Transactions on Systems, Man, and Cybernetics, Part A*, vol.34, no.1, pp.93–101.
23. Fanti, M.P., Zhou, M.C. (2004) Deadlock control methods in automated manufacturing systems. *IEEE Transactions on Systems, Man, and Cybernetics, Part A*, vol.34, no.1, pp.5–22.
24. Fanti, M.P., Zhou, M.C. (2005) Deadlock control methods in automated manufacturing systems. In *Deadlock Resolution in Computer-Integrated Systems*, New York: Marcel Dekker, pp.1–22.
25. García-Vallés, F., Colom, J.M. (1999) Implicit places in net systems. In *Proc. 8th Int. Workshop on Petri Nets and Performance Models*, pp.104–113.
26. Ghaffari, A., Rezg, N., Xie, X.L. (2003) Design of a live and maximally permissive Petri net controller using the theory of regions. *IEEE Transactions on Robotics and Automation*, vol.19, no.1, pp.137–142.
27. Gligor, V., Shattuck, S. (1980) On deadlock detection in distributed systems. *IEEE Transactions on Software Engineering*, vol.6, no.5, pp.435–440.

28. Gold, E.M. (1978) Deadlock predication: Easy and difficult cases. *SIAM Journal of Computing*, vol.7, no.3, pp.320–336.
29. Haberman, A. (1969) Prevention of system deadlocks. *Communications of the ACM*, vol.12, no.7, pp.373–377.
30. Hack, M.H.T. (1972) Analysis of Production Schemata by Petri Nets. Master Thesis, Massachusetts Institute of Technology, Cambridge, Massachusetts, USA.
31. Holloway, L.E., Krogh, B.H., Giua, A. (1997) A survey of Petri net methods for controlled discrete event systems. *Discrete Event Dynamic Systems: Theory and Applications*, vol.7, no.2, pp.151–190.
32. Holt, R. (1972) Some deadlock properties of computer systems. *ACM Computing Surveys*, vol.4, no.3, pp.179–196.
33. Hruz, B., Zhou, M.C (2007) *Modeling and Control of Discrete-Event Dynamic Systems: With Petri Nets and Other Tools*. London: Springer.
34. Hsieh, F.S., Chang, S.C. (1994) Dispatching-driven deadlock avoidance controller synthesis for flexible manufacturing systems. *IEEE Transactions on Robotics and Automation*, vol.10, no.2, pp.196–209.
35. Hsieh, F.S. (2004) Fault-tolerant deadlock avoidance algorithm for assembly processes. *IEEE Transactions Systems, Man, and Cybernetics, Part A*, vol.34, no.1, pp.65–79.
36. Huang, Y.S., Jeng, M.D., Xie, X.L., Chung, S.L. (2001) Deadlock prevention policy based on Petri nets and siphons. *International Journal of Production Research*, vol.39, no.2, pp.283–305.
37. Huang, Y.S., Jeng, M.D., Xie, X.L., Chung, S.L. (2001) A deadlock prevention policy for flexible manufacturing systems using siphons. In *Proc. IEEE Int. Conf. on Robotics and Automation*, pp.541–546.
38. Huang, Y.S., Jeng, M.D., Xie, X.L., Chung, D.H. (2006) Siphon-based deadlock prevention policy for flexible manufacturing systems. *IEEE Transactions on Systems, Man, and Cybernetics, Part A*, vol.36, no.6, pp.2152–2160.
39. Iordache, M.V. (2003) Methods for the Supervisory Control of Concurrent Systems Based on Petri Net Abstractions. Doctoral Dissertation, University of Notre Dame.
40. Isloor, S.S., Marsland, T.A. (1980) The deadlock problem: An overview. *Computer*, vol.13, no.9, pp.58–77.
41. Jeng, M.D., DiCesare, F. (1993) A review of synthesis techniques for Petri nets with applications to automated manufacturing systems. *IEEE Transactions on Systems, Man, and Cybernetics, Part A*, vol.23, no.1, pp.301–312.
42. Jeng, M.D., DiCesare, F. (1995) Synthesis using resource control nets for modeling shared-resource systems. *IEEE Transactions on Robotics and Automation*, vol.11, no.3, pp.317–327.
43. Jeng, M.D. (1997) A Petri net synthesis theory for modeling flexible manufacturing systems. *IEEE Transactions on Systems, Man and Cybernetics, Part B*, vol.27, no.2, pp.169–183.
44. Jeng, M.D., Xie, X.L. (1999) Analysis of modularly composed nets by siphons. *IEEE Transactions on Systems, Man, and Cybernetics, Part A*, vol.29, no.4, pp.399–406.
45. Jeng, M.D., Xie, X.L., Peng, M.Y. (2002) Process nets with resources for manufacturing modeling and their analysis. *IEEE Transactions on Robotics and Automation*, vol.18, no.6, pp.875–889.
46. Jeng, M.D., Xie, X.L., Chung, S.L. (2004) ERCN* merged nets for modeling degraded behavior and parallel processes in semiconductor manufacturing systems. *IEEE Transactions on Systems, Man, and Cybernetics, Part A*, vol.34, no.1, pp.102–112.
47. Jeng, M.D., Xie, X.L. (2005) Deadlock detection and prevention of automated manufacturing systems using Petri nets and siphons. In *Deadlock Resolution in Computer-Integrated Systems*, M. C. Zhou and M. P. Fanti (Eds.), pp.233-281, New York: Marcel Dekker.
48. Kumar, R. Garg, V. (1995) *Modeling and Control of Logical Discrete Event Systems*. Boston, MA: Kluwer.
49. Kumaran, T.K., Chang, W., Cho, H., Wysk, R.A. (1994) A structured approach to deadlock detection, avoidance and resolution in flexible manufacturing systems. *International Journal of Production Research*, vol.32, no.10, pp.2361–2379.

50. Lautenbach, K. (1987) Linear algebraic calculation of deadlocks and traps. In *Concurrency and Nets*, K. Voss, H. J. Genrich and G. Rozenberg (Eds.), pp.315–336.
51. Lautenbach, K., Ridder, H. (1993) Liveness in bounded Petri nets which are covered by T-invariants. In *Proc. 13th Int. Conf. on Applications and Theory of Petri Nets, Lecture Notes in Computer Science*, vol.815, R. Valette (Ed.), pp.358–375.
52. Lautenbach, K., Ridder, H. (1996) The linear algebra of deadlock avoidance–a Petri net approach. No.25-1996, Technical Report, Institute of Software Technology, University of Koblenz-Landau, Koblenz, Germany.
53. Li, Z.W., Zhou, M.C. (2004) Elementary siphons of Petri nets and their application to deadlock prevention in flexible manufacturing systems. *IEEE Transactions on Systems, Man, and Cybernetics, Part A*, vol.34, no.1, pp.38–51.
54. Li, Z.W., Zhou, M.C. (2006) Two-stage method for synthesizing liveness-enforcing supervisors for flexible manufacturing systems using Petri nets. *IEEE Transactions on Industrial Informatics*, vol.2, no.4, pp.313–325.
55. Li, Z.W., Zhou, M.C. (2006) Clarifications on the definitions of elementary siphons of Petri nets. *IEEE Transactions on Systems, Man, and Cybernetics, Part A*, vol.36, no.6, pp.1227–1229.
56. Li, Z.W., Hu, H.S., Wang, A.R. (2007) Design of liveness-enforcing supervisors for flexible manufacturing systems using Petri nets. *IEEE Transactions on Systems, Man, and Cybernetics, Part C*, vol.37, no.4, pp.517–526.
57. Li, Z.W., Zhou, M.., Jeng, M.D. (2008) A maximally permissive deadlock prevention policy for FMS based on Petri net siphon control and the theory of regions. *IEEE Transactions on Automation Science and Engineering*, vol.5, no.1, pp.182–188.
58. Li, Z.W. (2009) On systematic methods to remove redundant monitors from liveness-enforcing net supervisors. To appear in *Computer and Industrial Engineering*.
59. Moody, J.O., Antsaklis, P.J. (1998) *Supervisory Control of Discrete Event Systems Using Petri Nets*. Boston, MA: Kluwer.
60. Newton, G. (1979) Deadlock prevention, detection, and resolution: An annotated bibliography. *ACM SIGOPS Operating Systems Review*, vol.13, no.2, pp.33–44.
61. Pablo, J., Colom, J.M. (2006) Resource allocation systems: Some complexity results on the S^4PR class. In *Proc. IFIP International Federation for Information Processing, Lecture Notes in Computer Science*, vol.4229, E. Najm et al. (Eds.), pp.323–338.
62. Park, J., Reveliotis, S.A. (2001) Deadlock avoidance in sequential resource allocation systems with multiple resource acquisitions and flexible routings. *IEEE Transactions on Automatic Control*, vol.46, no.10, pp.1572–1583.
63. Peterson, J.L., Silberschatz, A. (1985) *Operating System Concepts*. Reading, MA: Addison-Wesley.
64. Porth, J.M., Xie, X.L. (1996) *Petri Nets, A Tool for Design and Management of Manufacturing Systems*. New York: John Wiley & Sons.
65. Ramadge, P., Wonham, W.M. (1987) Supervisory control of a class of discrete event processes. *SIAM Journal on Control and Optimization*, vol.25. no.1, pp.206–230.
66. Ramadge, P., Wonham, W.M. (1987) Modular feedback logic for discrete event systems. *SIAM Journal on Control and Optimization*, vol.25, no.5, pp.1202–1218.
67. Ramadge, P., Wonham, W.M. (1989) The control of discrete event systems. *Proceedings of the IEEE*, vol.77, no.1, pp.81–89.
68. Recalde, L., Teruel, E., Silva, M., (1997) Improving the decision power of rank theorems. In *Proc. IEEE Int. Conf. on Systems, Man, and Cybernetics*, pp.3768–3773.
69. Reveliotis, S.A. (2005) *Real-time Management of Resource Allocation Systems: A Discrete Event Systems Approach*. New York: Springer.
70. Roszkowska, E., Wojcik, R. (1993) Problems of process flow feasibility in FAS. In *CIM in Process and Manufacturing Industries*, Oxford: Pergamon Press, pp.115–120.
71. Roszkowska, E., Jentink, J. (1993) Minimal restrictive deadlock avoidance in FMSs. In *Proc. European Control Conf.*, J. W. Nieuwenhuis, C. Pragman, and H. L. Trentelman, (Eds.), vol.2, pp. 530–534.

72. Silva, M., Teruel, E. (1996) A systems theory perspective of discrete event dynamic systems: The Petri net paradigm. In *Symposium on Discrete Events and Manufacturing Systems*, IMACS Multiconference, P. Borne, J. C. Gentina, E. Craye, and S. El Khattabi, (Eds.), Lille, France, pp.1–12.
73. Silva, M., Teruel, E., Colom, J.M. (1998) Linear algebraic and linear programming techniques for the analysis of place/transition net systems. In *Lectures on Petri Nets I: Basic Models*, *Lectures Notes in Computer Science*, vol.1491, W. Reisig and G. Rozenberg (Eds.), pp.309–373.
74. Singhal, M. (1989) Deadlock detection in distributed systems. *IEEE Computer*, vol.22, no.11, pp.37–48.
75. Thistle, J.G. (1996) Supervisory control of discrete event systems. *Mathematical and Computer and Modeling*, vol.23, no.11–12, pp.25–53.
76. Tricas, F., Martinez, J. (1995) An extension of the liveness theory for concurrent sequential processes competing for shared resources. In *Proc. IEEE Int. Conf. on Systems, Man, and Cybernetics*, pp.3035–3040.
77. Tricas, F., García-Vallés, F., Colom, J.M., Ezpeleta, J. (1998) A structural approach to the problem of deadlock prevention in processes with shared resources. In *Proc. 4th Workshop on Discrete Event Systems*, pp.273–278.
78. Tricas, F., García-Vallés, F., Colom, J.M., Ezpeleta, J. (2000) An iterative method for deadlock prevention in FMS. In *Proc. 5th Workshop on Discrete Event Systems*, R. Boel and G.Stremersch (Eds.), pp.139–148.
79. Uzam, M. (2002) An optimal deadlock prevention policy for flexible manufacturing systems using Petri net models with resources and the theory of regions. *International Journal of Advanced Manufacturing Technology*, vol.19, no.3, pp.192–208.
80. Viswanadham, N., Narahari, Y., Johnson, T. (1990) Deadlock prevention and deadlock avoidance in flexible manufacturing systems using Petri net models. *IEEE Transactions on Robotics and Automation*, vol.6, no.6, pp.713–723.
81. Viswanadham, N., Narahari, Y. (1992) *Performance Modelling of Automated Manufacturing Systems*. Englewood Cliffs, NJ: Prentice Hall.
82. Wu, N. Q. (1999) Necessary and sufficient conditions for deadlock-free operation in flexible manufacturing systems using a colored Petri net model. *IEEE Transactions on Systems, Man, and Cybernetics, Part C*, vol.29, no.2, pp.192–204.
83. Wu, N.Q., Zhou, M.C. (2001) Avoiding deadlock and reducing starvation and blocking in automated manufacturing systems. *IEEE Transactions on Robotics and Automation*, vol.17, no.5, pp.658–669.
84. Wu, N.Q., Zhou, M.C. (2005) Modeling and deadlock avoidance of automated manufacturing systems with multiple automated guided vehicles. *IEEE Transactions on Systems, Man, and Cybernetics, Part B*, vol.35, no.6, pp.1193–1202.
85. Wysk, R.A., Yang, N.S., Joshi, S. (1994) Resolution of deadlocks in flexible manufacturing systems: avoidance and recovery approaches. *Journal of Manufacturing Systems*, vol.13, no.2, pp.128–138.
86. Xie, X.L., Jeng, M.D. (1999) ERCN-merged nets and their analysis using siphons. *IEEE Transactions on Robotics and Automation*, vol.15, no.4, pp.692–703.
87. Xing, K.Y., Hu, B.S., Chen, H.X. (1996) Deadlock avoidance policy for Petri-net modelling of flexible manufacturing systems with shared resources. *IEEE Transactions on Automatic Control*, vol.41, no.2, pp.289–295.
88. Zhou, M.C., DiCesare, F. (1991) Parallel and sequential exclusions for Petri net modeling for manufacturing systems. *IEEE Transactions on Robotics and Automation*, vol.7, no.4, pp.515–527.
89. Zhou, M.C., DiCesare, F. (1992) A hybrid methodology for synthesis of Petri nets for manufacturing systems. *IEEE Transactions on Robotics and Automation*, vol.8, no.3, pp.350–361.
90. Zhou, M.C., DiCesare, F. (1993) *Petri Net Synthesis for Discrete Event Control of Manufacturing Systems*. Boston, MA: Kluwer.
91. Zhou, M.C. (Ed.) (1995) *Petri Nets in Flexible and Agile Automation*. Norwell, MA: Kluwer.

92. Zhou, M.C., Venkatesh, K. (1998) *Modelling, Simulation and Control of Flexible Manufacturing Systems: A Petri Net Approach.* Singapore: World Scientific.
93. Zhou, M.C., Fanti, M.P. (Eds.) (2005) *Deadlock Resolution in Computer-Integrated Systems.* New York: MarcelDekker.
94. Zouari, B., Barkaoui, K. (2003) Parameterized supervisor synthesis for a modular class of discrete event systems. In *Proc. IEEE Int. Conf. on Systems, Man, and Cybernetics*, pp.1874–1879.

Chapter 2
Petri Nets

Abstract This chapter presents a mathematical treatment of Petri nets, including their formal definitions, structural and behavioral properties such as invariants, siphons, traps, reachability graphs, and state equations that are necessary to understand the subjects presented in this book. A number of important subclasses of Petri nets are introduced such as state machines and marked graphs. They are essential for the development of manufacturing-oriented Petri net models and the deadlock control strategies. The basics of automata are also covered in this chapter to facilitate the reader to understand well the deadlock prevention policy based on theory of regions. The concepts of a plant model, supervisor, and controlled system are defined.

2.1 Introduction

Though Petri nets and automata lack the full modeling and decision power of Turing machines, they still rank the top popular modeling tools for DES. As for Petri nets, this is partially attributed to their capability to provide the simple, direct, faithful, and convenient graphical representation of DES. Moreover, the well-established set of mathematical approaches employing linear matrix algebra makes them particularly useful for the modeling, analysis, and control of DES [44]. This chapter presents a mathematical treatment of Petri net theory. It is fundamental for understanding of the ideas presented in the following chapters.

2.2 Formal Definitions

A Petri net is a directed bipartite graph. It consists of two components: a net structure and an initial marking. A net (structure) contains two sorts of nodes: places and transitions. There are directed arcs from places to transitions and directed arcs from transitions to places in a net. Places are graphically represented by circles and transitions by boxes or bars. A place can hold tokens denoted by black dots, or a positive integer representing their number. The distribution of tokens over the places of a net is called a marking that corresponds to a state of the modeled system. The initial token distribution is hence called the initial marking. Let $\mathbb{N}$ denote the set of non-negative integers and $\mathbb{N}^+$ the set of positive integers.

Definition 2.1. A generalized Petri net (structure) is a 4-tuple $N = (P, T, F, W)$ where P and T are finite, non-empty, and disjoint sets. P is the set of places and T is the set of transitions with $P \cup T \neq \emptyset$ and $P \cap T = \emptyset$. $F \subseteq (P \times T) \cup (T \times P)$ is called a flow relation of the net, represented by arcs with arrows from places to transitions or from transitions to places. $W : (P \times T) \cup (T \times P) \rightarrow \mathbb{N}$ is a mapping that assigns a weight to an arc: $W(x,y) > 0$ iff $(x,y) \in F$, and $W(x,y) = 0$ otherwise, where $x, y \in P \cup T$.

Definition 2.2. A marking M of a Petri net N is a mapping from P to $\mathbb{N}$. $M(p)$ denotes the number of tokens in place p. A place p is marked by a marking M iff $M(p) > 0$. A subset $S \subseteq P$ is marked by M iff at least one place in S is marked by M. The sum of tokens of all places in S is denoted by $M(S)$, i.e., $M(S) = \sum_{p \in S} M(p)$. S is said to be empty at M iff $M(S) = 0$. (N, M_0) is called a net system or marked net and M_0 is called an initial marking of N.

We usually describe markings and vectors using a multiset (bag) or formal sum notation for economy of space. As a result, $\sum_{p \in P} M(p)p$ is used to denote vector M. For instance, a marking that puts four tokens in place p_2 and two tokens in place p_4 only in a net with $P = \{p_1 – p_6\}$ is denoted by $4p_2 + 2p_4$ instead of $(0,4,0,2,0,0)^T$.

In general, (N, M_0) is directly called a net where there is no confusion. $N = (P, T, F, W)$ is called an ordinary net, denoted by $N = (P, T, F)$, if $\forall f \in F, W(f) = 1$. Note that ordinary and generalized Petri nets have the same modeling power. The only difference is that the latter may have improved modeling efficiency and convenience for some systems. For convenience, (P, T, F, W, M_0) is sometimes used to denote a marked net. It is also called a net system.

Example 2.1. Figure 2.1a shows a simple Petri net with $P = \{p_1 – p_5\}$, $T = \{t_1 – t_3\}$, $F = \{(p_1,t_1), (t_3,p_1), (p_2,t_2), (t_1,p_2), (p_3,t_3), (t_2,p_3), (p_4,t_2), (t_3,p_4), (p_5,t_1), (p_5,t_2), (t_3,p_5)\}$, $W(p_1,t_1) = W(t_3,p_1) = W(p_2,t_2) = W(t_1,p_2) = W(p_3,t_3) = W(t_2, p_3) = W(p_4,t_2) = W(t_3,p_4) = W(p_5,t_1) = 1$, $W(p_5,t_2) = 2$, and $W(t_3,p_5) = 3$. Places are graphically represented by circles and transitions are represented by boxes. It is clear that the net is not ordinary because of the multiplicity of arcs (p_5,t_2) and (t_3,p_5).

Each of places p_1 and p_5 has three tokens, denoted by three black dots or number 3 inside. Place p_4 holds two tokens and there is no token in p_2 and p_3. This token distribution leads to the initial marking of the net with $M_0 = 3p_1 + 2p_4 + 3p_5$. The net's alternative graphical representation is given in Fig 2.1b, where multiple arcs are replaced with an arc with its weight and multiple tokens in a place can be replaced by a corresponding number for the sake of simplicity. For example, the number of tokens in place p_1 is denoted by number 3.

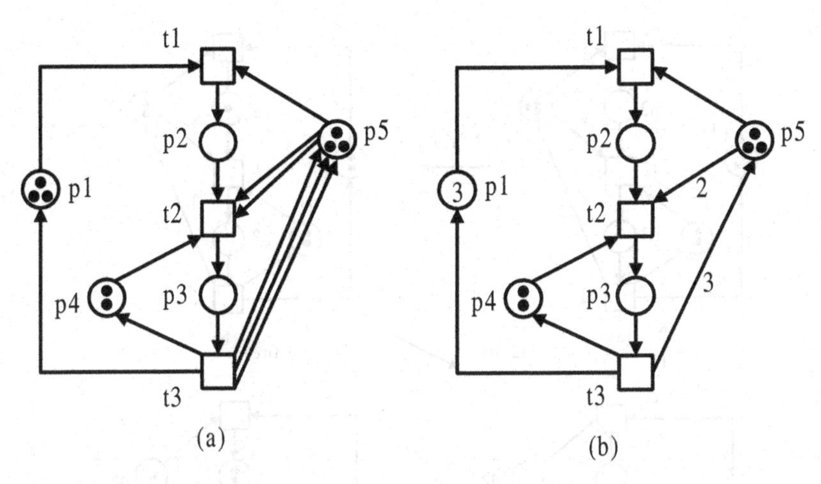

Fig. 2.1 A Petri net (N, M_0) with $M_0 = 3p_1 + 2p_4 + 3p_5$ represented by (a) multiplicity of arcs and (b) weight of arcs

Definition 2.3. Let $x \in P \cup T$ be a node of net $N = (P, T, F, W)$. The preset of x is defined as ${}^\bullet x = \{y \in P \cup T | (y, x) \in F\}$. While the postset of x is defined as $x^\bullet = \{y \in P \cup T | (x, y) \in F\}$. This notation can be extended to a set of nodes as follows: given $X \subseteq P \cup T$, ${}^\bullet X = \cup_{x \in X} {}^\bullet x$, and $X^\bullet = \cup_{x \in X} x^\bullet$. Given place p, we denote $max\{W(p,t) \mid t \in p^\bullet\}$ by $max_{p^\bullet}$.

For $t \in T$, $p \in {}^\bullet t$ is called an input place of t and $p \in t^\bullet$ is called an output place of t. For $p \in P$, $t \in {}^\bullet p$ is called an input transition of p and $t \in p^\bullet$ is called an output transition of p.

Example 2.2. In Fig. 2.1a, we have ${}^\bullet t_1 = \{p_1, p_5\}$, ${}^\bullet t_2 = \{p_2, p_4, p_5\}$, $t_2^\bullet = \{p_3\}$, $t_3^\bullet = \{p_1, p_4, p_5\}$, ${}^\bullet p_3 = \{t_2\}$, $p_3^\bullet = \{t_3\}$, ${}^\bullet p_5 = \{t_3\}$, and $p_5^\bullet = \{t_1, t_2\}$. Let $S = \{p_3, p_5\}$. Then, ${}^\bullet S = {}^\bullet p_3 \cup {}^\bullet p_5 = \{t_2, t_3\}$ and $S^\bullet = p_3^\bullet \cup p_5^\bullet = \{t_1, t_2, t_3\}$. It is easy to see that $max_{p_5^\bullet} = 2$ and $\forall p \in P \backslash \{p_5\}$, $max_{p^\bullet} = 1$.

Definition 2.4. A transition $t \in T$ is enabled at a marking M iff $\forall p \in {}^\bullet t$, $M(p) \geq W(p,t)$. This fact is denoted by $M[t\rangle$. Firing it yields a new marking M' such that $\forall p \in P$, $M'(p) = M(p) - W(p,t) + W(t,p)$, as denoted by $M[t\rangle M'$. M' is called an immediately reachable marking from M. Marking M'' is said to be reachable from M if there exists a sequence of transitions $\sigma = t_0 t_1 \cdots t_n$ and markings $M_1, M_2, \cdots$, and M_n such that $M[t_0\rangle M_1[t_1\rangle M_2 \cdots M_n[t_n\rangle M''$ holds. The set of markings reachable from M in N is called the reachability set of Petri net (N, M) and denoted by $R(N, M)$.

Example 2.3. In Fig. 2.2a, t_1 is enabled at initial marking $M_0 = 3p_1 + 2p_4 + 3p_5$ since ${}^\bullet t_1 = \{p_1, p_5\}$, $M_0(p_1) = 3 > W(p_1, t_1) = 1$, and $M_0(p_5) = 3 > W(p_5, t_1) = 1$. Firing t_1 leads to M_1 with $M_1(p_1) = M_0(p_1) - W(p_1, t_1) + W(t_1, p_1) = 2$, $M_1(p_2) = M_0(p_2) - W(p_2, t_1) + W(t_1, p_2) = 1$, $M_1(p_3) = M_0(p_3) - W(p_3, t_1) + W(t_1, p_3) = 0$,

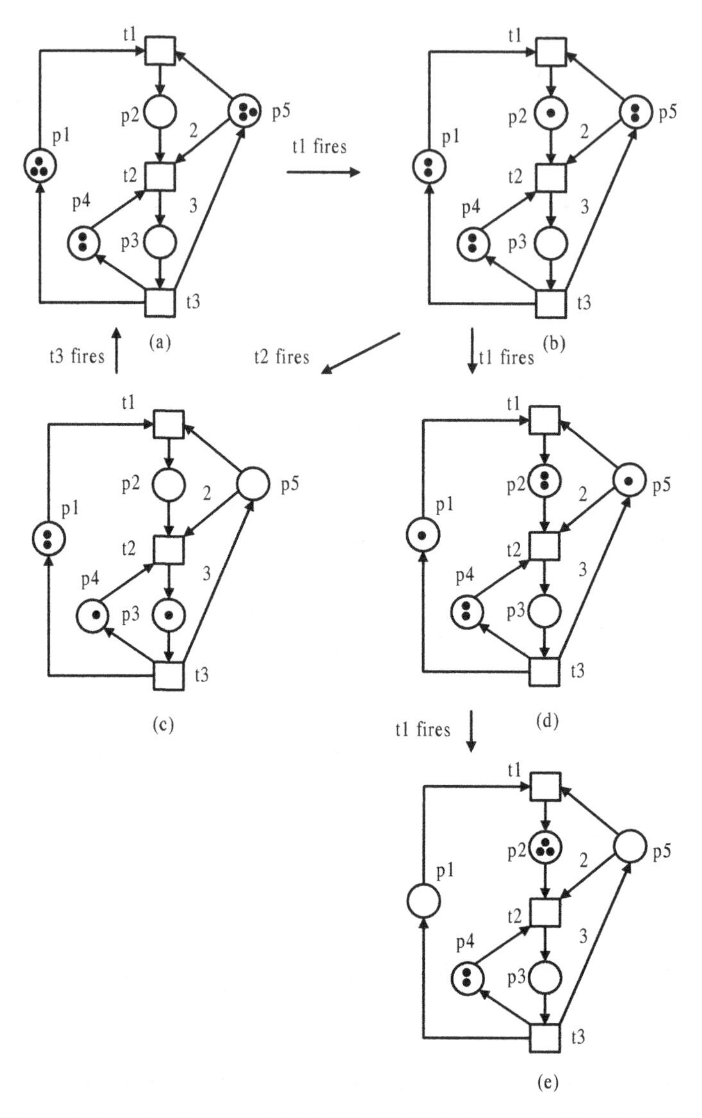

Fig. 2.2 The evolution of a Petri net: (a) (N,M_0), (b) (N,M_1), (c) (N,M_2), (d) (N,M_3), and (e) (N,M_4)

$M_1(p_4) = M_0(p_4) - W(p_4,t_1) + W(t_1,p_4) = 2$, and $M_1(p_5) = M_0(p_5) - W(p_5,t_1) + W(t_1,p_5) = 2$, as shown in Fig 2.2b.

In marking M_1, both t_1 and t_2 are enabled. Firing t_2 at M_1 leads to M_2 as shown in Fig. 2.2c. Firing t_1 at M_1 leads to M_3 as shown in Fig. 2.2d. Only t_1 is enabled at M_3. Figure 2.2e is the net after t_1 fires at M_3 and corresponds to M_4. At M_2, only t_3 is enabled. Firing it leads back to M_0. As a result, the reachability set of the net in Fig. 2.2a is $R(N,M_0) = \{M_0,M_1,M_2,M_3,M_4\}$, where $M_0 = 3p_1 + 2p_4 + 3p_5$,

$M_1 = 2p_1 + p_2 + 2p_4 + 2p_5$, $M_2 = 2p_1 + p_3 + p_4$, $M_3 = p_1 + 2p_2 + 2p_4 + p_5$, and $M_4 = 3p_2 + 2p_4$. Note that at M_4, no transition is enabled.

Definition 2.5. A Petri net (N, M_0) is safe if $\forall M \in R(N, M_0)$, $\forall p \in P$, $M(p) \leq 1$ is true. It is bounded if $\exists k \in \mathbb{N}^+$, $\forall M \in R(N, M_0)$, $\forall p \in P$, $M(p) \leq k$. It is said to be unbounded if it is not bounded. A net N is structurally bounded if it is bounded for any initial marking.

Note that a net is bounded iff its reachability set has a finite number of elements. The reachability set of a net (N, M_0) can be expressed by a reachability graph. A reachability graph is a directed graph whose nodes are markings in $R(N, M_0)$ and arcs are labeled by the transitions of N. An arc from M_1 to M_2 is labeled by t iff $M_1[t\rangle M_2$.

Example 2.4. Figure 2.3 shows the reachability graph of the Petri net depicted in Fig. 2.2a. The net is bounded and its reachability graph is finite.

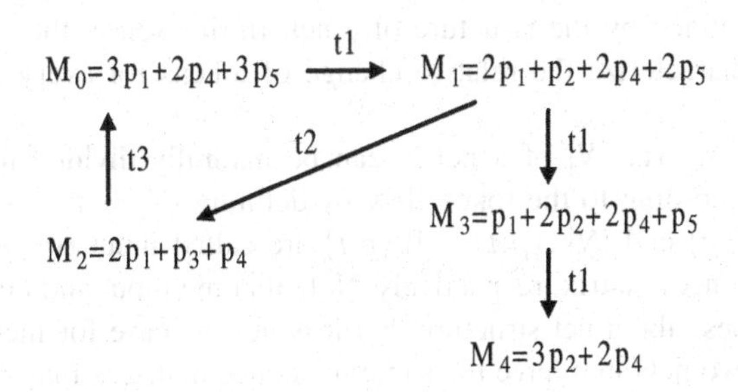

Fig. 2.3 The reachability graph of net (N, M_0) shown in Fig 2.2a

Definition 2.6. A net $N = (P, T, F, W)$ is pure (self-loop free) iff $\forall x, y \in P \cup T$, $W(x, y) > 0$ implies $W(y, x) = 0$.

Definition 2.7. A pure net $N = (P, T, F, W)$ can be represented by its incidence matrix $[N]$, where $[N]$ is a $|P| \times |T|$ integer matrix with $[N](p, t) = W(t, p) - W(p, t)$. For a place p (transition t), its incidence vector, a row (column) in $[N]$, is denoted by $[N](p, \cdot)$ ($[N](\cdot, t)$).

According to the definition, it is easy to see the physical meanings of an element in an incidence matrix of a Petri net N. Specifically, $[N](p, t)$ indicates that p receives (loses) $|[N](p, t)|$ tokens if $[N](p, t) > 0$ ($[N](p, t) < 0$) after t fires. The number of tokens in p does not change if $[N](p, t) = 0$ after t fires. Vector $[N](p, \cdot)$ shows the token variation in p with respect to the firing of each transition once in the net N. Let $S \subseteq P$ be a subset of places in net N. $[N](S, \cdot)$ is used to denote $\Sigma_{p \in S}[N](p, \cdot)$.

Example 2.5. The incidence matrix of the net in Fig. 2.1a is shown below:

$$[N] = \begin{pmatrix} -1 & 0 & 1 \\ 1 & -1 & 0 \\ 0 & 1 & -1 \\ 0 & -1 & 1 \\ -1 & -2 & 3 \end{pmatrix}.$$

$[N](p_1, t_1) = -1$ implies that p_1 loses a token after firing t_1. $[N](p_1, t_3) = 1$ indicates that p_1 gets a token after t_3 fires. $[N](p_1, t_2) = 0$ means that the number of tokens in p_1 does not change after t_2 fires. Note that $[N](p_5, \cdot) = (-1, -2, 3)$. It implies that firing t_1 removes one token from p_5, firing t_2 removes two tokens from p_5, and firing t_3 deposits three tokens into p_5.

Let $S = \{p_3, p_5\}$. $[N](S, \cdot) = [N](p_3, \cdot) + [N](p_5, \cdot) = (-1, -1, 2)$. It indicates that firing t_1 or t_2 removes one token from S, and firing t_3 puts two tokens into S.

It is important to note that the change of the number of tokens in a place p caused by firing some transition t does not depend on the current marking. Instead, it is completely determined by the structure of a net. In this sense, the incidence matrix suffices to characterize the relative change of tokens for every place when a transition fires.

The incidence matrix $[N]$ of a net N can be naturally divided into two parts $[N]^+$ and $[N]^-$ according to the token flow by defining $[N] = [N]^+ - [N]^-$, where $[N]^+(p,t) = W(t,p)$ and $[N]^-(p,t) = W(p,t)$ are called input (incidence) matrix and output (incidence) matrix, respectively. Note that the input and output matrices can completely describe a net structure, but it is not the case for incidence matrices in general. Two nets that have the same incidence matrices may have different net structures. This case likes an expression $a - b = c - d$ but $a = c$ and $b = d$ are not necessarily true. However, if there are no self-loops in a Petri net, its incidence matrix can completely determine its structure.

Example 2.6. For the net in Fig. 2.1a, its input matrix and output matrix are as follows:

$$[N]^+ = \begin{pmatrix} 0 & 0 & 1 \\ 1 & 0 & 0 \\ 0 & 1 & 0 \\ 0 & 0 & 1 \\ 0 & 0 & 3 \end{pmatrix}, \quad [N]^- = \begin{pmatrix} 1 & 0 & 0 \\ 0 & 1 & 0 \\ 0 & 0 & 1 \\ 0 & 1 & 0 \\ 1 & 2 & 0 \end{pmatrix}.$$

Accordingly, the enabling condition of a transition t can be rewritten as $M \geq [N]^-(\cdot, t)$.

Definition 2.8. Given a Petri net (N, M_0), $t \in T$ is live under M_0 iff $\forall M \in R(N, M_0)$, $\exists M' \in R(N, M)$, $M'[t\rangle$. (N, M_0) is live iff $\forall t \in T$, t is live under M_0. (N, M_0) is dead under M_0 iff $\nexists t \in T$, $M_0[t\rangle$. (N, M_0) is deadlock-free (weakly live or live-locked) iff $\forall M \in R(N, M_0)$, $\exists t \in T$, $M[t\rangle$.

Definition 2.9. Petri net (N, M_0) is quasi-live iff $\forall t \in T$, there exists $M \in R(N, M_0)$ such that $M[t\rangle$ holds.

A live Petri net guarantees deadlock-freedom no matter what firing sequence is chosen but the converse is not true. However, this property is costly to verify.

Example 2.7. The net shown in Fig. 2.4a is deadlock-free since transitions t_1 and t_2 are live, while the net in Fig 2.4b is live since all transitions are live. The net in Fig. 2.2e is dead since no transition is enabled under the current marking M_4.

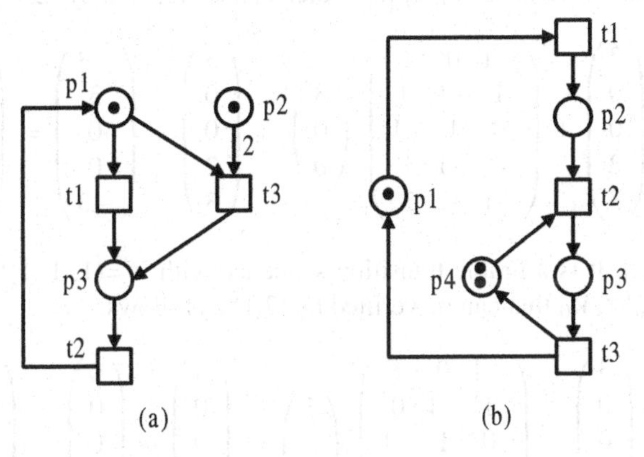

Fig. 2.4 Two Petri nets: (a) is deadlock-free and (b) is live

Definition 2.10. Let $N = (P, T, F, W)$ be a net and σ be a finite sequence of transitions. The Parikh vector of σ is $\vec{\sigma} : T \to \mathbb{N}$ which maps t in T to the number of occurrences of t in σ. Define $\vec{t_1} = (1, 0, \ldots, 0)^T$, $\vec{t_2} = (0, 1, 0, \ldots, 0)^T$, and $\vec{t_k} = (0, 0, \ldots, 0, 1)^T$ assuming $k = |T|$.

Example 2.8. Let $\sigma_1 = t_1 t_3 t_2 t_4 t_5 t_2$ and $\sigma_2 = t_1$ be two sequences of transitions of some net N with $|T| = 6$. Their Parikh vectors are $\vec{\sigma_1} = (1, 2, 1, 1, 1, 0)^T$ and $\vec{\sigma_2} = (1, 0, 0, 0, 0, 0)^T$, respectively. Clearly, we have $\vec{\sigma_2} = \vec{t_1} = (1, 0, 0, 0, 0, 0)^T$. For the transition sequence $\sigma = t_1 t_1 t_1$ in the net shown in Fig. 2.1a, $\vec{\sigma} = (3, 0, 0)^T$.

It is trivial that for each transition t, we have $[N](\cdot, t) = [N]\vec{t}$. Note that $M[t\rangle M'$ leads to $M' = M + [N](\cdot, t)$. Consequently, if $M[t\rangle M'$, we have $M' = M + [N]\vec{t}$. For an arbitrary finite transition sequence σ such that $M[\sigma\rangle M'$, we have

$$M' = M + [N]\vec{\sigma}. \tag{2.1}$$

Equation 2.1 is called the state equation of a Petri net (N, M), which presents an algebraic description of the marking change in a Petri net. In other words, it is a compact way to express the interrelation between markings and numbers of transition occurrences in a transition sequence. Such a linear algebraic expression is very helpful because it allows one to apply the concepts and results of linear algebra to the domain of Petri nets.

Any reachable marking fulfils the state equation but the converse is not true. In this sense, the state equation provides a necessary condition for a marking M to be reachable from an initial marking M_0. That is to say, if marking M is reachable from M_0, the state equation $M = M_0 + [N]\overrightarrow{\sigma}$ must have a vector solution for σ with its components in $\mathbb{N}$. Conversely, if the marking equation is not soluble, marking M is not reachable from M_0.

Example 2.9. In Fig. 2.1a, $\sigma = t_1t_1t_1$ is a firable transition sequence with $\overrightarrow{\sigma} = (3, 0, 0)^T$. From Fig. 2.2d, we have $M_0[\sigma\rangle M_4$, which can be verified by (2.1) as follows:

$$M_0 + [N]\overrightarrow{\sigma} = \begin{pmatrix} 3 \\ 0 \\ 0 \\ 2 \\ 3 \end{pmatrix} + \begin{pmatrix} -1 & 0 & 1 \\ 1 & -1 & 0 \\ 0 & 1 & -1 \\ 0 & -1 & 1 \\ -1 & -2 & 3 \end{pmatrix} \begin{pmatrix} 3 \\ 0 \\ 0 \end{pmatrix} = \begin{pmatrix} 3 \\ 0 \\ 0 \\ 2 \\ 3 \end{pmatrix} + \begin{pmatrix} -3 \\ 3 \\ 0 \\ 0 \\ -3 \end{pmatrix} = \begin{pmatrix} 0 \\ 3 \\ 0 \\ 2 \\ 0 \end{pmatrix} = M_4.$$

Let $\sigma = t_1t_2t_3$. It is a firable transition sequence with $\overrightarrow{\sigma}$=(1, 1, 1)^T. From Fig. 2.2, we have $M_0[\sigma\rangle M_0$ that can be verified by (2.1) as follows.

$$M_0 + [N]\overrightarrow{\sigma} = \begin{pmatrix} 3 \\ 0 \\ 0 \\ 2 \\ 3 \end{pmatrix} + \begin{pmatrix} -1 & 0 & 1 \\ 1 & -1 & 0 \\ 0 & 1 & -1 \\ 0 & -1 & 1 \\ -1 & -2 & 3 \end{pmatrix} \begin{pmatrix} 1 \\ 1 \\ 1 \end{pmatrix} = \begin{pmatrix} 3 \\ 0 \\ 0 \\ 2 \\ 3 \end{pmatrix} + \begin{pmatrix} 0 \\ 0 \\ 0 \\ 0 \\ 0 \end{pmatrix} = \begin{pmatrix} 3 \\ 0 \\ 0 \\ 2 \\ 3 \end{pmatrix} = M_0.$$

Definition 2.11. Let (N, M_0) be a net system. Its linearized reachability set by using the state equation over the real numbers is defined as $R^S(N, M_0) = \{M | M = M_0 + [N]Y, M \geq 0, Y \geq 0\}$.

We have $R(N, M_0) \subseteq R^S(N, M_0)$ since the state equation does not check whether there is a sequence of intermediate markings such that some transition sequence σ is actually firable. The markings in $R^S(N, M_0) \backslash R(N, M_0)$ are called spurious markings (with respect to the state equation).

Although the reachability set derived from the state equation may contain spurious markings, in some cases its linear description facilitates the analysis of a Petri net.

For example, the verification of predicate $min\{M(S) | M \in R(N, M_0)\} \geq k_1$ is difficult due to a potentially huge number of reachable markings in $R(N, M_0)$, where S is a subset of places and k_1 is a non-negative integer. However, the minimal number of tokens holding by S in $R^S(N, M_0)$ can be found by solving the following linear programming problem (LPP):

$MIN\ M(S)$

s.t.

$M = M_0 + [N]Y$

$M \geq 0$
$Y \geq 0$

It is known that an LPP can be solved in polynomial time. Let k_2 be a feasible solution of the above LPP. Obviously, we have $k_2 \leq min\{M(S)|M \in R(N,M_0)\}$. If $k_1 \leq k_2$, one gets $k_1 \leq k_2 \leq min\{M(S)|M \in R(N,M_0)\}$, leading to the truth of this predicate. Certainly, if $k_1 > k_2$, one cannot give a definite answer to the truth of this predicate.

Example 2.10. $S = \{p_1, p_3, p_4\}$ is a set of places in the net shown in Fig 2.5, where $M_0 = p_3$. A question is whether S can be always marked. By solving an LPP, we have $min\{M(S)|M = M_0 + [N]Y, M \geq 0, Y \geq 0\} = 1$. This leads to the fact that S can never be emptied, i.e., under any reachable marking, there is at least one place that is marked.

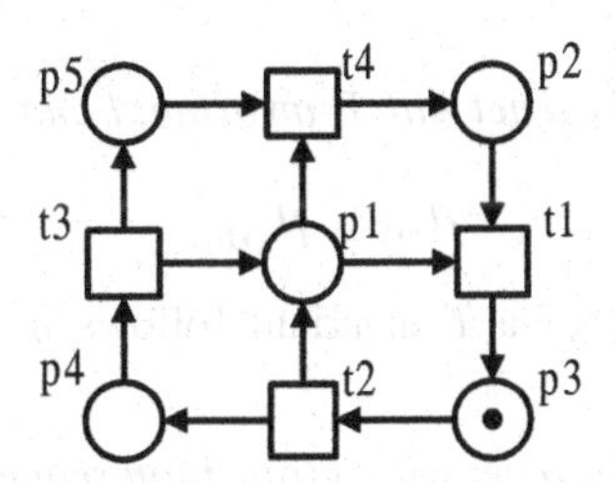

Fig. 2.5 A Petri net (N, M_0)

2.3 Structural Invariants

One important feature of Petri nets is that their structural properties can be obtained by linear algebraic techniques [13, 18, 42]. These properties that depend on only the topological structure of a Petri net and are independent of the initial marking are called invariants. Invariants are an important means for analyzing the behavior of a Petri net from a structural viewpoint.

Definition 2.12. A P-vector is a column vector $I : P \to \mathbb{Z}$ indexed by P and a T-vector is a column vector $J : T \to \mathbb{Z}$ indexed by T, where $\mathbb{Z}$ is the set of integers.

We denote column vectors where every entry equals 0(1) by $\mathbf{0}$($\mathbf{1}$). I^T and $[N]^T$ are the transposed versions of vector I and matrix $[N]$, respectively. A $P(T)$-vector is non-negative if no element in it is negative.

Definition 2.13. P-vector I is called a P-invariant (place invariant) iff $I \neq \mathbf{0}$ and $I^T[N] = \mathbf{0}^T$. T-vector J is called a T-invariant (transition invariant) iff $J \neq \mathbf{0}$ and $[N]J = \mathbf{0}$.

Definition 2.14. P-invariant I is a P-semiflow if every element of I is non-negative. $||I|| = \{p|I(p) \neq 0\}$ is called the support of I. $||I||^+ = \{p|I(p) > 0\}$ denotes the positive support of P-invariant I and $||I||^- = \{p|I(p) < 0\}$ denotes the negative support of I. I is called a minimal P-invariant if $||I||$ is not a superset of the support of any other one and its components are mutually prime.

Definition 2.15. T-invariant J is a T-semiflow if every element of J is non-negative. $||J|| = \{t|J(t) \neq 0\}$ is called the support of J. $||J||^+ = \{t|J(t) > 0\}$ denotes the positive support of T-invariant J and $||J||^- = \{t|J(t) < 0\}$ denotes the negative support of J. J is called a minimal T-invariant if $||J||$ is not a superset of the support of any other one and its components are mutually prime.

Note that a set of numbers is mutually prime if their common divisor is one. For example, 4, 7, and 16 are mutually prime. But 4, 6, and 16 are not since 2 is their common divisor. A P-invariant corresponds to a set of places whose weighted token count is a constant for any reachable marking. It follows immediately from the state equation.

Theorem 2.1. *Let (N, M_0) be a net with P-invariant I and M be a reachable marking from M_0. Then*

$$I^T M = I^T M_0.$$

A fundamental property of a T-invariant follows immediately from the state equation.

Theorem 2.2. *Let (N, M_0) be a net with a transition sequence σ such that $M_0[\sigma\rangle M$. $M = M_0$ iff $\vec{\sigma}$ is a T-invariant of N.*

Note that for a specific marked net, the existence of a T-invariant does not imply that there exists a transition sequence whose Parikh vector is the T-vector such that it is firable and its firing leads the net from the initial marking back to it. Furthermore, it is easy to see that any linear combination of $P(T)$-invariants of a net is still a $P(T)$-invariant of the net.

Property 2.1. If I is a P-semiflow of a net, ${}^\bullet||I|| = ||I||^\bullet$.

Example 2.11. In the net shown in Fig. 2.1a, there are three minimal P-invariants: $I_1 = p_1 + p_2 + p_3$, $I_2 = p_3 + p_4$, and $I_3 = p_2 + 3p_3 + p_5$, since $\forall i \in \{1,2,3\}$, $I_i^T[N]=\mathbf{0}^T$. $\forall M \in R(N, M_0)$, $I_1^T M = I_1^T M_0 = M_0(p_1) + M_0(p_2) + M_0(p_3) = 3$. This indicates that the token count in places p_1, p_2, and p_3 keeps three under any reachable marking, which can be verified from the reachability graph, which is identical with the one shown in Fig 2.3.

The net has a unique T-invariant $J = t_1 + t_2 + t_3$ and the transition sequence $\sigma = t_1 t_2 t_3$ is firable. As a result, $M_0[t_1\rangle M_1[t_2\rangle M_2[t_3\rangle M_0$.

Since I_1 and I_2 are P-invariants, $I = I_1 - I_2 = p_1 + p_2 - p_4$ is a P-invariant as well. Note that I is not a P-semiflow due to its negative component. Moreover, one can get $||I_1|| = \{p_1, p_2, p_3\}$, $||I||^+ = \{p_1, p_2\}$, and $||I||^- = \{p_4\}$. It is easy to see that ${}^\bullet||I_1|| = {}^\bullet p_1 \cup {}^\bullet p_2 \cup {}^\bullet p_3 = \{t_3\} \cup \{t_1\} \cup \{t_2\} = \{t_1, t_2, t_3\}$ and $||I_1||^\bullet = p_1^\bullet \cup p_2^\bullet \cup p_3^\bullet = \{t_1\} \cup \{t_2\} \cup \{t_3\} = \{t_1, t_2, t_3\}$. $||I_1||^\bullet = {}^\bullet ||I_1||$ will not be surprising since I_1 is a P-semiflow.

A Petri net is strongly connected if $\forall x, y \in P \cup T$, there is a sequence of nodes x, a, b, ..., c, y such that (x,a), (a,b), ..., $(c,y) \in F$, where $\{a,b,\ldots,c\} \subseteq P \cup T$. A string $x_1 \ldots x_n$ is called a path of N iff $\forall i \in \mathbb{N}_{n-1}$, $x_{i+1} \in x_i^\bullet$, where $\forall x \in \{x_1, \cdots, x_n\}$, $x \in P \cup T$. An elementary path from x_1 to x_n is a path whose nodes are all different (except, perhaps, x_1 and x_n). A path $x_1 \cdots x_n$ is called a circuit iff it is an elementary path and $x_1 = x_n$.

The liveness of a Petri net is close to its connectedness. A result is given in [17]: Each connected net with a live and bounded marking is strongly connected. A result that establishes a bridge between strong connectedness and invariants is given as follows owing to [18]:

Theorem 2.3. *Each connected net with a positive place invariant and positive transition invariant is strongly connected.*

2.4 Siphons and Traps

P-invariants that can be derived from the state equation of a Petri net are marking invariants. The token count in their corresponding places stays constant, i.e., the invariant law associated with a P-invariant holds for any reachable marking. In a Petri net, siphons and traps are also structural objects that involve marking invariants. However, the invariant laws associated with them do not hold under any reachable marking, but once they become true they remain true for any subsequently reachable markings. A siphon remains empty once it loses all tokens. A trap remains marked once it has any token in it. Siphons and traps have been extensively investigated and used for the structural analysis of a Petri net. They also play an important role in the liveness analysis of a net, particularly in ordinary ones.

Definition 2.16. A non-empty set $S \subseteq P$ is a siphon iff ${}^\bullet S \subseteq S^\bullet$. $S \subseteq P$ is a trap iff $S^\bullet \subseteq {}^\bullet S$. A siphon (trap) is minimal iff there is no siphon (trap) contained in it as a proper subset. A minimal siphon S is said to be strict if ${}^\bullet S \subsetneq S^\bullet$.

Property 2.2. Let S_1 and S_2 are two siphons (traps). Then, $S_1 \cup S_2$ is a siphon (trap).

Example 2.12. In the net shown in Fig. 2.1a, $S_1 = \{p_1, p_2, p_3\}$, $S_2 = \{p_4, p_3\}$, $S_3 = \{p_2, p_3, p_5\}$, and $S_4 = \{p_3, p_5\}$ are siphons, among which S_1, S_2, and S_4 are minimal since the removal of any place from each of these sets leads to the fact that the resultant set is not a siphon any more. Note that ${}^\bullet S_1 = S_1^\bullet$, ${}^\bullet S_2 = S_2^\bullet$, and ${}^\bullet S_3 = S_3^\bullet$. S_1, S_2, and S_3 are also traps. By ${}^\bullet S_4 = \{t_2, t_3\}$ and $S_4^\bullet = \{t_1, t_2, t_3\}$, we have ${}^\bullet S_4 \subset S_4{}^\bullet$. $S_4 = \{p_3, p_5\}$ is therefore a strict minimal siphon.

Corollary 2.1. *If I is a P-semiflow, then $||I||$ is both a siphon and trap.*

Note that the converse of Corollary 2.1 is not true since a P-invariant depends on not only the topological structure of a net but also the weights attached to the arcs. However, a siphon or trap depends on the topological structure only. For example,

$S = \{p_1, p_2\}$ in Fig. 2.6 is both a siphon and trap. However, it is not the support of a P-semiflow. In this sense, the converse of Corollary 2.1 is true in the domain of ordinary nets.

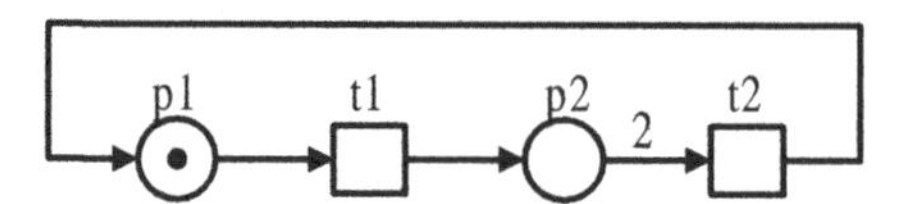

Fig. 2.6 A siphon and trap in a net without P-semiflow

If a siphon contains the support of a P-semiflow and the support is initially marked, then it can never be emptied. In addition, traps and siphons have the following marking invariant laws.

Property 2.3. Let $M \in R(N, M_0)$ be a marking of net (N, M_0) and S a trap. If $M(S) > 0$, then $\forall M' \in R(N, M)$, $M'(S) > 0$.

This property implies that once a trap is marked under a marking, it is always marked under the subsequent markings that are reachable from the current one.

Property 2.4. Let $M \in R(N, M_0)$ be a marking of net (N, M_0) and S a siphon. If $M(S) = 0$, then $\forall M' \in R(N, M)$, $M'(S) = 0$.

Property 2.4 indicates that once a siphon loses all its tokens, it remains unmarked under any subsequent markings that are reachable from the current marking. An empty siphon S causes that no transition in $S^\bullet$ is enabled. Due to the definition of siphons, all transitions connected to S can never be enabled once it is emptied. The transitions are therefore dead, leading to the fact that the net containing these transitions is not live.

As a result, deadlock-freedom and liveness of a Petri net are closely related to its siphons, which is shown by the following known results [16].

Theorem 2.4. *Let (N, M_0) be an ordinary net and Π the set of its siphons. The net is deadlock-free if* $\forall S \in \Pi$, $\forall M \in R(N, M_0)$, $M(S) > 0$.

This theorem states that an ordinary Petri net is deadlock-free if no (minimal) siphon eventually becomes empty.

Theorem 2.5. *Let (N, M) be an ordinary net that is in a deadlock state. Then, $\{p \in P | M(p) = 0\}$ is a siphon.*

This result means that if an ordinary net is dead, i.e., no transition is enabled, then the unmarked places form a siphon.

Example 2.13. The net shown in Fig. 2.7a is a famous example as first discussed by Zhou et al. [61, 62] and later by Chu and Xie [12] and many other researchers [6]. It has four minimal siphons $S_1 = \{p_1, p_2, p_3, p_4\}$, $S_2 = \{p_3, p_5\}$, $S_3 = \{p_2, p_4, p_6\}$, and $S_4 = \{p_4, p_5, p_6\}$. S_1, S_2 and S_3 are also traps that are initially marked. Note that S_4 is a strict minimal siphon since ${}^\bullet S_4 = \{t_2, t_3, t_4\}$ and $S_4{}^\bullet = \{t_1, t_2, t_3, t_4\}$, leading to the truth of ${}^\bullet S_4 \subsetneq S_4{}^\bullet$.

In Fig. 2.7a, $\sigma = t_1 t_2 t_1$ is a firable transition sequence whose firing leads to a new marking as shown in Fig. 2.7b. The net in Fig. 2.7b is dead since no transition is enabled in the current marking. The unmarked places p_1, p_4, p_5, and p_6 form a siphon $S = \{p_1, p_4, p_5, p_6\}$ that is not minimal since it contains S_4. The emptiness of S disables every transition in $S^\bullet$ such that no transition in this net is enabled. As a result, the net is dead.

Based on Theorem 2.5, we can achieve the following results.

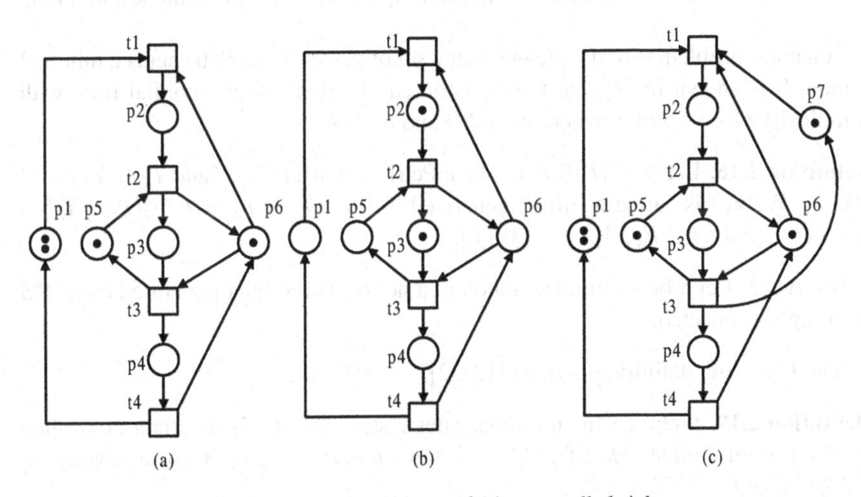

Fig. 2.7 (a) A Petri net [61], (b) a dead marking, and (c) a controlled siphon

Corollary 2.2. *A deadlocked ordinary Petri net contains at least one empty siphon.*

Corollary 2.3. *Let $N = (P, T, F, W)$ be a deadlocked net under marking M. Then, it has at least one siphon S such that $\forall p \in S$, $\exists t \in p^\bullet$ such that $W(p,t) > M(p)$.*

Definition 2.17. A siphon S is said to be controlled in a net system (N, M_0) iff $\forall M \in R(N, M_0)$, $M(S) > 0$.

Clearly, any siphon that contains a marked trap is controlled since it can never be emptied. In an ordinary Petri net, a siphon that is controlled does not imply a deadlock. This is not the case in a generalized Petri net. For example, there are two minimal siphons in the generalized Petri net shown in Fig 2.4a. They are $S_1 = \{p_1, p_3\}$ and $S_2 = \{p_2\}$. Both of them can never be unmarked. However, the insufficient number of tokens in S_2 disables t_3. In fact, t_3 is a dead transition in the net. Hence, the net is not live even though each siphon is always marked. Chapter 3 shows that a siphon in a generalized Petri net does not lead to dead transitions if it is max-controlled [4].

For a siphon that can be emptied in a net, some external control mechanism can be exerted on the net such that it becomes controlled. In Fig. 2.7a, S_4 is a strict minimal siphon whose emptiness leads to the deadlock of the net. To prevent S_4 from being unmarked, a place p_7 is added with ${}^\bullet p_7 = \{t_3\}$ and $p_7^\bullet = \{t_1\}$, as shown in Fig. 2.7c. The initial marking of p_7 is one. Such an additional place is called a *monitor* or control place in terms of its role. In Fig. 2.7c, the addition of p_7 leads to an extra minimal siphon $S_5 = \{p_2, p_3, p_7\}$ that is a marked trap. As a result, no siphon can be emptied in the net and it is deadlock-free (actually, live). This example motivates one to explore the mechanism to make a siphon controlled by adding a monitor.

When we talk about siphon control, we are usually concerned with minimal siphons since the controllability of a minimal siphon implies that of those containing it.

A natural problem is to decide whether a set of places S in a Petri net is a minimal siphon. It is shown in [2] that the decision can be done in polynomial time with complexity $O(m^2 + mn^2)$, where $m = |S^\bullet|$ and $n = |S|$.

Definition 2.18. Let $N = (P, T, F, W)$ be a Petri net with $P_X \subseteq P$ and $T_X \subseteq T$. $N_X = (P_X, T_X, F_X, W_X)$ is called a subnet generated by $P_X \cup T_X$ if $F_X = F \cap [(P_X \times T_X) \cup (T_X \times P_X)]$ and $\forall f \in F_X$, $W_X(f) = W(f)$.

Property 2.5. Let S be a minimal siphon in a net N. The subnet generated by $S \cup {}^\bullet S$ is strongly connected.

The following definition is from [12, 32].

Definition 2.19. Siphon S in an ordinary net system (N, M_0) is invariant-controlled by P-invariant I under M_0 iff $I^T M_0 > 0$ and $\forall p \in P \backslash S, I(p) \leq 0$, or equivalently, $I^T M_0 > 0$ and $||I||^+ \subseteq S$.

If S is controlled by P-invariant I under M_0, S cannot be emptied, i.e., $\forall M \in R(N, M_0)$, S is marked under M.

Example 2.14. In Fig. 2.7c, one can verify that $I_1 = p_3 + p_5$, $I_2 = p_2 + p_4 + p_6$, and $I_3 = p_2 + p_3 + p_7$ are P-invariants. As a result, $I = I_1 + I_2 - I_3 = p_4 + p_5 + p_6 - p_7$ is a P-invariant as well. It is easy to see that siphon $S_4 = \{p_4, p_5, p_6\}$ is controlled by P-invariant I since $||I||^+ = \{p_4, p_5, p_6\} = S_4$ and $I^T M_0 = M_0(p_4) + M_0(p_5) + M_0(p_6) - M_0(p_7) = 2 - 1 = 1 > 0$. The controllability of $S_4 = \{p_4, p_5, p_6\}$ implies that of siphon $S = \{p_1, p_4, p_5, p_6\}$ that is not minimal. Note that ${}^\bullet S_4 = \{t_2, t_3, t_4\}$. The subset generated by $S_4 \cup {}^\bullet S_4$ is shown in Fig. 2.8. It is clearly strongly connected since $S_4 = \{p_4, p_5, p_6\}$ is a minimal siphon.

In essence, the controllability of siphon S by adding a monitor is ensured by the fact that the number of tokens leaving S is limited by a marking invariant law imposed on the Petri net, which is implemented by a P-invariant whose support contains the monitor.

In order to test whether a siphon S is controlled by a P-invariant I, it is sufficient to solve the following system of linear homogeneous inequalities and equations:

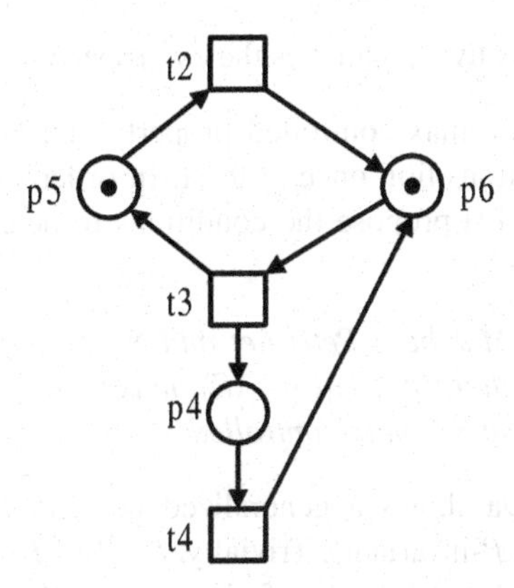

Fig. 2.8 A subnet generated by a minimal siphon and its preset

$$I^T[N] = \mathbf{0}^T$$
$$I^T M_0 > 0$$
$$I(p) \leq 0, \forall p \in P \backslash S$$

For the above system, the existence of a solution can be proved through *Phase I* of the simplex algorithm applied to the following LPP:

$$\text{maximize } \mathbf{0}^T I$$
$$\text{s. t.}$$
$$I^T[N] = \mathbf{0}^T$$
$$I^T M_0 > 0$$
$$I(p) \leq 0, \forall p \in P \backslash S$$

Phase I of the simplex algorithm computes a *basic feasible solution* of the set of constraints of the LPP if it exists.

An empty or insufficiently marked siphon in a Petri net can cause some transitions not to be enabled. A siphon in an ordinary Petri net can be made invariant-controlled as defined above. The case in a generalized Petri net is much more complicated and is treated as follows.

Definition 2.20. Let (N, M_0) be a net system and S be a siphon of N. S is said to be max-marked at a marking $M \in R(N, M_0)$ iff $\exists p \in S$ such that $M(p) \geq max_{p^\bullet}$.

Definition 2.21. A siphon is said to be max-controlled iff it is max-marked at any reachable marking.

Definition 2.22. (N, M_0) satisfies the maximal cs-property (maximal controlled-siphon property) iff each minimal siphon of N is max-controlled.

The following results are owing to [4]. In case of no confusion, maximal cs-property is called cs-property for the sake of simplification.

Property 2.6. If (N, M_0) satisfies the cs-property, it is deadlock-free.

Property 2.7. If (N, M_0) is live, it satisfies the cs-property.

A siphon satisfying the max-controlled property can be always marked sufficiently to allow firing a transition once at least. In order to check and use the cs-property, Barkaoui et al. [4] propose the conditions to determine whether a given siphon is max-controlled.

Proposition 2.1. *Let (N, M_0) be a Petri net and S be a siphon of N. If there exists a P-invariant I such that $\forall p \in (||I||^- \cap S)$, $max_{p^\bullet} = 1$, $||I||^+ \subseteq S$ and $I^T M_0 > \sum_{p \in S} I(p)(max_{p^\bullet} - 1)$, then S is max-controlled.*

Example 2.15. Figure 2.9a shows a generalized net and $I_1 = p_2 + p_6$ and $I_2 = p_2 + 3p_3 + p_5$ are its two P-invariants. Trivially, $I = I_2 - I_1 = 3p_3 + p_5 - p_6$ is also a P-invariant. Let $S = \{p_3, p_5\}$ be a set of places. Since ${}^\bullet S \subset S^\bullet$, S is a strict minimal siphon. Next we show that it is max-controlled by P-invariant I. It is clear that $||I||^- \cap S = \emptyset$ and $||I||^+ = S$. We then check the truth of $I^T M_0 > \sum_{p \in S} I(p)(max_{p^\bullet} - 1)$. $I^T M_0 = M_0(p_5) + 3M_0(p_3) - M_0(p_6) = 3 - 1 = 2$. $\sum_{p \in S} I(p)(max_{p^\bullet} - 1) = I(p_3)(max_{p_3^\bullet} - 1) + I(p_5)(max_{p_5^\bullet} - 1)$. Considering $max_{p_3^\bullet} = 1$ and $max_{p_5^\bullet} = 2$, we have $\sum_{p \in S} I(p)(max_{p^\bullet} - 1) = 1$. Therefore, $I^T M_0 > \sum_{p \in S} I(p)(max_{p^\bullet} - 1)$ and S is max-controlled. Figure 2.9b shows the reachability graph of the net in Fig. 2.9a.

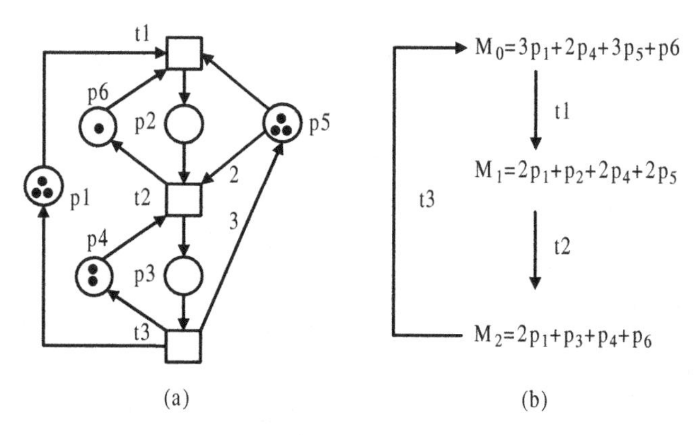

Fig. 2.9 A max-controlled siphon in a net (N, M_0) (a) Petri net model and (b) reachability graph

By comparing the net in Fig. 2.9a with Fig 2.1a as well as their reachability graphs as shown in Fig. 2.9b and Fig. 2.3, respectively, one concludes that the addition of p_6 removes two markings M_3 and M_4 in Fig. 2.3: one is a deadlock marking and the other is a marking that inevitably leads the system to deadlock.

Remark 2.1. The number of siphons (minimal siphons) grows fast with respect to the size of a Petri net and in the worst case grows exponentially with a net size. However, many deadlock control approaches depend on the complete or partial enumeration of siphons in a plant net model [23, 33, 34, 40, 50–52, 58, 59]. It is well known that the complete siphon enumeration is time-consuming. Extensive studies have been conducted on the siphon computation, leading to a variety of methods [1, 14, 22, 31, 35, 53, 54]. A recent work [15] by Cordone et al. claims that their proposed siphon computation method can find more than 2×10^7 siphons in less than one hour.

2.5 Subclasses of Petri Nets

There are a number of interesting subclasses of ordinary Petri nets. The reasons that they are interesting are twofold. First, they play an important role in the development of certain application of Petri nets [11, 12]. Second, some relevant analysis problems in these classes can be solved in polynomial time [3, 21, 36]. For example, the problem of deciding whether a free-choice Petri net is live and bounded can be solved in $O(nm)$ [21], where n and m are the number of places and transitions of the net, respectively. In turn, many analysis problems of live and bounded free-choice nets are also shown to have polynomial time complexity [16].

Definition 2.23. A Petri net $N = (P, T, F)$ is called a state machine iff $\forall\, t \in T$, $|{}^\bullet t| = |t^\bullet| = 1$.

In a state machine, each transition has exactly one input place and exactly one output place. Each transition allows tokens to flow from one place to another. Multiple transitions may allow tokens to flow from their respective places to the same place. In addition, a single token in a place p enables all transitions in $p^\bullet$. Firing any of them disables the others. This is called a conflict. Note that all finite automata can be described as the state machines of Petri nets.

Theorem 2.6. *A state machine (N, M_0) is live iff N is strongly connected and M_0 marks at least one place.*

Definition 2.24. A Petri net $N = (P, T, F)$ is said to be a marked graph iff $\forall p \in P$, $|{}^\bullet p| = |p^\bullet| = 1$.

In a marked graph, each place has exactly one input transition and exactly one output transition. A transition may have multiple input places and output places. In this sense, a marked graph allows concurrent and synchronization structure. A state machine admits no synchronization and a marked graph allows no conflict.

Theorem 2.7. *A marked graph (N, M_0) is live iff M_0 places at least one token on each circuit in N.*

Definition 2.25. A Petri net is a free-choice net iff $\forall p_1, p_2 \in P$, $p_1^\bullet \cap p_2^\bullet \neq \emptyset \Rightarrow |p_1^\bullet| = |p_2^\bullet| = 1$.

In a free-choice net, every arc from a place is either a unique outgoing arc or a unique incoming arc to a transition. A free-choice net allows both conflict and synchronization, i.e., state machines and marked graphs fall under the class of free-choice nets.

Theorem 2.8. *A free-choice net (N, M_0) is live iff every siphon in it contains a marked trap.*

Definition 2.26. A Petri net is an extended free-choice net iff $\forall p_1, p_2 \in P$, $p_1^\bullet \cap p_2^\bullet \neq \emptyset \Rightarrow p_1^\bullet = p_2^\bullet$.

Definition 2.27. A Petri net is an asymmetric choice net iff $\forall p_1, p_2 \in P$, $p_1^\bullet \cap p_2^\bullet \neq \emptyset \Rightarrow p_1^\bullet \subseteq p_2^\bullet$ or $p_2^\bullet \subseteq p_1^\bullet$.

Theorem 2.9. *An asymmetric choice net (N, M_0) is live if (but not only if) every siphon in N contains a marked trap.*

Example 2.16. Figure 2.10 shows some subclasses of Petri nets.

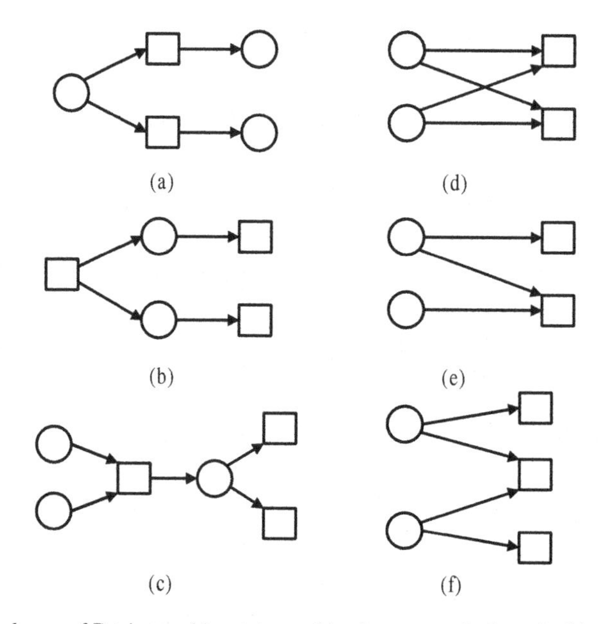

Fig. 2.10 Subclasses of Petri nets: (a) a state machine but not marked graph, (b) a marked graph but not state machine, (c) a free-choice net, (d) an extended free-choice net, (e) an asymmetric net, and (f) a Petri net

2.6 Petri Nets and Automata

Since the reachability graph of a Petri net is an automaton, this section presents some basics of finite-state automata [26], which are helpful to understand what is presented in this book.

Definition 2.28. A (deterministic) finite-state automaton is a 5-tuple $G = (Q, \Sigma, \delta, q_0, Q_m)$, where Q is a finite set of states, Σ is a finite alphabet of symbols that we refer to as event labels, δ: $Q \times \Sigma$ the (partial) transition function, q_0 the initial state, and $Q_m \subseteq Q$ the set of marker states.

δ is a partial function since $\delta(q, \alpha)$ may not be defined for all $(q, \alpha) \in Q \times \Sigma$. When $\delta(q, \alpha)$ is defined, it implies that $\exists\, q' \in Q$ and $\alpha \in \Sigma$, the occurrence of event α transits the automaton from states q to q'.

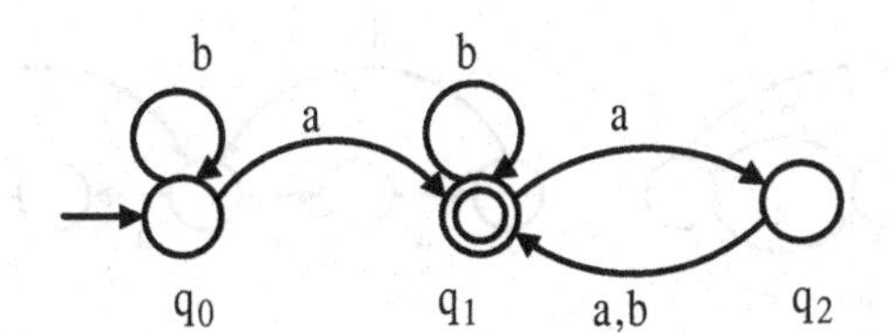

Fig. 2.11 An automaton

The operation of a finite-state automaton is always illustrated in a state diagram. Graphically, the initial state is marked with an input arrow and the marker states are denoted by double circles. For instance, Figure 2.11 shows an automaton G, where $Q = \{q_0, q_1, q_2\}$, $\Sigma = \{a, b\}$, the initial state is q_0, $Q_m = \{q_1\}$, $\delta(q_0, b) = q_0$, $\delta(q_0, a) = q_1$, $\delta(q_1, b) = q_1$, $\delta(q_1, a) = q_2$, $\delta(q_2, a) = q_1$, and $\delta(q_2, b) = q_1$.

The behavior of a system modeled by an automaton can be characterized by the language that the automaton speaks, i.e., a set of sequences of symbols of events from Σ, which are physically possible. For example, $\sigma = abab$ is a possible sequence of events in the automaton in Fig. 2.11. The set of all finite sequences over Σ is denoted by Σ^*, which includes the empty string whose length is zero and which is denoted by ε.

Definition 2.29. A labeled Petri net is a net with a labeling function $l : T \rightarrow 2^{\Sigma} \cup \{\varepsilon\}$, where Σ is the set of events and ε is a null event. A net is said to be free-labeled if each transition $t \in T$ is labeled by a single event $a \in \Sigma$ and different transitions bear different labels.

The reachability graph of a free-labeled Petri net corresponds to a deterministic automaton. A finite automaton can easily be converted into a labeled Petri net by inserting a transition that is labeled by the symbol between two connected states. The states in the automaton are differently numbered by places. Figure 2.12 is the equivalent labeled Petri net of the automaton depicted in Fig. 2.11.

For supervisory control of DES in a Petri net formalism, we are more concerned with a free-labeled Petri net representation. Unfortunately, it is shown that not all finite automata admit a free-labeled Petri net representation. It remains unanswered what finite automata do have a free-labeled Petri net realization. Figure 2.13 shows two finite automata that have no such realizations.

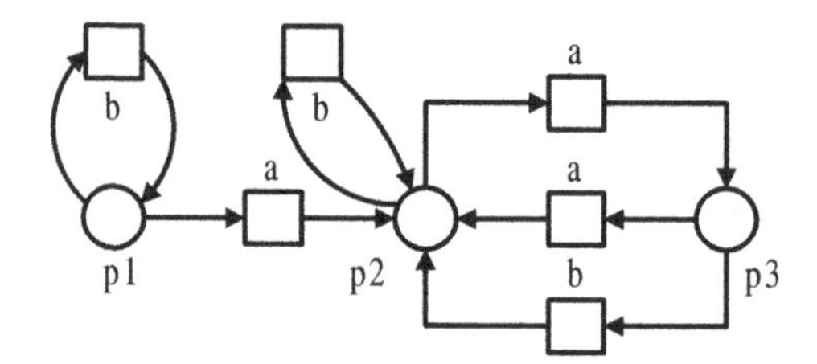

Fig. 2.12 The equivalent labeled Petri net of a finite automaton (a) Petri net A and (b) Petri net B

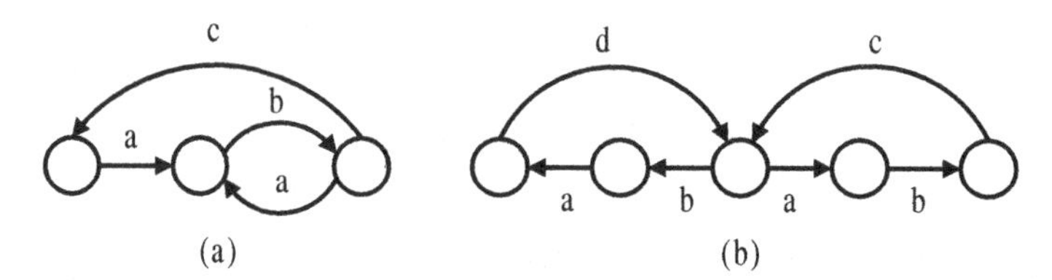

Fig. 2.13 Two finite automata without Petri net realizations

2.7 Plants, Supervisors, and Controlled Systems

In traditional supervisory control theory of DES, a system to be controlled is called a plant or a plant net model if Petri nets are used as a formalism. The external agent that forces the system to behave to satisfy given control specifications and requirements is usually called a supervisor. In a Petri net formalism, a supervisor is a Petri net that usually consists of a set of *monitors*, sometimes called control places, and a set of transitions of the plant net model. There are no places of the plant model in its supervisor. The role of the monitors in a supervisor is to supervise the plant such that its behavior satisfies the control specifications. The compound of a plant net model and its Petri net supervisor is called the controlled (net) system of the plant, whose behavior does not violate the given control specifications and requirements. To formally define a controlled system, it is necessary to first define a class of compositions of two Petri nets via shared transitions. This composition is also called synchronous synthesis of Petri nets.

Definition 2.30. Let (N_1, M_1) and (N_2, M_2) be two nets with $N_i = (P_i, T_i, F_i, W_i)$, $i = 1,2$, satisfying $P_1 \cap P_2 = \emptyset$. (N,M) with $N = (P,T,F,W)$ is said to be a synchronous synthesis net resulting from the merge of (N_1,M_1) and (N_2,M_2), denoted by $(N_1,M_1) \otimes (N_2,M_2)$, iff

1. $P = P_1 \cup P_2$
2. $T = T_1 \cup T_2$
3. $F = F_1 \cup F_2$
4. $W(f) = W_i(f)$ if $f \in F_i$, $i = 1,2$
5. $M(p) = M_i(p)$ if $p \in P_i$, $i = 1,2$.

Definition 2.31. Let (N_1,M_1), (N_2,M_2), ..., and (N_k,M_k) be k nets satisfying $P_i \cap P_j = \emptyset$, $\forall i,j \in \mathbb{N}_k$, $i \neq j$. The synchronous synthesis of the k Petri nets (N_1,M_1), (N_2,M_2), ..., and (N_k,M_k) is defined as $(N,M) = (N_k,M_k) \otimes (\otimes_{i=1}^{k-1}(N_i,M_i))$.

In a Petri net formalism, a supervisor is a Petri net that usually consists of a set of monitors and a set of transitions, which is a subset of the set of transitions in the plant net model. The controlled system is the synchronous synthesis of a plant net model and its supervisor via shared transitions.

Definition 2.32. Let (N_p,M_p) with $N_p = (P,T,F,W)$ be a plant model and (N_{sup}, M_{sup}) with $N_{sup} = (P_V,T_V,F_V,W_V)$ its supervisor, where $P \cap P_V = \emptyset$ and $T_V \subseteq T$. The controlled system of the plant model is $(N_p,M_p) \otimes (N_{sup},M_{sup})$.

Example 2.17. The Petri net shown in Fig. 2.14a is a plant model. The control specification is that the number of tokens in place p_2 is not greater than one at any reachable marking. The net depicted in Fig. 2.14b is a supervisor that can implement this control specification, where p_3 is a monitor, $P_V = \{p_3\}$, and $T_V = \{t_1,t_2\}$. Figure 2.14c shows the controlled system that can be obtained by synchronous synthesis of the nets in Fig. 2.14a, b. It is easy to verify that the number of tokens in p_2 can never be greater than one.

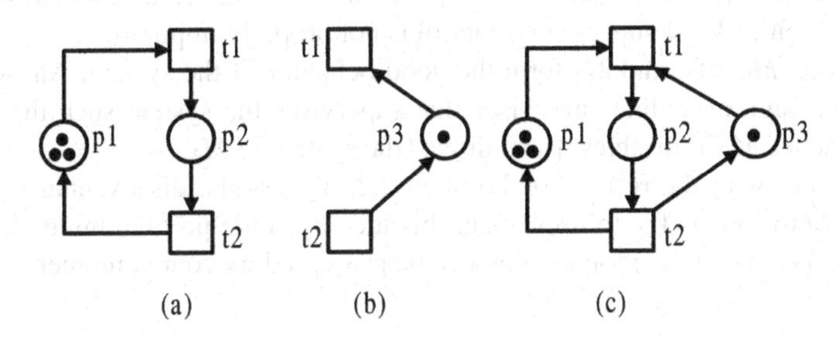

Fig. 2.14 (a) a plant net model, (b) the supervisor, and (c) the controlled system

2.8 Bibliographical Remarks

All the material covered in this chapter can be found in standard books [16, 39, 43] and survey papers [37, 38]. A good paper on siphons is [2], which presents an effective characterization of minimal siphons and traps from the viewpoint of graph theory. The algorithms calculating siphons and traps can be found in [5, 9, 10, 14, 15, 22, 29, 31, 48, 54, 57, 60]. For a general introduction to the subclasses of Petri nets, we refer readers to [37]. Good surveys of Petri nets from a system theory view can be found in [24, 46].

For a more extensive discussion of the original framework of DES supervisory control based on formal languages and automata, we refer readers to the tutorial surveys, papers and books [7, 8, 25, 27, 30, 41, 49].

Problems

2.1. It is known that the siphons are closely related to the deadlock or the existence of dead transitions in a Petri net. Suppose that (N, M_0) is a net without siphons. Is it live? Results can be found in [55].

2.2. INA [47] is a widely used tool that supports the behavioral and structural analysis of Petri nets. Let us define the size of a net (N, M_0) as $||\mathscr{N}|| = |P| + |T| + \sum_{p \in P} M_0(p)$. By using INA, compute the reachability graphs for a number of Petri nets with different sizes 5, 10, 20, ..., and 100, and observe the relationship between the CPU-time and the size of a Petri net.

2.3. Figure 2.15 shows the reduced version of the reachability graph of the net in Fig. 2.1a. It is clear that M_4 is a deadlock marking and M_3 is a marking that definitely leads the system to a deadlock state. These are "bad" states , which the system is not allowed to enter. M_1 is called a *dangerous marking* since, at this marking, the system may enter M_3 if supervisory control is not properly imposed.

Therefore, M_0, M_1, and M_2 form the good behavior of the system. An intuitive idea is to design an online supervisor that supervises the system such that if the system reaches M_1, it disables t_1 and directs the system to M_2.

Combining with the results for Problem 2.2, discuss the disadvantages of this intuitive control idea. Try to implement this idea by some programming language and check the size of the problem that can be processed by your computer.

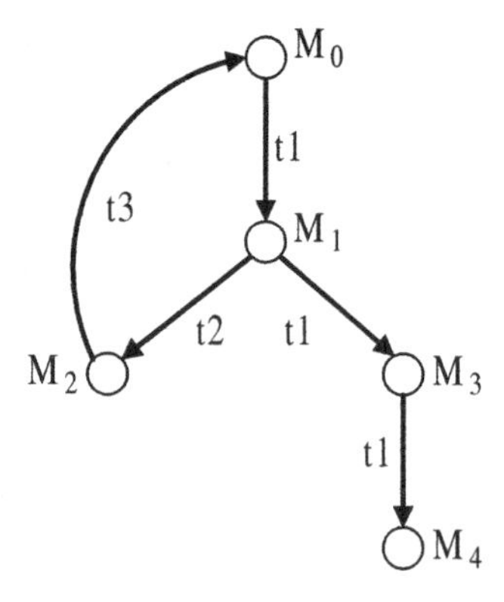

Fig. 2.15 A reachability graph

2.4. Prove Corollary 2.3.

2.5. Find and compare the strict minimal siphons in Fig 2.16a, b. Change the initial markings of places p_1 and p_4 and verify the liveness of the two nets by INA. This verification may find an interesting problem about deadlocks and siphons in a generalized Petri net.

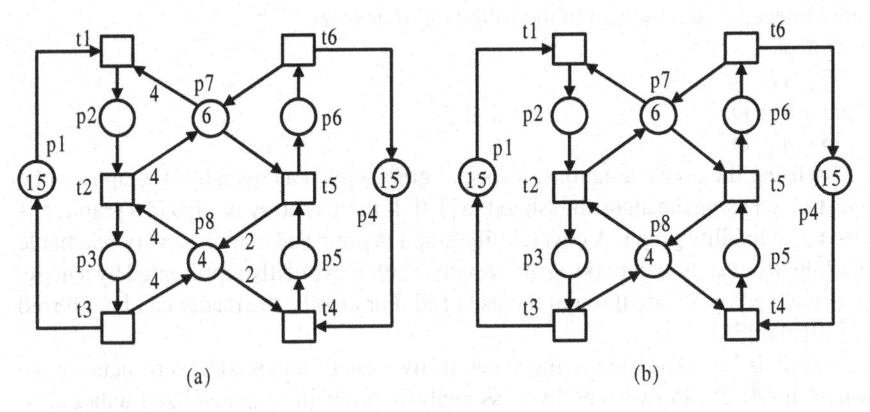

Fig. 2.16 Two Petri nets (a) a generalized one and (b) an ordinary one

2.6. The reachability graph of a Petri net (N, M_0) can be constructed using the following algorithm that terminates in a finite number of steps if its reachability set is finite. Starting with M_0, all the enabled transitions can be fired. These firings can lead to new markings that may enable other transitions. Taking each of the new markings as a new root, we can recursively generate all the reachable markings. The following reachability graph generation algorithm can be found in [37, 39].

Algorithm 2.1 Reachability graph

1: The root node is M_0. This node has initially no label.
2: **while** There are nodes with no label **do**
3: Consider a node M with no label.
4: (a) For each transition t enabled at M:
5: Let $M' = M + [N](\cdot, t)$.
6: **if** There does not exist a node M' in the graph **then**
7: Add it.
8: Add an arc t from M to M'.
9: **end if**
10: (b) Label the node M "old".
11: **end while**
12: Remove all labels from nodes.

Implement the algorithm in a programming language and find the reachability graphs for all the Petri nets in Chap. 2 and those used in Problem 2.2. Check the maximal size of a reachability graph that your computer can process in reasonable time. Compare the CPU-times needed by your program and INA.

Let M, $M' \in R(N, M_0)$ be two reachable markings of a Petri net (N, M_0) with $N = (P, T, F, W)$. M is said to cover M' if $M \geq M'$, i.e., $\forall p \in P$, $M(p) \geq M'(p)$.

For an unbounded Petri net, its reachability graph can grow indefinitely. To reduce and keep the size of the graph finite, a special symbol $\overline{\omega}$ is usually introduced, which represents a number of tokens that can be made arbitrarily large. For any finite integer a, $\overline{\omega}$ is subject to the following four rules:

$a \leq \overline{\omega}$,
$\overline{\omega} \leq \overline{\omega}$,
$\overline{\omega} + a = \overline{\omega}$,
$\overline{\omega} - a = \overline{\omega}$.

By using the above notations, a special graph called a coverability graph can be constructed using the algorithm stated in [19]. If there is no symbol $\overline{\omega}$ in a graph, it is also a reachability graph. A coverability graph is finite and contains every reachable marking from an initial marking M_0, which is either explicitly represented by a node, or is covered by a node through the use of $\overline{\omega}$. For details, the reader can be referred to [19] and [37].

Additional work related to the check of liveness of unbounded Petri nets can be found in [20, 28, 45, 56]. The liveness analysis problem of generalized unbounded Petri nets remains open.

References

1. Abdallah, I.B., ElMaraghy, H.A., ElMekkawy, T. (1997) A logic programming approach for finding minimal siphons in S^{3}PR nets applied to manufacturing systems. In *Proc. IEEE Int. Conf. on Systems, Man, and Cybernetics*, pp.1710–1715.
2. Barkaoui, K., Lemaire, B. (1989) An effective characterization of minimal deadlocks and traps in Petri nets based on graph theory. In *Proc. 10th Int. Conf. on Applications and Theory of Petri Nets*, pp.1–21.
3. Barkaoui, K., Minoux, M. (1992) A polynomial time graph algorithm to decide liveness of some basic classes of Petri nets. In *Proc. 13th Int. Conf. on Applications and Theory of Petri Nets, Lecture Notes in Computer Science*, vol.616, pp.62–75.
4. Barkaoui, K., Pradat-Peyre, J.F. (1996) On liveness and controlled siphons in Petri nets. In *Proc. 17th Int. Conf. on Applications and Theory of Petri Nets Lecture Notes in Computer Science*, vol.1091, pp.57–72.
5. Boer, E.R., Murata, T. (1994) Generating basis siphons and traps of Petri nets using the sign incidence matrix. *IEEE Transactions on Circuits and Systems I–Fundamental Theory and Applications*, vol.41, no.4, pp.266–271.
6. Bogdan, S., Lewis, F.L., Kovacic, Z., Mireles, J. (2006) *Manufacturing Systems Control Design*. London: Springer.
7. Cassandras, C.G., Lafortune, S. (1999) *Introduction to Discrete Event Systems*. Boston, MA: Kluwer.
8. Cassandras, C.G., Lafortune, S. (2008) *Introduction to Discrete Event Systems*. Springer.
9. Chao, D.Y. (2006) Computation of elementary siphons in Petri nets for deadlock control. *Computer Journal*, vol.49, no.4, pp.470–479.
10. Chao, D.Y. (2006) Searching strict minimal siphons for SNC-based resource allocation systems, *Journal of Information Science and Engineering*, vol.23, no.3, pp.853–867.

11. Cheung, K.S. (2004) New characterization for live and reversible augmented Petri nets. *Information Processing Letters*, vol.92, no.5, pp.239–243.
12. Chu, F., Xie, X.L. (1997) Deadlock analysis of Petri nets using siphons and mathematical programming. *IEEE Transactions on Robotics and Automation*, vol.13, no.6, pp.793–804.
13. Colom, J.M., Campos, J., Silva, M. (1990) On liveness analysis through linear algebraic techniques. In *Proc. of Annual General Meeting of ESPRIT Basic Research Action 3148 Design Methods Based on Nets DEMON.*
14. Cordone, R., Ferrarini, L., Piroddi, L. (2003) Some results on the computation of minimal siphons in Petri nets. In *Proc. 42nd IEEE Conf. on Decision and Control*, pp.3754–3759.
15. Cordone, R., Ferrarini, L., Piroddi, L. (2005) Enumeration algorithms for minimal siphons in Petri nets based on place constraints. *IEEE Transactions on Systems, Man and Cybernetics, Part A*, vol.35, no.6, pp.844–854.
16. Desel, J., Esparza, J. (1995) *Free Choice Petri Nets*. London: Cambridge University Press.
17. Desel, J., Reisig, W. (1998) Place/transition Petri nets. In *Lectures on Petri Nets I: Basic Models, Lecture Notes in Computer Science*, vol.1491, W. Reisig and G. Rozenberg (Eds.), pp.122–174.
18. Desel, J. (1998) Basic linear algebraic techniques for place/transition nets. In *Lectures on Petri Nets I: Basic Models, Lecture Notes in Computer Science*, vol.1491, W. Reisig and G. Rozenberg (Eds.), pp.257–308.
19. Desrocher, A.A., AI-Jaar, R.Y. (1995) *Applications of Petri Nets in Manufacturing Systems: Modeling, Control, and Performance Analysis*, Piscataway, NJ: IEEE Press.
20. Ding, Z.J., Jiang, C.J., Zhou, M.C. (2008) Deadlock checking for one-place unbounded Petri nets based on modified reachability trees. *IEEE Transactions on Systems, Man, and Cybernetics, Part B*, vol.38, no.3, pp.881–882.
21. Esparza, J., Silva, M. (1992) A polynomial-time algorithm to decide liveness of bounded free choice nets. *Theoretical Computer Sciences*, vol.102, no.1, pp.185–205.
22. Ezpeleta, J., Couvreur, J.M., Silva, M. (1993) A new technique for finding a generating family of siphons, traps, and st-components: Application to colored Petri nets. In *Advances in Petri Nets, Lecture Notes in Computer Science*, vol.674, G. Rozenberg (Ed.), pp.126–147.
23. Ezpeleta, J, Colom, J.M., Martinez, J. (1995) A Petri net based deadlock prevention policy for flexible manufacturing systems. *IEEE Transactions on Robotics and Automation*, vol.11, no.2, pp.173–184.
24. Giua, A., Seatzu, C. (2007) A systems theory view of Petri nets. In *Advances in Control Theory and Applications, Lecture Notes in Control and Information Science*, vol.353, C. Bonivento et al. (Eds.), pp.99–127.
25. Holloway, L.E., Krogh, B.H., Giua, A. (1997) A survey of Petri net methods for controlled discrete event systems. *Discrete Event Dynamic Systems: Theory and Applications*, vol.7, no.2, pp.151–190.
26. Hopcroft, J.E., Motwani, R., Ullman, J.D. (2000) *Introduction to Automata Theory, Languages, and Computation*, 2nd ed., New York: Addison-Wesley.
27. Hruz, B., Zhou, M.C (2007) *Modeling and Control of Discrete-Event Dynamic Systems: With Petri Nets and Other Tools*. London: Springer.
28. Jeng, M.D, Peng, M.Y. (1999) Augmented reachability trees for 1-place-unbounded generalized Petri nets. *IEEE Transactions on Systems, Man, and Cybernetics, Part A*, vol.29, no.2, pp.173–183.
29. Jeng, M.D., Peng, M.Y., Huang, Y.S. (1999) An algorithm for calculating minimal siphons and traps in Petri nets. *International Journal of Intelligent Control and Systems*, vol.3, no.3, pp.263–275.
30. Kumar, R. Garg, V. (1995) *Modeling and Control of Logical Discrete Event Systems*. Boston, MA: Kluwer.
31. Lautenbach, K. (1987) Linear algebraic calculation of deadlocks and traps. In *Concurrency and Nets*, K. Voss, H. J. Genrich and G. Rozenberg (Eds.), pp.315–336.
32. Lautenbach, K., Ridder, H. (1993) Liveness in bounded Petri nets which are covered by T-invariants. In *Proc. 13th Int. Conf. on Applications and Theory of Petri Nets, Lecture Notes in Computer Science*, vol.815, R. Valette (Ed.), pp.358–375.

33. Li, Z.W., Zhou, M.C. (2004) Elementary siphons of Petri nets and their application to deadlock prevention in flexible manufacturing systems. *IEEE Transactions on Systems, Man, and Cybernetics, Part A*, vol.34, no.1, pp.38–51.
34. Li, Z.W., Wei, N. (2007) Deadlock control of flexible manufacturing systems via invariant-controlled elementary siphons of Petri nets. *International Journal of Advanced Manufacturing Technology*, vol.33, no.1–2, pp.24–35.
35. Li, Z.W., Zhou, M.C. (2008) On siphon computation for deadlock control in a class of Petri nets. *IEEE Transactions on Systems, Man, and Cybernetics, A.*, vol.38, no.3, pp.667–679.
36. Minoux, M., Barkaoui, K. (1990) Deadlocks and traps in Petri nets as horn-satisfiability solutions and some related polynomially solvable problems. *Discrete Mathematics*, vol.29, no.2–3, pp.195–210.
37. Murata, T. (1989) Petri nets: Properties, analysis, and applications. *Proceedings of the IEEE*, vol.77, no.4, pp.541–580.
38. Peterson, J.L. (1977) Petri nets. *Computing Surveys*, vol.9, no.3, pp.223–252.
39. Peterson, J.L. (1981) *Petri Net Theory and the Modeling of Systems*. Englewood Cliffs, NJ: Prentice-Hall.
40. Piroddi, L., Cordone, R., Fumagalli, I. (2008) Selective siphon control for deadlock prevention in Petri nets, *IEEE Transactions on Systems, Man, and Cybernetics, Part A*, vol. 38, no. 6, pp.1337–1348.
41. Ramadge, P., Wonham, W.M. (1989) The control of discrete event systems. *Proceedings of the IEEE*, vol.77, no.1, pp.81–89.
42. Recalde, L., Teruel, E., Silva, M., (1998) On linear algebraic techniques for liveness analysis of P/T systems. *Journal of Circuits, Systems, and Computers*, vol.8, no.1, pp.223–265.
43. Reisig, W. (1985) *Petri Nets: An Introduction*. New York: Springer.
44. Reutenauer, C. (1990) *The Mathematics of Petri Nets*. Translated by I. Varig, Englewood Cliffs, NJ: Prentice-Hall.
45. Ru, Y., Wu, W.M., Hadjicostis, C.N. (2006) Comments on "A modified reachability tree approach to analysis of unbounded Petri nets". *IEEE Transactions on Systems, Man, and Cybernetics, Part B*, vol.36, no.5, p.1210.
46. Silva, M., Teruel, E. (1996) A systems theory perspective of discrete event dynamic systems: The Petri net paradigm. In P. Borne, J. C. Gentina, E. Craye, and S. El Khattabi, (Eds.), Symposium on *Discrete Events and Manufacturing Systems*, IMACS Multiconference, Lille, France, pp.1–12.
47. Starke, P. H. (2003) *INA: Integrated Net Analyzer*. http://www2.informatik.hu-berlin.de/~starke/ina.html.
48. Tanimoto, S., Yamauchi, M., Watanabe, T. (1996) Finding minimal siphons in general Petri nets. *IEICE Transactions on Fundamentals*, vol.E79-A, no.11, pp.1817–1824.
49. Thistle, J.G. (1996) Supervisory control of discrete event systems. *Mathematical and Computer and Modeling*, vol.23, no.11–12, pp.25–53.
50. Tricas, F., Garacía-Vallés, F., Colom, J.M., Ezpeleta, J. (1998) A partial approach to the problem of deadlocks in processes with resources. Research Report, GISI-RR-97-05, Departamento de Informática e Ingeniería de Sistemas, Universidad de Zaragoza, Spain.
51. Tricas, F., García-Vallés, F., Colom, J.M., Ezpeleta, J. (1998) A structural approach to the problem of deadlock prevention in processes with shared resources. In *Proc. 4th Workshop on Discrete Event Systems*, pp.273–278.
52. Tricas, F., Ezpeleta, J. (1999) A Petri net solution to the problem of deadlocks in systems of processes with resources. In *Proc. IEEE Int. Conf. on Emerging Technologies and Factory Automation*, pp.1047–1056.
53. Tricas, F., Ezpeleta, J. (2003) Some results on siphon computation for deadlock prevention in resource allocation systems modeled with Petri nets. In *Proc. IEEE Int. Conf. on Emerging Technologies and Factory Automation*, pp.322–329.
54. Tricas, F., Ezpeleta, J. (2006) Computing minimal siphons in Petri net models of resource allocation systems: A parallel solution. *IEEE Transactions on Systems, Man, and Cybernetics, Part A*, vol.36, no.3, pp.532–539.

55. Tsuji, K., Murata, T. (1993) On reachability conditions for unrestricted Petri nets. In *Proc. IEEE Int. Symp. on Circuits and Systems*, pp.2713–2716.
56. Wang, F.Y., Gao, Y.Q., Zhou, M.C. (2004) A modified reachability tree approach to analysis of unbounded Petri nets. *IEEE Transactions on Systems, Man, and Cybernetics, Part B*, vol.34, no.1, pp.303–308.
57. Watanabe, T., Yamauchi, M., Tanimoto, S. (1998) Extracting siphons containing specified set of places in Petri nets. In *Proc. IEEE Int. Conf. on Systems, Man, and Cybernetics*, pp.142–147.
58. Xing, K.Y., Hu, B.S., Chen, H.X. (1996) Deadlock avoidance policy for Petri-net modelling of flexible manufacturing systems with shared resources. *IEEE Transactions on Automatic Control*, vol.41, no.2, pp.289–295.
59. Xing, K.Y., Hu, B.S. (2005) Optimal liveness Petri net controllers with minimal structures for automated manufacturing systems. In *Proc. IEEE Int. Conf. on Systems, Man and Cybernetics*, pp.282–287.
60. Yamauchi, M., Watanabe, T. (1999) Algorithms for extracting minimal siphons containing specified places in a general Petri net. *IEICE Transactions on Fundamentals*, vol.E82-A, no.11, pp.2566–2575.
61. Zhou, M.C., DiCesare, F. (1991) Parallel and sequential exclusions for Petri net modeling for manufacturing systems. *IEEE Transactions on Robotics and Automation*, vol.7, no.4, pp.515–527.
62. Zhou, M.C., DiCesare, F. (1993) *Petri Net Synthesis for Discrete Event Control of Manufacturing Systems*. Boston, MA: Kluwer.

Chapter 3
Elementary Siphons of Petri Nets

Abstract This chapter first presents the concepts of poor, rich, and equivalent siphons and the related invariant-control theory. It then discusses elementary and dependent siphons in a Petri net, which play a key role in the development of structurally simple liveness-enforcing (Petri net) supervisors. Dependent siphons are further divided into strongly and weakly dependent ones. Next, a more general result on the control of a siphon is given. Later, it is shown that the controllability of a dependent siphon can be ensured by supervising its elementary siphons via properly setting their control depth variables. For well-initially-marked Petri nets that cover most of the manufacturing-oriented Petri net models in the literature, it is shown that strongly and weakly dependent siphons have identical controllability conditions. Finally, an elementary siphon identification algorithm for a deadlock control purpose is presented.

3.1 Introduction

This chapter presents the concept of elementary and dependent siphons in a Petri net as well as their controllability. Elementary siphons are a special class of structural objects. Their development is motivated by the observation that the change of the number of tokens in a siphon linearly depends on the token variation of a set of other siphons. It is shown that a dependent siphon can be implicitly controlled by properly supervising the number of tokens staying in its elementary siphons. Elementary siphons play an important role in the development of deadlock prevention approaches that lead to structurally simple liveness-enforcing monitor-based supervisors.

3.2 Equivalent Siphons

In this section, we present the concept of equivalent siphons.

Definition 3.1. Let $S \subseteq P$ be a subset of places of Petri net $N = (P, T, F, W)$. P-vector λ_S is called the characteristic P-vector of S iff $\forall p \in S$, $\lambda_S(p) = 1$; otherwise $\lambda_S(p) = 0$.

Definition 3.2. $\eta_S = [N]^T \lambda_S$ is called the characteristic T-vector of S, where $[N]^T$ is the transpose of incidence matrix $[N]$.

By the above definitions, $\eta_S^T = \sum_{p \in S} [N](p, \cdot)$ is trivially true. The physical implication of the T-vector of a subset of places is clear: $\eta_S(t) > 0$ means that $\eta_S(t)$ tokens are put into S after t fires; $\eta_S(t) = 0$ indicates that the number of tokens in S does not change after t fires; and $\eta_S(t) < 0$ implies that $|\eta_S(t)|$ tokens are removed from S after t fires.

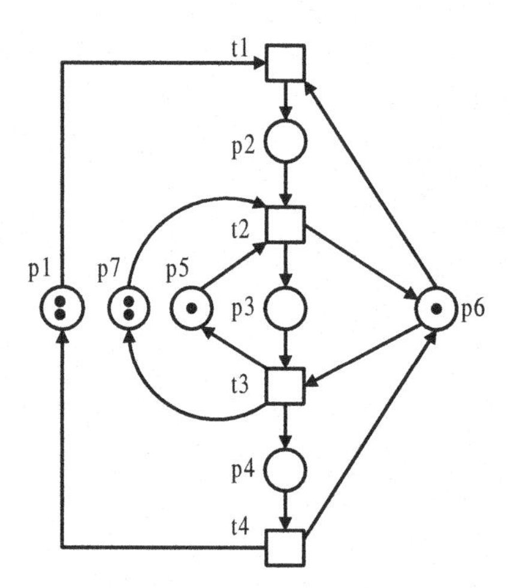

Fig. 3.1 A Petri net (N, M_0)

Example 3.1. Let $S_1 = \{p_1 - p_4\}$, $S_2 = \{p_2, p_4, p_6\}$, $S_3 = \{p_4 - p_6\}$, and $S_4 = \{p_4, p_6, p_7\}$ be subsets of places of the Petri net shown in Fig. 3.1. We have $\lambda_{S_1} = p_1 + p_2 + p_3 + p_4$ (i.e., the space-saving version of vector $(1\ 1\ 1\ 1\ 0\ 0\ 0)^T$), $\lambda_{S_2} = p_2 + p_4 + p_6$, $\lambda_{S_3} = p_4 + p_5 + p_6$, and $\lambda_{S_4} = p_4 + p_6 + p_7$. We have $\eta_{S_1} = \mathbf{0}^T$ and $\eta_{S_2} = \mathbf{0}^T$. This is not surprising since both λ_{S_1} and λ_{S_2} are P-semiflows. Firing a transition does not change their token count. In addition, it is easy to verify that $\eta_{S_3} = \eta_{S_4} = -t_1 + t_3$. Firing t_1 removes a token from S_3 and S_4, respectively, while firing t_3 adds a token to S_3 and S_4, respectively.

More specifically, take $S_3 = \{p_4, p_5, p_6\}$ as an example. Firing t_1 removes a token from p_6 and does not change the number of tokens in p_4 and p_5. Therefore, S_3 loses a token if t_1 fires. Firing t_2 removes a token from p_5 and puts a token into p_6, keeping the constant number of tokens in S_3. Firing t_3 removes a token from p_6, and puts a token into p_4 and p_5, respectively, and thereby S_3 gains a token. Firing t_4 removes a token from p_4 and puts a token into p_6, thus keeping the token count in S_3 constant. These facts are well reflected by $\eta_{S_3} = -t_1 + t_3$.

Definition 3.3. Let S_1 and S_2 be two siphons in a net. S_1 and S_2 are said to be equivalent, denoted by $S_1 \cong S_2$, if $\eta_{S_1} = \eta_{S_2}$.

Definition 3.4. Let Π be a set of siphons in a net. $\langle S \rangle \subseteq \Pi$ is called a set of equivalent siphons if (1) $S \in \langle S \rangle$, (2) $\forall S', S'' \in \langle S \rangle$, $\eta_{S'} = \eta_{S''}$, and (3) $\forall S' \in \langle S \rangle$, $\forall S'' \in \Pi \backslash \langle S \rangle$, $\eta_{S'} \neq \eta_{S''}$.

Proposition 3.1. *Let $R = \{(S', S'') \mid S', S'' \in \langle S \rangle\}$. Thus R is an equivalent relationship on $\langle S \rangle$.*

Proof. We have to prove that R is reflexive, symmetric, and transitive. It is easy to see that $\forall S^{\nabla}, S'$, and $S'' \in \langle S \rangle$, one can get (a) $\eta_{S^{\nabla}} = \eta_{S^{\nabla}}$, (b) $\eta_{S^{\nabla}} = \eta_{S'} \Rightarrow \eta_{S'} = \eta_{S^{\nabla}}$, and (c) $\eta_{S^{\nabla}} = \eta_{S'} \wedge \eta_{S'} = \eta_{S''} \Rightarrow \eta_{S^{\nabla}} = \eta_{S''}$. Hence, this proposition holds. $\square$

It is easy to see that $\langle S \rangle$ is an equivalent class of Π.

Corollary 3.1. *Let S and $S' \in \Pi$ be two siphons of a net. We have (1)$\langle S \rangle \neq \emptyset$ and $\langle S \rangle \subseteq \Pi$; (2)$\langle S \rangle = \langle S' \rangle$ if $(S, S') \in R$; (3)$\langle S \rangle \cap \langle S' \rangle = \emptyset$ if $(S, S') \notin R$; (4)$\cup_{S \in \Pi} \langle S \rangle = \Pi$; (5) Let π be the set of equivalent classes of the elements in Π. π is a partition of Π.*

In what follows, $\langle S \rangle$ is used to denote a set of equivalent siphons in net system (N, M_0), which contains the siphon S.

Example 3.2. The net shown in Fig. 3.2 is a Petri net (not strongly connected). $S_1 = \{p_1\}$, $S_2 = \{p_2\}$, $S_3 = \{p_3\}$, $S_4 = \{p_4\}$, and $S_5 = \{p_5\}$ are minimal siphons with $\eta_{S_1} = \eta_{S_2} = \eta_{S_3} = -t_1$ and $\eta_{S_4} = \eta_{S_5} = -t_2 - t_3$. Therefore, S_1, S_2, and S_3 are equivalent. So are S_4 and S_5. These equivalent relationships lead to the fact that $\langle S_1 \rangle = \langle S_2 \rangle = \langle S_3 \rangle = \{S_1, S_2, S_3\}$ and $\langle S_4 \rangle = \langle S_5 \rangle = \{S_4, S_5\}$.

Take another example from Fig. 3.1. Consider siphons $S_3 = \{p_4, p_5, p_6\}$ and $S_4 = \{p_4, p_6, p_7\}$. S_3 and S_4 are equivalent since their characteristic T-vectors are identical.

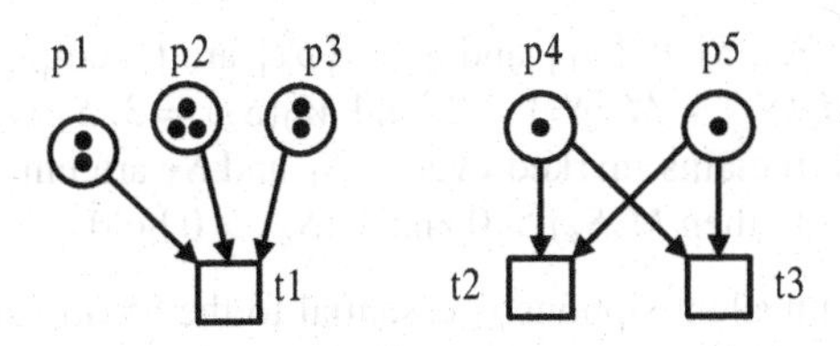

Fig. 3.2 A Petri net (N, M_0) with equivalent siphons

Definition 3.5. Let $\langle S \rangle$ be a set of equivalent siphons in a Petri net (N, M_0). Siphon $S' \in \langle S \rangle$ is said to be token-poor in $\langle S \rangle$ if $\nexists S'' \in \langle S \rangle$ such that $M_0(S'') < M_0(S')$ holds. It is called token-rich otherwise.

A token-poor siphon holds the smallest number of tokens at initial marking M_0 among a set of equivalent siphons. Clearly, among these equivalent siphons, all token-poor siphons have the same token count at M_0.

Theorem 3.1. *$M_0(S') = M_0(S'')$ if siphons S' and S'' are token-poor in $\langle S \rangle$.*

If siphons S_1 and S_2 are both token-rich in $\langle S \rangle$, $M_0(S_1) = M_0(S_2)$ is not necessarily true. Recalling that a siphon is said to be controlled if it is always marked at any reachable marking, we immediately have the following result.

Theorem 3.2. *A token-rich siphon is controlled.*

Proof. If S is token-rich, there must exist a siphon $S' \in \langle S \rangle$ such that $\eta_{S'} = \eta_S$ and $M_0(S) > M_0(S')$ are true. Hence, firing any transition removes the same number of tokens from S and S'. Let $M \in R(N, M_0)$ be any reachable marking. We have two subcases: $M(S') > 0$ and $M(S') = 0$.

If $M(S') > 0$, then $M(S) > M(S') > 0$. Thus S is marked.

If $M(S') = 0$, then $M(S) = M_0(S) - M_0(S') > 0$. It is easy to see, in this case, that at any reachable marking M, no output transitions of S' can be enabled and fire to remove tokens from S and S'. Thus S remains to be marked. □

Theorem 3.3. *Let S' and S'' be two token-poor siphons in $\langle S \rangle$. S' is controlled iff S'' is controlled.*

Proof. Assume that S' and S'' are token-poor siphons in (N, M_0), where $N = (P, T, F, W)$. Therefore, they have the identical characteristic T-vector and the same number of initial tokens. $\forall t \in T$, the number of tokens removed from S' equals that from S'' if t fires. The controllability of S' means that the least number of tokens in S' is greater than 0. Hence, the least number of tokens staying in S'' is also greater than 0. S'' is therefore controlled.

Similarly, we can prove that S' is controlled if S'' is so. □

Theorem 3.4. *If a token-poor siphon in $\langle S \rangle$ is controlled, all siphons in $\langle S \rangle$ are controlled.*

Theorem 3.4 indicates that the controllability of a token-poor siphon in $\langle S \rangle$ is sufficient for that of all others in $\langle S \rangle$.

Example 3.3. In Fig. 3.2, $S_1 = \{p_1\}$ and $S_3 = \{p_3\}$ are token-poor and $S_2 = \{p_2\}$ is token-rich since $M_0(S_1) = M_0(S_3) = 2$, and $M_0(S_2) = 3$. S_2 is controlled due to Theorem 3.2 since it remains marked even if S_1 and S_3 are emptied. Also, $\forall M \in R(N, M_0)$, if $M(S_1) > 0$, then $M(S_3) > 0$ and $M(S_2) > 0$ hold.

The concept of equivalent siphons is essential to the identification of the set of elementary siphons in a net system for deadlock control purposes. The motivation to propose the concept of elementary siphons is to control dependent siphons by explicitly controlling their elementary siphons only.

3.3 Elementary and Dependent Siphons

Theorems 3.3 and 3.4 show the controllability relations among equivalent siphons. The concept of elementary and dependent siphons [20, 21] reveals the relations among siphons belonging to different sets of equivalent siphons.

Definition 3.6. Let $N = (P,T,F,W)$ be a net with $|P| = m, |T| = n$ and $\Pi = \{S_1, S_2, \ldots, S_k\}$ be a set of siphons of N ($m,n,k \in \mathbb{N}^+$). Let $\lambda_{S_i}(\eta_{S_i})$ be the characteristic $P(T)$-vector of siphon $S_i, i \in \mathbb{N}_k$. $[\lambda]_{k\times m} = [\lambda_{S_1}|\lambda_{S_2}|\cdots|\lambda_{S_k}]^T$ and $[\eta]_{k\times n} = [\lambda]_{k\times m} \times [N]_{m\times n} = [\eta_{S_1}|\eta_{S_2}|\cdots|\eta_{S_k}]^T$ are called the characteristic P- and T-vector matrices of the siphons in N, respectively.

Definition 3.7. Let η_{S_α}, η_{S_β}, ..., and η_{S_γ} ($\{\alpha,\beta,\ldots,\gamma\} \subseteq \mathbb{N}_k$) be a linearly independent maximal set of matrix $[\eta]$. Then $\Pi_E = \{S_\alpha, S_\beta, \ldots, S_\gamma\}$ is called a set of elementary siphons in N.

Definition 3.8. $S \notin \Pi_E$ is called a strongly dependent siphon if $\eta_S = \sum_{S_i\in\Pi_E} a_i\eta_{S_i}$, where $a_i \geq 0$.

Definition 3.9. $S \notin \Pi_E$ is called a weakly dependent siphon if $\exists A, B \subset \Pi_E$, such that $A \neq \emptyset$, $B \neq \emptyset$, $A \cap B = \emptyset$, and $\eta_S = \sum_{S_i\in A} a_i\eta_{S_i} - \sum_{S_i\in B} a_i\eta_{S_i}$, where $a_i > 0$.

Let $\Gamma^+(S) = \sum_{S_i\in A} a_i\eta_{S_i}$ and $\Gamma^-(S) = \sum_{S_i\in B} a_i\eta_{S_i}$ for a weakly dependent siphon S. We have $\eta_S = \Gamma^+(S) - \Gamma^-(S)$. If S is strongly dependent, we define $\Gamma^-(S) = 0$.

Definition 3.10. Dependent siphons S_1 and S_2 are said to be quasi-equivalent iff $\Gamma^+(S_1) = \Gamma^+(S_2)$.

Lemma 3.1. *The number of elements in any set of elementary siphons in net N equals the rank of* $[\eta]$.

Let Π_E denote a set of the elementary siphons in a Petri net. Since the rank of $[\eta]$ is at most the smaller of $|P|$ and $|T|$, Lemma 3.1 leads to the following important conclusion.

Theorem 3.5. $|\Pi_E| \leq min\{|P|, |T|\}$.

This result indicates that the number of elementary siphons in a Petri net is bounded by the smaller of place count and transition count.

Let S be a (strongly or weakly) dependent siphon. In sequel, if η_S can be linearly represented by elementary siphons' characteristic T-vectors η_{S_1}, η_{S_2}, ..., and η_{S_n} with non-zero coefficients, we say that S_1, S_2, ..., and S_n are the elementary siphons of S. Let Π be the set of siphons in which we are interested given a net, and Π_D be the set of dependent ones within the scope of Π. Obviously, we have $\Pi = \Pi_E \cup \Pi_D$.

Example 3.4. The Petri net shown in Fig. 3.3(a) has 10 minimal siphons: $S_1 = \{p_5, p_9, p_{12}, p_{13}\}$, $S_2 = \{p_4, p_6, p_{13}, p_{14}\}$, $S_3 = \{p_6, p_9, p_{12}–p_{14}\}$, $S_4 = \{p_2, p_{15}\}$, $S_5 = \{p_7, p_{11}\}$, $S_6 = \{p_1–p_3, p_5–p_7\}$, $S_7 = \{p_4, p_8\text{-}p_{10}\}$, $S_8 = \{p_3, p_9, p_{12}\}$,

$S_9 = \{p_4, p_5, p_{13}\}$, and $S_{10} = \{p_6, p_8, p_{14}\}$. Each S_4–S_{10} is both a siphon and trap. Such a siphon cannot be emptied once it is initially marked. Now we consider the strict minimal siphons S_1, S_2, and S_3. We have

$\lambda_{S_1} = p_5 + p_9 + p_{12} + p_{13}$,
$\lambda_{S_2} = p_4 + p_6 + p_{13} + p_{14}$,
$\lambda_{S_3} = p_6 + p_9 + p_{12} + p_{13} + p_{14}$,
$\eta_{S_1} = -t_2 + t_3 - t_9 + t_{10}$,
$\eta_{S_2} = -t_3 + t_4 - t_8 + t_9$,
$\eta_{S_3} = -t_2 + t_4 - t_8 + t_{10}$.

Accordingly, $[\lambda]$ and $[\eta]$ are shown as follows:

$$[\lambda] = \begin{pmatrix} 0\,0\,0\,0\,1\,0\,0\,0\,1\,0\,0\,1\,1\,0\,0 \\ 0\,0\,0\,1\,0\,1\,0\,0\,0\,0\,0\,0\,1\,1\,0 \\ 0\,0\,0\,0\,0\,1\,0\,0\,1\,0\,0\,1\,1\,1\,0 \end{pmatrix},$$

$$[\eta] = \begin{pmatrix} 0 & -1 & 1 & 0\,0\,0\,0 & 0 & -1 & 1\,0 \\ 0 & 0 & -1 & 1\,0\,0\,0 & -1 & 1 & 0\,0 \\ 0 & -1 & 0 & 1\,0\,0\,0 & -1 & 0 & 1\,0 \end{pmatrix}.$$

It is easy to verify that $\eta_{S_3} = \eta_{S_1} + \eta_{S_2}$ and the rank of $[\eta]$ is two, i.e., $rank([\eta]) = |\Pi_E| = 2$. This means that there are two elementary siphons. If S_1 and S_2 are selected as elementary siphons, S_3 is a strongly dependent one. If S_1 and S_3 are selected as elementary ones, S_2 becomes weakly dependent. If S_2 and S_3 are selected as elementary ones, S_1 is weakly dependent.

3.4 Controllability of Dependent Siphons in Ordinary Petri Nets

This section mainly focuses on the development of conditions under which a dependent siphon can be always marked if their elementary ones are controlled. The results presented are useful in the design of liveness-enforcing net supervisors for resource allocation systems.

Invariant-based siphon control is a well-established technique in the literature. The concept of general-invariant-controlled siphons is an extension to the known results in [9, 15, 19, 32].

Let (N, M_0) be a net system. S is a siphon and I is a P-invariant of N, where $S \cap ||I||^+ \neq \emptyset$ and $S \cap ||I||^- = \emptyset$. Let $S \cap ||I||^+ = A$, $||I||^+ \setminus S = B$, and $||I||^- = C$. Clearly, we have $||I|| = ||I||^+ \cup ||I||^- = A \cup B \cup C$.

Definition 3.11. S is called general-invariant-controlled if

$$I^T M_0 - max\{\sum_{p \in B} I(p)M(p)\} + min\{\sum_{p \in C} |I(p)|M(p)\} > 0,$$

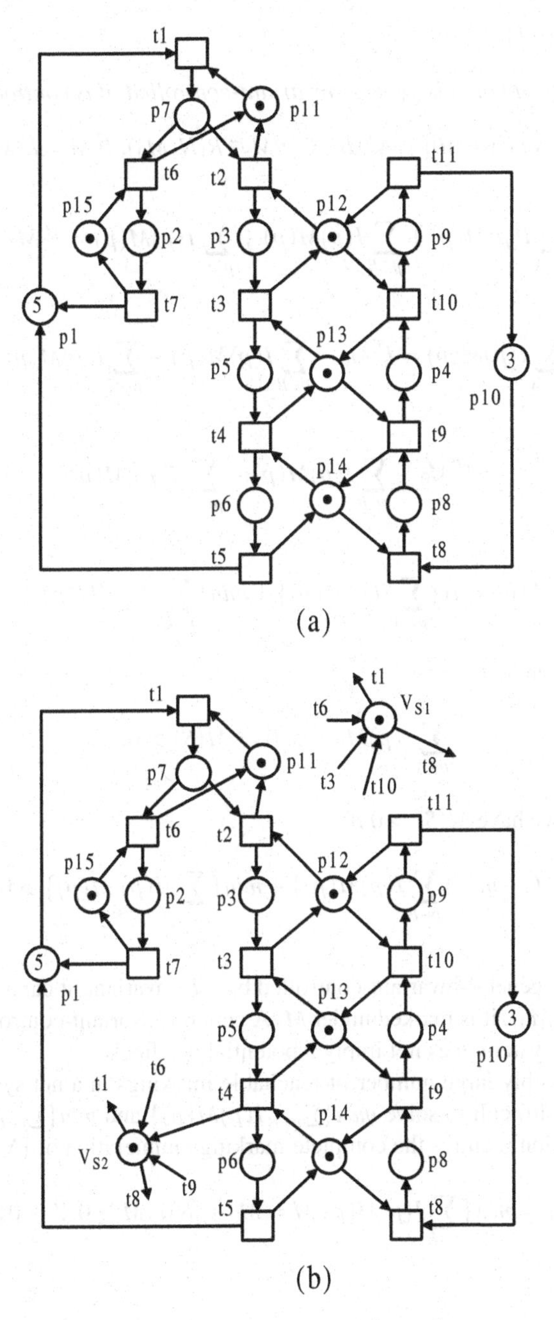

Fig. 3.3 (a) A Petri net (N_1, M_1) and (b) its augmented net (N_2, M_2)

where $M \in R(N, M_0)$.

Theorem 3.6. *If siphon S is general-invariant-controlled, it is controlled.*

Proof. Clearly, we have $||I|| = A \cup B \cup C$. $\forall M \in R(N, M_0)$, $I^T M = I^T M_0$. As a result, we have

$$\sum_{p \in A} I(p)M(p) + \sum_{p \in B} I(p)M(p) + \sum_{p \in C} I(p)M(p) = I^T M_0.$$

$$\sum_{p \in A} I(p)M(p) = I^T M_0 - \sum_{p \in B} I(p)M(p) - \sum_{p \in C} I(p)M(p)$$

$$= I^T M_0 - \sum_{p \in B} I(p)M(p) + \sum_{p \in C} |I(p)|M(p)$$

$$\geq I^T M_0 - max\{\sum_{p \in B} I(p)M(p)\} + min\{\sum_{p \in C} |I(p)|M(p)\} > 0.$$

It is easy to see that

$$\sum_{p \in A} I(p)M(p) > 0 \Rightarrow M(S) > 0.$$

As a result, we have $M(S) > 0$ if

$$I^T M_0 - max\{\sum_{p \in B} I(p)M(p)\} + min\{\sum_{p \in C} |I(p)|M(p)\} > 0.$$

□

If a siphon is general-invariant-controlled by a *P*-invariant, it cannot be emptied, i.e., $\forall M \in R(N, M_0)$, it is marked under M. A general-invariant-controlled siphon in an ordinary net system does not imply a potential deadlock.

Due to a possibly large number of reachable markings in a net system (N, M_0), in general, it is difficult to solve $max\{\sum_{p \in B} I(p)M(p)\}$ and $min\{\sum_{p \in C} |I(p)|M(p)\}$ since their solution requires the complete marking enumeration of (N, M_0). Define

$$m_B = max\{\sum_{p \in B} I(p)M(p) | M = M_0 + [N]Y, M \geq 0, Y \geq 0\}$$

and

$$m_C = min\{\sum_{p \in C} |I(p)|M(p) | M = M_0 + [N]Y, M \geq 0, Y \geq 0\},$$

where $[N]$ is the incidence matrix and M and Y are vectors of real numbers. From the basic theory of Petri nets, any reachable marking fulfils the state equation but the converse is not true. Hence, we have

$$m_B \geq max\{\sum_{p \in B} I(p)M(p)|M \in R(N, M_0)\},$$

and

$$m_C \leq min\{\sum_{p \in C} |I(p)|M(p)|M \in R(N, M_0)\}.$$

Corollary 3.2. *S is controlled if*

$$I^T M_0 - m_B + m_C > 0.$$

Proof. It is known that

$$m_B \geq max\{\sum_{p \in B} I(p)M(p)\}$$

and

$$m_C \leq min\{\sum_{p \in C} |I(p)|M(p)\}.$$

Hence, $I^T M_0 - m_B + m_C > 0$ implies the truth of

$$I^T M_0 - max\{\sum_{p \in B} I(p)M(p)\} + min\{\sum_{p \in C} |I(p)|M(p)\} > 0.$$

□

The problem to find m_B and m_C is an LPP and can be solved in polynomial time [26, 30]. As shown in [9], imposing integrity constraints to M and Y as it is usually done in the literature significantly increases the computational complexity.

If $B = ||I||^+ \setminus S = \emptyset$, then $||I||^+ \subseteq S$. This immediately leads to the following result.

Corollary 3.3. *If $||I||^+ \setminus S = \emptyset$, S is controlled if either*

$$I^T M_0 + min\{\sum_{p \in C} |I(p)|M(p)\} > 0$$

or

$$I^T M_0 + m_C > 0.$$

Corollary 3.4. *If* $||I||^+ \setminus S = \emptyset$, *S is controlled if* $I^T M_0 > 0$.

Proof. Since $min\{\sum_{p \in C} |I(p)|M(p)\} \geq 0$, $I^T M_0 > 0$ leads to the truth of

$$I^T M_0 + min\{\sum_{p \in C} |I(p)|M(p)\} > 0.$$

By Corollary 3.3, S is controlled. □

The concept of general-invariant-controlled siphons is useful to develop controllability conditions for dependent siphons since it is more general than invariant-controlled siphons defined in Definition 2.19.

Let

$$M_{min}(S) = min\{M(S)|M \in R(N, M_0)\} \tag{3.1}$$

and

$$M_{max}(S) = max\{M(S)|M \in R(N, M_0)\}. \tag{3.2}$$

The linear dependency of the T-vector of a dependent siphon on those T-vectors of its elementary siphons reveals the relationship between its and their token changes. That is to say, the number of tokens in a dependent siphon can be controlled by supervising the token flow in its elementary siphons.

Without loss of generality, we assume that in a Petri net (N, M_0), a strongly dependent S satisfies $\eta_S = \sum_{i=1}^{n} a_i \eta_{S_i}$, $a_i > 0$, $i \in \mathbb{N}_n$, and a weakly dependent siphon S satisfies $\eta_S = \sum_{i=1}^{n} a_i \eta_{S_i} - \sum_{j=n+1}^{m} a_j \eta_{S_j}$, $a_i > 0$, $i \in \mathbb{N}_m$. Concerning the controllability of a dependent siphon, we have the following results.

Theorem 3.7. *A weakly dependent siphon S is controlled if*

$$M_0(S) > \sum_{i=1}^{n} a_i(M_0(S_i) - M_{min}(S_i)) - \sum_{j=n+1}^{m} a_j(M_0(S_j) - M_{max}(S_j)). \tag{3.3}$$

Proof. Since

$$\eta_S = \sum_{i=1}^{n} a_i \eta_{S_i} - \sum_{j=n+1}^{m} a_j \eta_{S_j},$$

we have

$$\lambda_S^T[N] = \sum_{i=1}^{n} a_i \lambda_{S_i}^T[N] - \sum_{j=n+1}^{m} a_j \lambda_{S_j}^T[N],$$

i.e.,

$$(\lambda_S - \sum_{i=1}^{n} a_i\lambda_{S_i} + \sum_{j=n+1}^{m} a_j\lambda_{S_j})^T[N] = \mathbf{0}^T.$$

Let $I = \lambda_S - \sum_{i=1}^{n} a_i\lambda_{S_i} + \sum_{j=n+1}^{m} a_j\lambda_{S_j}$. There are hence two subcases: (a) I is a P-invariant and (b) $I = \mathbf{0}$.

First we consider that

$$I = \lambda_S - \sum_{i=1}^{n} a_i\lambda_{S_i} + \sum_{j=n+1}^{m} a_j\lambda_{S_j}$$

is a P-invariant of N. $\forall M \in R(N,M_0)$, we have $I^T M = I^T M_0$. Therefore, the following equation holds by noting $M(S') = \lambda_{S'}^T M$, $\forall S' \in \Pi$, $M \in R(N,M_0)$.

$$M(S) - \sum_{i=1}^{n} a_iM(S_i) + \sum_{j=n+1}^{m} a_jM(S_j) = M_0(S) - \sum_{i=1}^{n} a_iM_0(S_i) + \sum_{j=n+1}^{m} a_jM_0(S_j). \tag{3.4}$$

From (3.4), we have:

$$M(S) = M_0(S) - \sum_{i=1}^{n} a_iM_0(S_i) + \sum_{i=1}^{n} a_iM(S_i) + \sum_{j=n+1}^{m} a_jM_0(S_j) - \sum_{j=n+1}^{m} a_jM(S_j). \tag{3.5}$$

Obviously, S is controlled, i.e., $M(S) > 0$ if

$$M_0(S) > \sum_{i=1}^{n} a_iM_0(S_i) - \sum_{i=1}^{n} a_iM(S_i) - \sum_{j=n+1}^{m} a_jM_0(S_j) + \sum_{j=n+1}^{m} a_jM(S_j). \tag{3.6}$$

Noticing $\forall M \in R(N,M_0)$, $M_{min}(S) \leq M(S)$ and $M_{max}(S) \geq M(S)$, (3.6) is true if the following inequality holds.

$$M_0(S) > \sum_{i=1}^{n} a_i(M_0(S_i) - M_{min}(S_i)) - \sum_{j=n+1}^{m} a_j(M_0(S_j) - M_{max}(S_j)). \tag{3.7}$$

As a result, S is controlled if

$$M_0(S) > \sum_{i=1}^{n} a_i(M_0(S_i) - M_{min}(S_i)) - \sum_{j=n+1}^{m} a_j(M_0(S_j) - M_{max}(S_j)).$$

Second, we consider the case of $I = \mathbf{0}$. $\forall M \in R(N,M_0)$, the following equation is true.

$$M(S) - \sum_{i=1}^{n} a_iM(S_i) + \sum_{j=n+1}^{m} a_jM(S_j) = 0.$$

Thus, one concludes that $M(S) > 0$ if

$$\sum_{i=1}^{n} a_i M(S_i) - \sum_{j=n+1}^{m} a_j M(S_j) > 0.$$

Since

$$M_0(S) - \sum_{i=1}^{n} a_i M_0(S_i) + \sum_{j=n+1}^{m} a_j M_0(S_j) = 0,$$

$M(S) > 0$ if

$$M_0(S) - \sum_{i=1}^{n} a_i M_0(S_i) + \sum_{j=n+1}^{m} a_j M_0(S_j) + \sum_{i=1}^{n} a_i M(S_i) - \sum_{j=n+1}^{m} a_j M(S_j) > 0.$$

i.e.,

$$M_0(S) > \sum_{i=1}^{n} a_i (M_0(S_i) - M_{min}(S_i)) - \sum_{j=n+1}^{m} a_j (M_0(S_j) - M_{max}(S_j)).$$

Subcases (a) and (b) lead to the truth of the theorem.

□

Corollary 3.5. *A strongly dependent siphon S is controlled if*

$$M_0(S) > \sum_{i=1}^{n} a_i (M_0(S_i) - M_{min}(S_i)).$$

Definition 3.12. (N, M_0) is said to be well-initially-marked iff $\forall S \in \Pi$, $M_{max}(S) = M_0(S)$.

This definition indicates that a siphon in a well-initially-marked Petri net has the maximal number of tokens at the initial marking. For example, the net in Fig. 3.3(a) is well-initially-marked.

Corollary 3.6. *Let (N, M_0) be a well-initially-marked net system. A (strongly or weakly) dependent siphon S is controlled if*

$$M_0(S) > \sum_{i=1}^{n} a_i (M_0(S_i) - M_{min}(S_i)).$$

Proof. It follows immediately due to Corollary 3.5 if S is a strongly dependent siphon.

Let S be a weakly dependent siphon. From Theorem 3.7, it is controlled if

$$M_0(S) > \sum_{i=1}^{n} a_i(M_0(S_i) - M_{min}(S_i)) - \sum_{j=n+1}^{m} a_j(M_0(S_j) - M_{max}(S_j)).$$

Since $\forall S \in \Pi$, $M_{max}(S) = M_0(S)$, we have

$$\sum_{j=n+1}^{m} a_j(M_0(S_j) - M_{max}(S_j)) = 0.$$

This leads to the truth of the corollary.

□

The condition ($\forall S \in \Pi$, $M_{max}(S) = M_0(S)$) in a well-initially-marked net is reasonable and meaningful in practice. As far as the authors know, this is true for application-oriented Petri net subclasses in the literature, which can model flexible manufacturing systems, e.g., PPN [2], augmented marked graphs [9], S^3PR [11], L-S^3PR [12], S^4R [1], S^4PR [28], ES^3PR [14], WS^3PSR [27], S^*PR [13], PNR [17], RCN-merged nets [16], ERCN-merged nets [31], ERCN*-merged nets [18], S^2LSPR [24], S^3PGR^2 [25], G-task [4], and G-system [33]. However, the verification of the well-initially-markedness seems a formidable job given an arbitrary Petri net.

By Theorem 3.7, in order to verify the controllability of a dependent siphon, we need to compute $M_{min}(S)$ and $M_{max}(S)$, $S \in \Pi_E$. $M_{min}(S)$ and $M_{max}(S)$ can be obtained by solving problems (3.1) and (3.2), respectively. Due to a large number of reachable markings, $M_{min}(S)$ and $M_{max}(S)$ are difficult to find. For this, we consider $M^{min}(S)$ and $M^{max}(S)$ defined as follows by relaxing M and Y to any non-negative real numbers.

Let

$$M^{min}(S) = min\{M(S)|M = M_0 + [N]Y, M \geq 0, Y \geq 0\}$$

and

$$M^{max}(S) = max\{M(S)|M = M_0 + [N]Y, M \geq 0, Y \geq 0\}.$$

Obviously, we have

$$M^{min}(S) \leq M_{min}(S)$$

and

$$M^{max}(S) \geq M_{max}(S).$$

Corollary 3.7. *A weakly dependent siphon S is controlled if*

$$M_0(S) > \sum_{i=1}^{n} a_i(M_0(S_i) - M^{min}(S_i)) - \sum_{j=n+1}^{m} a_j(M_0(S_j) - M^{max}(S_j)).$$

Proof. Since

$$M^{min}(S) \leq M_{min}(S)$$

and

$$M^{max}(S) \geq M_{max}(S),$$

it is easy to see that

$$M_0(S) > \sum_{i=1}^{n} a_i(M_0(S_i) - M^{min}(S_i)) - \sum_{j=n+1}^{m} a_j(M_0(S_j) - M^{max}(S_j))$$

implies the truth of

$$M_0(S) > \sum_{i=1}^{n} a_i(M_0(S_i) - M_{min}(S_i)) - \sum_{j=n+1}^{m} a_j(M_0(S_j) - M_{max}(S_j)).$$

□

Corollary 3.8. *A strongly dependent siphon S is controlled if*

$$M_0(S) > \sum_{i=1}^{n} a_i(M_0(S_i) - M^{min}(S_i)).$$

Proof. It follows immediately from Corollary 3.5 by considering $M^{min}(S) \leq M_{min}(S)$. □

Corollary 3.9. *Let (N, M_0) be a well-initially-marked net. A (strongly or weakly) dependent siphon S is controlled if*

$$M_0(S) > \sum_{i=1}^{n} a_i(M_0(S_i) - M^{min}(S_i)).$$

Proof. It follows immediately from Corollary 3.6 by considering $M^{min}(S) \leq M_{min}(S)$. □

Corollary 3.9 is very useful when we deal with a subclass of Petri nets modeling manufacturing systems. Corollaries 3.7–3.9 can be employed to verify whether a dependent siphon is controlled. To use them, in the worst case, we need to solve $2|\Pi_E|$ LPP to find $M^{min}(S)$ and $M^{max}(S)$.

Define

$$D_1 = min\{\sum_{i=1}^{n} a_i M(S_i) | M = M_0 + [N]Y, M \geq 0, Y \geq 0\}$$

and

$$D_2 = max\{\sum_{j=n+1}^{m} a_j M(S_j) | M = M_0 + [N]Y, M \geq 0, Y \geq 0\}.$$

Next we present weaker conditions under which a dependent siphon can be controlled.

Corollary 3.10. *A weakly dependent siphon S is controlled if*

$$M_0(S) > \sum_{i=1}^{n} a_i M_0(S_i) - D_1 - \sum_{j=n+1}^{m} a_j M_0(S_j) + D_2.$$

Proof. From (3.6), S is controlled if

$$M_0(S) > \sum_{i=1}^{n} a_i M_0(S_i) - \sum_{i=1}^{n} a_i M(S_i) - \sum_{j=n+1}^{m} a_j M_0(S_j) + \sum_{j=n+1}^{m} a_j M(S_j).$$

Note that

$$D_1 = min\{\sum_{i=1}^{n} a_i M(S_i) | M = M_0 + [N]Y, M \geq 0, Y \geq 0\} \leq \sum_{i=1}^{n} a_i M(S_i)$$

and

$$D_2 = max\{\sum_{j=n+1}^{m} a_j M(S_j) | M = M_0 + [N]Y, M \geq 0, Y \geq 0\} \geq \sum_{j=n+1}^{m} a_j M(S_j).$$

As a result, the truth of

$$M_0(S) > \sum_{i=1}^{n} a_i M_0(S_i) - D_1 - \sum_{j=n+1}^{m} a_j M_0(S_j) + D_2$$

implies that of

$$M_0(S) > \sum_{i=1}^{n} a_i M_0(S_i) - \sum_{i=1}^{n} a_i M(S_i) - \sum_{j=n+1}^{m} a_j M_0(S_j) + \sum_{j=n+1}^{m} a_j M(S_j).$$

□

Corollary 3.11. *Let S be a strongly dependent siphon in a net system* (N, M_0). *S is controlled if*

$$M_0(S) > \sum_{i=1}^{n} a_i M_0(S_i) - D_1.$$

Corollary 3.12. *Let S be a (weakly or strongly) dependent siphon in a well-initially-marked net (N, M_0). S is controlled if*

$$M_0(S) > \sum_{i=1}^{n} a_i M_0(S_i) - D_1.$$

The controllability condition stated in Corollary 3.10 is weaker than that in Corollary 3.7. However, its usage requires the computation of D_1 and D_2 for all dependent siphons, which is time-consuming since the number of dependent siphons is exponential with respect to the size of a net. For practical purposes, we can first use Corollary 3.7 to verify the controllability of a dependent siphon. If it is not controlled due to Corollary 3.7, Corollary 3.10 can be then employed. Our experimental studies show that most controlled dependent siphons satisfy the condition in Corollary 3.7.

Example 3.5. It is easy to verify that in net (N_2, M_2) as shown in Fig. 3.3(b), $I_1 = p_5 - p_7 - p_8 + p_9 + p_{12} + p_{13} - V_{S_1}$ and $I_2 = -p_3 + p_4 + p_6 - p_7 + p_{13} + p_{14} - V_{S_2}$ are P-invariants of N_2. Clearly, $S_1 = \{p_5, p_9, p_{12}, p_{13}\}$ is invariant-controlled by I_1 since $\{p | I_1(p) > 0\} = S_1$ and $I_1^T M_2 = M_2(S_1) - M_2(p_7) - M_2(p_8) - M_2(V_{S_1}) = 2 - 1 > 0$, where M_2 is an initial marking. Likewise, S_2 is controlled by P-invariant I_2. Now we check the controllability of S_3, a strongly dependent siphon with respect to S_1 and S_2 with $\eta_{S_3} = \eta_{S_1} + \eta_{S_2}$.

By solving LPP, we have $M^{min}(S_1) = M^{min}(S_2) = 1$. Note that $M_2(S_1) = 2$, $M_2(S_2) = 2$, $M_2(S_3) = 3$, and $\sum_{i=1}^{2}(M_2(S_i) - M^{min}(S_i)) = 2$. Due to Corollary 3.9, S_3 is controlled.

Remark 3.1. Compared with Fig. 3.3(a), places V_{S_1} and V_{S_2} in Fig. 3.3(b) are used to constrain the firing of transitions associated with siphons S_1 and S_2 respectively such that $M^{min}(S_1) = M^{min}(S_2) = 1$, leading to the satisfaction of inequality $M_2(S_3) > \sum_{i=1}^{2}(M_2(S_i) - M^{min}(S_i))$.

When designing a liveness-enforcing supervisor for a plant, emptiable minimal siphons in the plant model can be divided into elementary and dependent ones. Then we can make the elementary siphons properly controlled by using an active siphon control method, e.g., adding extra places like V_{S_1} and V_{S_2} in Fig. 3.3(b). Therefore, the dependent siphons can be implicitly controlled. In other words, if a plant Petri net model contains emptiable siphons, it is possible for us to control only those elementary ones. Since the number of elementary siphons is much smaller than that of dependent siphons in most of large Petri nets, the concept of elementary siphons can provide an effective way to prevent a large number of siphons from being emptied by controlling only a small number of them.

3.5 Controllability of Dependent Siphons in Generalized Petri Nets

This section discusses the controllability of dependent siphons in a generalized Petri net. Let (N, M_0) be a net system and S be a siphon of N. As stated in Chap. 2, S is said to be max-marked at a marking M iff $\exists p \in S$ such that $M(p) \geq max_{p^\bullet}$. It is said to be max-controlled iff it is max-marked at any reachable marking. (N, M_0) satisfies the cs-property iff each minimal siphon of N is max-controlled [3].

The cs-property is an important concept in liveness-enforcement to a generalized Petri net. A siphon satisfying the max-controlled property can be always marked sufficiently to allow firing a transition at least once. In order to check and use the cs-property, Barkaoui et al. [3] propose the conditions to determine whether a given siphon is max-controlled as stated in Proposition 2.1.

Lemma 3.2. *Let S be a siphon in net (N, M_0) and $M \in R(N, M_0)$ be a marking. S is max-marked under M if $M(S) > \omega(S)$, where $\omega(S) = \sum_{p \in S}(max_{p^\bullet} - 1)$.*

Proof. Let $S = \{p_1, p_2, \ldots, p_n\}$. By contradiction, suppose that S is not max-marked. This implies that $\forall i \in \mathbb{N}_n$, $M(p_i) \leq max_{p_i^\bullet} - 1$. Therefore, we have

$$\sum_{i=1}^{n} M(p_i) \leq \sum_{i=1}^{n}(max_{p_i^\bullet} - 1),$$

i.e.,

$$M(S) \leq \sum_{p \in S}(max_{p^\bullet} - 1),$$

which contradicts

$$M(S) > \sum_{p \in S}(max_{p^\bullet} - 1).$$

□

This lemma plays an important role in developing the controllability condition for dependent siphons in a generalized Petri net.

Example 3.6. The net shown in Fig. 3.4 is a generalized Petri net with three strict minimal siphons $S_1 = \{p_3, p_6, p_9, p_{13}, p_{14}\}$, $S_2 = \{p_2, p_5, p_{10}, p_{12}, p_{13}\}$, and $S_3 = \{p_3, p_6, p_{10}, p_{12}\text{-}p_{14}\}$. Thus, we have $\omega(S_1) = 0$, $\omega(S_2) = 1$, and $\omega(S_3) = 1$.

Without loss of generality, we assume that in a Petri net (N, M_0), a strongly dependent S satisfies $\eta_S = \sum_{i=1}^{n} a_i \eta_{S_i}$, $a_i > 0$, $i \in \mathbb{N}_n$, and a weakly dependent siphon S satisfies $\eta_S = \sum_{i=1}^{n} a_i \eta_{S_i} - \sum_{j=n+1}^{m} a_j \eta_{S_j}$, $a_i > 0$, $i \in \mathbb{N}_m$. Concerning the controllability of a dependent siphon in generalized Petri nets, we have the following results.

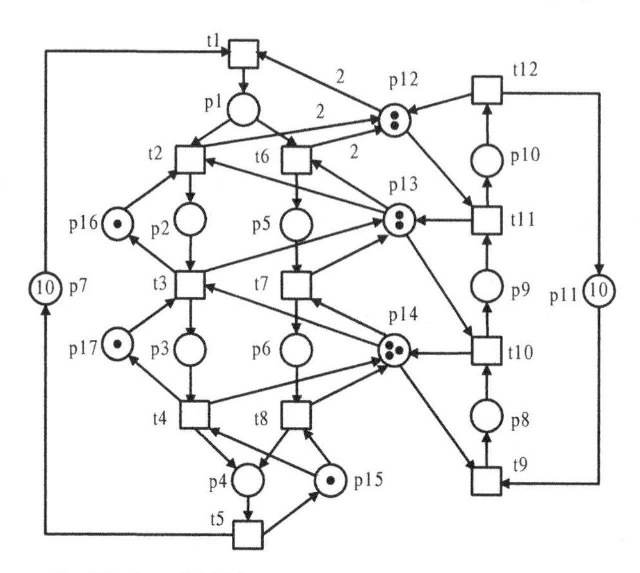

Fig. 3.4 A generalized Petri net (N, M_0)

Theorem 3.8. *A strongly dependent siphon S is max-controlled if*

$$M_0(S) > \sum_{i=1}^{n} a_i M_0(S_i) - \sum_{i=1}^{n} a_i M_{min}(S_i) + \omega(S).$$

Proof. Due to

$$\eta_S = \sum_{i=1}^{n} a_i \eta_{S_i},$$

we have

$$\lambda_S^T[N] = \sum_{i=1}^{n} a_i \lambda_{S_i}^T[N].$$

As a result,

$$(\lambda_S - \sum_{i=1}^{n} a_i \lambda_{S_i})^T[N] = \mathbf{0}^T.$$

Let

$$I = \lambda_S - \sum_{i=1}^{n} a_i \lambda_{S_i}.$$

$I = \mathbf{0}$ or I is a P-invariant of N. According to the proof of Theorem 3.7, this theorem is true in the case of $I = \mathbf{0}$. Next we consider that I is a P-invariant of N.

$\forall M \in R(N, M_0)$, $I^T M = I^T M_0$. Therefore, we have

$$M_0(S) - \sum_{i=1}^{n} a_i M_0(S_i) = M(S) - \sum_{i=1}^{n} a_i M(S_i),$$

$$M(S) = M_0(S) - \sum_{i=1}^{n} a_i M_0(S_i) + \sum_{i=1}^{n} a_i M(S_i),$$

and

$$M(S) \geq M_0(S) - \sum_{i=1}^{n} a_i M_0(S_i) + \sum_{i=1}^{n} a_i M_{min}(S_i).$$

Note that

$$M_0(S) > \sum_{i=1}^{n} a_i M_0(S_i) - \sum_{i=1}^{n} a_i M_{min}(S_i) + \omega(S).$$

It implies the truth of

$$M_0(S) - \sum_{i=1}^{n} a_i M_0(S_i) + \sum_{i=1}^{n} a_i M_{min}(S_i) > \omega(S).$$

Hence, $M(S) > \omega(S)$. By Lemma 3.2, S is max-marked under M. Therefore, $\forall M \in R(N, M_0)$, S is max-marked under M. In other words, S is max-controlled if

$$M_0(S) > \sum_{i=1}^{n} a_i M_0(S_i) - \sum_{i=1}^{n} a_i M_{min}(S_i) + \omega(S).$$

□

Example 3.7. There are three strict minimal siphons in Fig. 3.4. It is easy to verify that S_3 is a strongly dependent siphon with $\eta_{S_3} = \eta_{S_1} + \eta_{S_2}$.

In Fig. 3.5, we have $M_{min}(S_1) = M_{min}(S_2) = 2$. Considering that $M_1(S_1) = 5$, $M_1(S_2) = 4$, $M_1(S_3) = 7$, and $\omega(S_3) = 1$, $M_1(S_3) > \sum_{i=1}^{2}(M_1(S_i) - M_{min}(S_i)) + \omega(S_3)$ holds. As a result, S_3 is max-controlled in Fig. 3.5 due to the existence of monitors V_{S_1} and V_{S_2}.

For the controllability of a weakly dependent siphon in a net system (N, M_0), we have the following results.

Theorem 3.9. *A weakly dependent siphon S is max-controlled if*

$$M_0(S) > \sum_{i=1}^{n} a_i (M_0(S_i) - M_{min}(S_i)) - \sum_{j=n+1}^{m} a_j (M_0(S_j) - M_{max}(S_j)) + \omega(S).$$

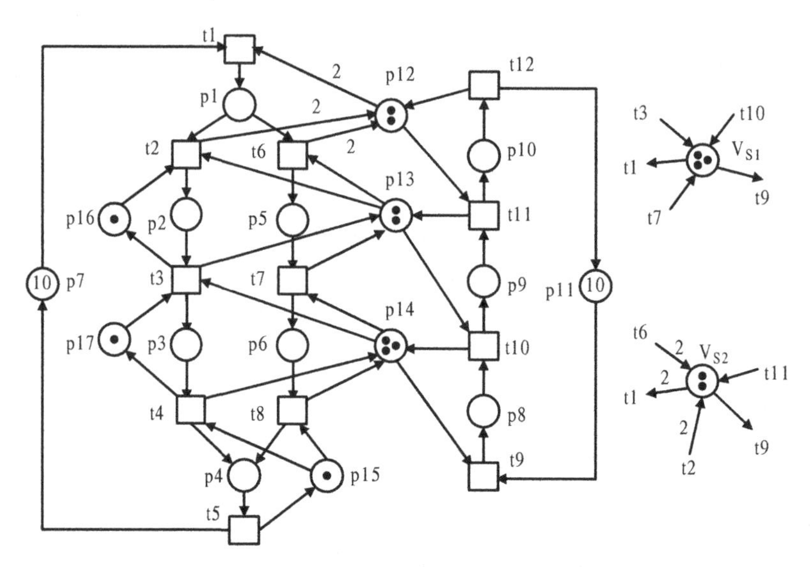

Fig. 3.5 A max-controlled siphon in a generalized Petri net (N_1, M_1)

Proof. Since

$$\eta_S = \sum_{i=1}^{n} a_i \eta_{S_i} - \sum_{j=n+1}^{m} a_j \eta_{S_j},$$

we have

$$\lambda_S^T[N] = \sum_{i=1}^{n} a_i \lambda_{S_i}^T[N] - \sum_{j=n+1}^{m} a_j \lambda_{S_j}^T[N],$$

and

$$(\lambda_S - \sum_{i=1}^{n} a_i \lambda_{S_i} + \sum_{j=n+1}^{m} a_j \lambda_{S_j})^T[N] = \mathbf{0}^T.$$

Let

$$I = \lambda_S - \sum_{i=1}^{n} a_i \lambda_{S_i} + \sum_{j=n+1}^{m} a_j \lambda_{S_j}.$$

We have that $I = \mathbf{0}$ or I is a P-invariant of N. According to the proof of Theorem 3.7, this theorem is true in the case of $I = \mathbf{0}$. Next we consider that I is a P-invariant of N.

$\forall M \in R(N, M_0)$, $I^T M = I^T M_0$. We have

$$M(S) - \sum_{i=1}^{n} a_i M(S_i) + \sum_{j=n+1}^{m} a_j M(S_j)$$

$$= M_0(S) - \sum_{i=1}^{n} a_i M_0(S_i) + \sum_{j=n+1}^{m} a_j M_0(S_j). \tag{3.8}$$

From (3.8), we have

$$M(S) = M_0(S) - \sum_{i=1}^{n} a_i (M_0(S_i) - M(S_i)) + \sum_{j=n+1}^{m} a_j (M_0(S_j) - M(S_j)). \tag{3.9}$$

Clearly, the following inequality is true.

$$M(S) \geq M_0(S) - \sum_{i=1}^{n} a_i (M_0(S_i) - M_{min}(S_i)) + \sum_{j=n+1}^{m} a_j (M_0(S_j) - M_{max}(S_j)). \tag{3.10}$$

From

$$M_0(S) > \sum_{i=1}^{n} a_i (M_0(S_i) - M_{min}(S_i)) - \sum_{j=n+1}^{m} a_j (M_0(S_j) - M_{max}(S_j)) + \omega(S),$$

we have

$$M_0(S) - \sum_{i=1}^{n} a_i (M_0(S_i) - M_{min}(S_i)) + \sum_{j=n+1}^{m} a_j (M_0(S_j) - M_{max}(S_j)) > \omega(S). \tag{3.11}$$

From (3.10) and (3.11), we have $M(S) > \omega(S)$ implying that S is max-marked under M. Since M is any marking in $R(N, M_0)$, S is max-controlled. □

Corollary 3.13. *A dependent siphon S in a well-initially-marked net (N, M_0) is max-controlled if*

$$M_0(S) > \sum_{i=1}^{n} a_i (M_0(S_i) - M_{min}(S_i)) + \omega(S).$$

Proof. It is trivial if S is strongly dependent. Let S be weakly dependent in (N, M_0). Due to Theorem 3.9, S is max-controlled if

$$M_0(S) > \sum_{i=1}^{n} a_i (M_0(S_i) - M_{min}(S_i)) - \sum_{j=n+1}^{m} a_j (M_0(S_j) - M_{max}(S_j)) + \omega(S).$$

By Definition 3.12, $\forall S \in \Pi$, $M_{max}(S) = M_0(S)$. As a result,

$$\sum_{j=n+1}^{m} a_j(M_0(S_j) - M_{max}(S_j)) = 0.$$

The controllability condition of a weakly dependent siphon coincides with that of a strongly dependent one. □

Corollary 3.14. *A strongly dependent siphon S is max-controlled if*

$$M_0(S) > \sum_{i=1}^{n} a_i M_0(S_i) - \sum_{i=1}^{n} a_i M^{min}(S_i) + \omega(S).$$

Proof. Since $M^{min}(S) \leq M_{min}(S)$,

$$M_0(S) > \sum_{i=1}^{n} a_i M_0(S_i) - \sum_{i=1}^{n} a_i M^{min}(S_i) + \omega(S)$$

$$\Rightarrow M_0(S) > \sum_{i=1}^{n} a_i M_0(S_i) - \sum_{i=1}^{n} a_i M_{min}(S_i) + \omega(S).$$

□

Corollary 3.15. *A weakly dependent siphon S is max-controlled if*

$$M_0(S) > \sum_{i=1}^{n} a_i(M_0(S_i) - M^{min}(S_i)) - \sum_{j=n+1}^{m} a_j(M_0(S_j) - M^{max}(S_j)) + \omega(S).$$

Corollary 3.16. *A dependent siphon S in a well-initially-marked net system is max-controlled if*

$$M_0(S) > \sum_{i=1}^{n} a_i(M_0(S_i) - M^{min}(S_i)) + \omega(S).$$

From the above discussion, in order to verify the controllability of dependent siphons, $\forall i \in \mathbb{N}_{|\Pi_E|}$, we need to compute $M^{min}(S_i)$ and $M^{max}(S_i)$ for S_i. In the worst case, we have to solve $2|\Pi_E|$ LPP, where $|\Pi_E|$ is bounded by the structural size of a net.

Recalling that

$$D_1 = min\{\sum_{i=1}^{n} a_i M(S_i) | M = M_0 + [N]Y, M \geq 0, Y \geq 0\}$$

and

$$D_2 = max\{\sum_{j=n+1}^{m} a_j M(S_j) | M = M_0 + [N]Y, M \geq 0, Y \geq 0\},$$

we present a weaker condition under which a dependent siphon can be max-controlled.

Corollary 3.17. *A weakly dependent siphon S is max-controlled if*

$$M_0(S) > \sum_{i=1}^{n} a_i M_0(S_i) - D_1 - \sum_{j=n+1}^{m} a_j M_0(S_j) + D_2 + \omega(S).$$

Proof.

$$M_0(S) > \sum_{i=1}^{n} a_i M_0(S_i) - D_1 - \sum_{j=n+1}^{m} a_j M_0(S_j) + D_2 + \omega(S)$$

implies the truth of

$$M_0(S) - \sum_{i=1}^{n} a_i M_0(S_i) + D_1 + \sum_{j=n+1}^{m} a_j M_0(S_j) - D_2 > \omega(S). \tag{3.12}$$

$\forall M \in R(N, M_0)$, (3.4) leads to the truth of

$$M(S) = M_0(S) - \sum_{i=1}^{n} a_i M_0(S_i) + \sum_{i=1}^{n} a_i M(S_i) + \sum_{j=n+1}^{m} a_j M_0(S_j) - \sum_{j=n+1}^{m} a_j M(S_j). \tag{3.13}$$

By the definitions of D_1 and D_2, $\forall M \in R(N, M_0)$, we have

$$D_1 \leq \sum_{i=1}^{n} a_i M(S_i)$$

and

$$D_2 \geq \sum_{j=n+1}^{m} a_j M(S_j).$$

As a result,

$$M(S) \geq M_0(S) - \sum_{i=1}^{n} a_i M_0(S_i) + D_1 + \sum_{j=n+1}^{m} a_j M_0(S_j) - D_2.$$

Considering (3.12), we have $\forall M \in R(N, M_0)$, $M(S) > \omega(S)$. S is hence max-controlled. □

Corollary 3.18. *A strongly dependent siphon S is max-controlled if*

$$M_0(S) > \sum_{i=1}^{n} a_i M_0(S_i) - D_1 + \omega(S).$$

Corollary 3.19. *A dependent siphon S in a well-initially-marked net* (N, M_0) *is max-controlled if*

$$M_0(S) > \sum_{i=1}^{n} a_i M_0(S_i) - D_1 + \omega(S).$$

A max-controlled siphon may satisfy Corollaries 3.17–3.19 but does not satisfy Theorem 3.9 and its related corollaries. However, the computation of D_1 (D_2) for all dependent siphons is expensive since in the worst case their number remains to be exponential with respect to the size of a net.

Corollary 3.20. $\omega(S) = 0$ *if S is a minimal siphon in an ordinary net.*

Proof. We claim that $\forall p \in S$, $max_{p^\bullet} = 1$. To prove this, we need to show that $\forall p \in S$, $p^\bullet \neq \emptyset$. By contradiction, we suppose that $\exists p \in S$, $p^\bullet = \emptyset$. Then $\forall t \in {}^\bullet p$, $\exists p' \in S$, $t \in (p')^\bullet$ since S is a siphon. As a result, we have ${}^\bullet(S \backslash \{p\}) \subseteq {}^\bullet S \subseteq S^\bullet = (S \backslash \{p\})^\bullet$. $S \backslash \{p\}$ is a siphon as well, which contradicts the minimality of S. In other words, $\forall p \in S$, $max_{p^\bullet} = 1$ is true. Hence, we have $\omega(S) = 0$. □

The controllability results of dependent siphons in an ordinary Petri net are given by Theorem 3.7 and the corollaries induced from it. They can also be derived from those in generalized nets by $\omega(S) = 0$ if the considered siphon S is minimal.

3.6 An Elementary Siphon Identification Algorithm

According to the definition of elementary siphons, their set is generally not unique in a given net structure unless its $[\eta]$ is a row full-rank matrix. Our investigation reveals that, in order to prevent deadlocks from occurring, it is crucial to choose a proper set of elementary siphons in a net system. For deadlock control purposes, an algorithm is presented in this section to determine elementary siphons within the scope of a given set of siphons Π that is not necessarily the set of all minimal siphons of a net. They are chosen such that the resulting dependent siphons can be easily marked sufficiently. When choosing elementary siphons, we should try to minimize the right sides of the controllability conditions such as those stated in Theorems 3.7 and 3.9, and their related corollaries. As a result, the number of tokens in an elementary siphon at an initial marking should be as small as possible. We present an algorithm below to identify such a set of elementary siphons.

Let Π be a set of interesting minimal siphons in a Petri net (N, M_0). Suppose that Π has n sets of equivalent siphons $\langle S_1^{\alpha}\rangle$-$\langle S_n^{\alpha}\rangle$. $\forall i \in \mathbb{N}_n$, a token-poor siphon S_i is selected from $\langle S_i^{\alpha}\rangle$. Clearly, $|\Pi_E| \leq n$ and Π_E can be found from these n token-poor siphons. Without loss of generality, let $\Pi = \{S_1, S_2, \ldots, S_n\}$. Note that the controllability of a token-poor siphon implies that of its equivalent siphons.

Algorithm 3.1 Identification of elementary siphons for deadlock control

Input: $\Pi = \{S_1, S_2, \ldots, S_n\}$

Output: Π_E, a set of elementary siphons

1: Find $[\eta]$ from Π
2: Compute $rank([\eta])$, $m := rank([\eta])$
3: Compute the initial number in tokens of siphons in Π and get a marking sequence $M_0(S_1), M_0(S_2), \ldots$, and $M_0(S_n)$
4: By the *merge sort* algorithm, sort the marking sequence to be $M_0(S_{k_1})$, $M_0(S_{k_2})$, $\ldots$, and $M_0(S_{k_n})$ in an ascending order, where $\{k_1, k_2, \ldots, k_n\} = \mathbb{N}_n$
5: **for** $j = 1$ to n **do**
6: $\quad \chi_j := \eta_{S_{k_j}}$
7: **end for**
8: $\Pi_E := \{S_{k_1}\}$
9: $A := [\chi_1]^T$
10: $i := 2$
11: **while** $(m \neq 1)$ **do**
12: $\quad$ **for** $i = 2$ to n **do**
13: $\quad\quad A_E := [A^T | \chi_i]^T$
14: $\quad\quad \Pi_E := \Pi_E \cup \{S | \eta_S = \chi_i\}$
15: $\quad\quad$ **while** $(rank(A_E) \neq m)$ **do**
16: $\quad\quad\quad$ **if** $rank(A_E) - rank(A) = 0$ **then**
17: $\quad\quad\quad\quad \Pi_E := \Pi_E \setminus \{S | \eta_S = \chi_i\}$
18: $\quad\quad\quad\quad i := i + 1$
19: $\quad\quad\quad$ **else**
20: $\quad\quad\quad\quad A := A_E$
21: $\quad\quad\quad\quad i := i + 1$
22: $\quad\quad\quad$ **end if**
23: $\quad\quad$ **end while**
24: $\quad\quad i := n + 1$
25: $\quad$ **end for**
26: **end while**
27: Output Π_E

We explain this algorithm as follows. If there is only one minimal siphon in Π, it is certainly elementary and this algorithm can terminate at once. Suppose that at some step, $rank(A_E) < m$ since $rank(A_E) > m$ is impossible and $rank(A_E) = m$ leads to the termination of this algorithm. We need to check the ranks of A_E and A. If $rank(A_E) = rank(A)$, then the siphon corresponding to χ_i is dependent on the siphons in Π_E resulting from the last step, i.e., it cannot be put into Π_E and become elementary. If $rank(A_E) \neq rank(A)$, it indicates $rank(A_E) > rank(A)$ since

$rank(A_E) < rank(A)$ is impossible by matrix A's construction in the algorithm. This implies that the siphon corresponding to χ_i is elementary. This process is repeated until $rank(A_E) = m$. In this case, we assign i to be $n+1$ so that the algorithm terminates at once.

Example 3.8. In Fig. 3.3(a), there are three strict minimal siphons S_1, S_2, and S_3 with $M_1(S_1) = M_1(S_2) = 2$ and $M_1(S_3) = 3$. It is easy to find that m, the number of elementary siphons, is two. Then, we have an ascending order sequence $M_1(S_1)$, $M_1(S_2)$, and $M_1(S_3)$. S_1 is first put into Π_E. Next we process S_2. Clearly, η_{S_2} is linearly independent of the characteristic T-vector of the siphon in Π_E ($\eta_{S_2} \neq \eta_{S_1}$). As a result, S_2 is put into Π_E. Now we have $\Pi_E = \{S_1, S_2\}$. Since $|\Pi_E| = m = 2$, the algorithm terminates and outputs $\Pi_E = \{S_1, S_2\}$. It is certain that η_{S_3} is linearly dependent on the characteristic T-vectors of the siphons in Π_E. Siphon S_3 is in fact a strongly dependent siphon with $\eta_{S_3} = \eta_{S_1} + \eta_{S_2}$.

Next we discuss its complexity. The complexity of the *merge sort* algorithm for a sequence with n elements is $O(n \lg n)$. In addition, we have to compute the ranks of matrices A and A_E. However, as variable i indicates, the number of times of computing $rank(A)$ and $rank(A_E)$ is bounded by n. The number of rows (columns) of A or A_E is bounded by $|P|$ ($|T|$). It is known that the complexity of computing the rank of a $k \times m$ matrix is $O(km^2)$ based on the singular value decomposition method. Thus, the complexity of the algorithm is $O(2n|P||T|^2 + n \lg n)$, which is independent of initial markings for a given net structure.

The development of this algorithm is motivated by the observation that different sets of elementary siphons usually lead to liveness-enforcing supervisors with different permissive behavior if elementary siphons are explicitly controlled only by some particular deadlock prevention policies in the literature. The algorithm outputs such a set of elementary siphons that the resultant supervisors have more permissive behavior.

To informally illustrate this, we take the controllability of a strongly dependent siphon as an example. Let S be a strongly dependent siphon with $\eta_S = \sum_{i=1}^{n} a_i \eta_{S_i}$ in an ordinary Petri net (N, M_0). It is shown that S is controlled if $M_0(S) > \sum_{i=1}^{n} a_i (M_0(S_i) - M_{min}(S_i))$. Note that a larger $M_{min}(S_i)$ results in more restrictive behavior. As a result, to easily guarantee the inequality for the controllability of siphon S, we want, for all $i \in \mathbb{N}_n$, $M_0(S_i)$ to be as small as possible. Otherwise, we have to enlarge $M_{min}(S_i)$.

For the net in Fig. 3.3(a), its initial marking is $M_1 = 5p_1 + 3p_{10} + p_{11} + p_{12} + p_{13} + p_{14} + p_{15}$. By the proposed elementary siphon identification algorithm, we have $\Pi_E = \{S_1, S_2\}$. Using the siphon control approach proposed in [11] to control elementary siphons only, we can have a liveness-enforcing supervisor that allows the controlled system to have 28 permissive reachable states, as shown in Fig. 3.3(b). If S_1 and S_3 (S_2 and S_3) are selected to be elementary siphons, the controlled system has 28 (28) reachable states as well.

However, for the same net structure (depicted in Fig. 3.3(a)) with initial marking $M_1 = 6p_1 + 4p_{10} + p_{11} + p_{12} + 2p_{13} + p_{14} + p_{15}$, the algorithm outputs $\Pi_E = \{S_1, S_2\}$ as the set of elementary siphons, which leads to a controlled system that has

110 reachable states. If S_1 and S_3 (S_2 and S_3) are selected to be elementary siphons, controlled system has 96 (96) reachable states only.

3.7 Existence of Dependent Siphons

This section focuses on the existence of dependent siphons within Π of a Petri net from the algebraic point of view. Note that Π is not necessarily the set of all minimal siphons of a Petri net N.

Theorem 3.10. *There is no dependent siphon in Π if $[\eta]$ derived from Π is a row full-rank matrix.*

Proof. This is trivial since all siphons are elementary ones in this case. □

The next result presents a necessary condition for the existence of a set of elementary siphons such that all other siphons are strongly dependent in Π.

Without loss of generality, suppose that $\Pi = \{S_1, S_2, \ldots, S_n\}$ and $[\eta] = [\eta_1|\eta_2|\cdots|\eta_n]^T$, where $\forall i, j \in \mathbb{N}_n$, $\eta_i \neq \eta_j$. Let $m = rank([\eta])$, $K = \{\eta_i | \eta_i$ cannot be linearly represented by the other vectors in $[\eta]$ with non-negative coefficients only$\}$, $k = |K|$, and r be the rank of the matrix consisting of all vectors in K.

Theorem 3.11. *If there is a set of elementary siphons such that all other siphons in Π are strongly dependent on them, then $k = m = r$.*

Proof. Suppose that $[\eta]$ has the following form, where η_α, η_β, $\ldots$, and η_γ correspond to the set of elementary siphons $\{S_\alpha, S_\beta, \ldots, S_\gamma\}$.

$$[\eta] = \begin{pmatrix} \eta_\alpha^T \\ \eta_\beta^T \\ \vdots \\ \eta_\gamma^T \\ \eta_u^T \\ \eta_v^T \\ \vdots \\ \eta_w^T \end{pmatrix}.$$

All other siphons corresponding to η_u, η_v, $\ldots$, and η_w are strongly dependent ones. Clearly, we have $\{\alpha, \beta, \ldots, \gamma\} \cap \{u, v, \ldots, w\} = \emptyset$ and $\{\alpha, \beta, \ldots, \gamma\} \cup \{u, v, \ldots, w\} = \mathbb{N}_n$.

By the definition of strongly dependent siphons, $\forall i \in \{u, v, \ldots, w\}$, we have $\eta_i = \sum_{j \in \{\alpha, \beta, \ldots, \gamma\}} a_j \eta_j$, where $a_j \geq 0$. This implies two facts: (1) $rank([\eta])$ depends on η_α, η_β, $\ldots$, and η_γ only; and (2) $\forall \eta \in \{\eta_\alpha, \eta_\beta, \ldots, \eta_\gamma\}$, η cannot be linearly represented by η_u, η_v, $\ldots$, and η_w with non-negative coefficients only (note that in $[\eta] = [\eta_1|\eta_2|\cdots|\eta_n]^T$, $\forall i \neq j, \eta_i \neq \eta_j$).

Since $\{S_\alpha, S_\beta, \ldots, S_\gamma\}$ is a set of elementary siphons, η_α, η_β, ..., and η_γ are linearly independent. That is to say, $\forall \eta \in \{\eta_\alpha, \eta_\beta, \ldots, \eta_\gamma\}$, η cannot be linearly represented by the vectors in $\{\eta_\alpha, \eta_\beta, \ldots, \eta_\gamma\} \backslash \{\eta\}$, not to speak of using non-negative coefficients. In conclusion, each of η_α, η_β, ..., and η_γ cannot be linearly represented by other vectors in $[\eta]$ with non-negative coefficients. This implies $K = \{\eta_\alpha, \eta_\beta, \ldots, \eta_\gamma\}$ and $|\{\alpha, \beta, \ldots, \gamma\}| = k$. Since η_α, η_β, ..., and η_γ are linearly independent, we have $k = r$.

As known, $\forall i \in \mathbb{N}_n \backslash \{\alpha, \beta, \ldots, \gamma\}$, η_i can be linearly represented by η_α, η_β, ..., and η_γ with non-negative coefficients. Therefore, m equals the rank of the matrix consisting of η_α, η_β, ..., and η_γ. Note that η_α, η_β, ... and η_γ are linearly independent. We have $m = |\{\alpha, \beta, \ldots, \gamma\}| = k$. Considering $k = r$, we have $k = m = r$.

□

Example 3.9. For the net in Fig. 3.3(a), we have $\eta_{S_3} = \eta_{S_1} + \eta_{S_2}$ implying that there is no weakly dependent siphon if S_1 and S_2 are chosen to be elementary siphons. η_{S_1} cannot be linearly represented by the two others with non-negative coefficients only. Neither can η_{S_2}. Therefore, $k = 2$. Since $\eta_{S_1} \neq \eta_{S_2}$ and $rank([\eta]) = 2$, we have $k = m = r = 2$.

Corollary 3.21. *If $k \neq m$, there is no such a set of elementary siphons that all other siphons are strongly dependent.*

Consider a case where there are four siphons $S_1 - S_4$ in a net with $\eta_{S_1} + \eta_{S_2} = \eta_{S_3} + \eta_{S_4}$. Clearly, we have $k = 4$ since $\forall i \in \{1,2,3,4\}$, η_{S_i} cannot be represented by others with non-negative coefficients. However, $m = 3$ since η_{S_1}, η_{S_2}, η_{S_3}, and η_{S_4} are linearly dependent. That is to say, there must exist a weakly dependent siphon whatever siphons are chosen to be elementary ones.

A natural question is to ask whether η_i can be linearly represented by η_1, η_2, ..., η_{i-1}, η_{i+1}, ..., and η_n with non-negative coefficients. This can be answered by solving the following LPP, where X is an $(n-1)$-dimensional vector:

$$\text{maximize} \quad \mathbf{0}^T X \tag{3.14}$$

$$\text{s.t.} \quad [\eta^i]X = \eta_i^T, \quad X \geq \mathbf{0}$$

where $[\eta^i] = [\eta_1|\eta_2|\cdots|\eta_{i-1}|\eta_{i+1}|\cdots|\eta_n]$. Note that if n vectors η_1, η_2, ..., and η_n are linearly dependent, solution $X = \mathbf{0}$ is impossible. As a result, problem (3.14) has either a feasible solution or none. The reason to maximize $\mathbf{0}^T X$ in the above problem is just to find the feasible solution when it exists.

Based on the above results, an intuitively sufficient condition under which there exists a set of elementary siphons such that all others are strongly dependent can be derived.

Suppose that $[\eta] = [\eta_1|\eta_2|\cdots|\eta_n]^T$ is known. Find $K = \{\eta | \eta$ cannot be linearly represented by other vectors in $[\eta]$ with non-negative coefficients$\}$ by using LPP (3.14). Let $K' = \{\eta_1, \eta_2, \ldots, \eta_n\} \backslash K$. As stated previously, $k = |K|$, $m = rank([\eta])$, and r be the rank of the matrix consisting of all vectors in K. Check whether every vector in K' can be linearly represented by vectors in K with non-negative coefficients. If this is true and $k = m = r$, we can conclude that the siphons corresponding to the vectors in K form the set of elementary siphons such that all other ones are strongly dependent. Certainly, if $|K| = m$ and the vectors in K are linearly independent, by Definition 3.7, $\{S|\eta_S \in K\}$ is a set of elementary siphons. Hence, we have the following result.

Corollary 3.22. *If $k = m = r$ and every vector in $K' = \{\eta_1, \eta_2, \ldots, \eta_n\} \setminus K$ can be linearly represented by those in K with non-negative coefficients, then $\{S|\eta_S \in K\}$ is the set of elementary siphons such that all other siphons are strongly dependent.*

3.8 Bibliographical Remarks

The material of this chapter consists mainly of results of elementary and dependent siphons in ordinary and generalized Petri nets. Most of them can be found in [20–23]. The concept of elementary and dependent siphons is due to [21]. For a class of Petri nets called LS^3PR (linear system of simple sequential processes with resources), an algorithm with polynomial complexity is developed in [29] to find the number of elementary siphons without the complete siphon enumeration. By using the concepts of handles and bridges [10], Chao proposes some novel methods to compute a set of elementary siphons for S^3PR [5, 8] and BS^3PR [7].

Problems

3.1. It is shown in the literature that the max-controllability of siphons in a generalized Petri net is too conservative [6]. Discuss the possibility to relax the conditions in the cs-property. This relaxation can be significant even in some subclasses of Petri nets.

3.2. Suppose that in a Petri net there exists a set of elementary siphons such that all others are strongly dependent. Develop an algorithm that identifies such a set of elementary siphons. Trivially, there are at most $n!/(m!(n-m)!)$ different sets of elementary siphons, where $n = |\Pi|$ and $m = |\Pi_E|$. They can be checked one by one. Discuss the existence of other efficient algorithms to find such a set of elementary siphons.

3.3. Develop an algorithm to find a set of elementary siphons that minimize the number of resultant weakly dependent siphons in a Petri net. Then, discuss its time complexity provided that all siphons are known.

References

1. Abdallah, I.B., ElMaraghy, H.A. (1998) Deadlock prevention and avoidance in FMS: A Petri net based approach. *International Journal of Advanced Manufacturing Technology*, vol.14, no.10, pp.704–715.
2. Banaszak, Z., Krogh, B.H. (1990) Deadlock avoidance in flexible manufacturing systems with concurrently competing process flows. *IEEE Transactions on Robotics and Automation*, vol.6, no.6, pp.724–734.
3. Barkaoui, K., Pradat-Peyre, J.F. (1996) On liveness and controlled siphons in Petri nets. In *Proc. 17th Int. Conf. on Applications and Theory of Petri Nets Lecture Notes in Computer Science*, vol.1091, pp.57–72.
4. Barkaoui, K., Chaoui, A., Zouari, B. (1997) Supervisory control of discrete event systems based on structure theory of Petri nets. In *Proc. IEEE Int. Conf. on Systems, Man, and Cybernetics*, pp.3750–3755.
5. Chao, D.Y. (2006) Computation of elementary siphons for deadlock control. *The Computer Journal*, vol.49, no.4, pp.470–479.
6. Chao, D.Y. (2007) Max′-controlled siphons for liveness of S^3PGR^2. *IET Control Theory and Applications*, vol.1, no.4, pp.933–936.
7. Chao, D.Y. (2007) A graphic-algebraic computation of elementary siphons of BS^3PR. *Journal of Information Science and Engineering*, vol.23 no.6, pp.1817–1831.
8. Chao, D.Y. (2007) Incremental approach to computation of elementary siphons for arbitrary S^3PR. *IET Control Theory and Applications*, vol.2, no.2, pp.168–179.
9. Chu, F., Xie, X.L. (1997) Deadlock analysis of Petri nets using siphons and mathematical programming. *IEEE Transactions on Robotics and Automation*, vol.13, no.6, pp.793–804.
10. Esparza, J., Silva, M. (1990) Circuits, handles, bridges, and nets. In *Advances in Petri Nets 1990, Lecture Notes in Computer Science*, vol.483, G. Rozenberg (Ed.), pp.210–242.
11. Ezpeleta, J, Colom, J.M., Martinez, J. (1995) A Petri net based deadlock prevention policy for flexible manufacturing systems. *IEEE Transactions on Robotics and Automation*, vol.11, no.2, pp.173–184.
12. Ezpeleta, J., García-Vallés, F., Colom, J.M. (1998) A class of well structured Petri nets for flexible manufacturing systems. In *Proc. 19th Int. Conf. on Applications and Theory of Petri Nets, Lecture Notes in Computer Science*, vol.1420, J. Desel and M. Silva (Eds.), pp.64–83.
13. Ezpeleta, J., Tricas, F., García-Vallés, F., Colom, J.M. (2002) A banker's solution for deadlock avoidance in FMS with flexible routing and multiresource states. *IEEE Transactions on Robotics and Automaton*, vol.18. no.4, pp.621–625.
14. Huang, Y.S., Jeng, M.D., Xie, X.L., Chung, S.L. (2001) A deadlock prevention policy for flexible manufacturing systems using siphons. In *Proc. IEEE Int. Conf. on Robotics and Automation*, pp.541–546.
15. Iordache, M.V., Moody, J.O., Antsaklis, P.J. (2002) Synthesis of deadlock prevention supervisors using Petri nets. *IEEE Transactions on Robotics and Automation*, vol.18, no.1, pp.59–68.
16. Jeng, M.D., Xie, X.L. (1999) Analysis of modularly composed nets by siphons. *IEEE Transactions on Systems, Man, and Cybernetics, Part A*, vol.29, no.4, pp.399–406.
17. Jeng, M.D., Xie, X.L., Peng, M.Y. (2002) Process nets with resources for manufacturing modeling and their analysis. *IEEE Transactions on Robotics and Automation*, vol.18, no.6, pp.875–889.
18. Jeng, M.D., Xie, X.L., Chung, S.L. (2004) ERCN* merged nets for modeling degraded behavior and parallel processes in semiconductor manufacturing systems. *IEEE Transactions on Systems, Man, and Cybernetics, Part A*, vol.34, no.1, pp.102–112.
19. Lautenbach, K., Ridder, H. (1996) The linear algebra of deadlock avoidance–a Petri net approach. No.25-1996, Technical Report, Institute of Software Technology, University of Koblenz-Landau, Koblenz, Germany.
20. Li, Z.W., Zhou, M.C. (2004) Elementary siphons of Petri nets and their application to deadlock prevention in flexible manufacturing systems. *IEEE Transactions on Systems, Man, and Cybernetics, Part A*, vol.34, no.1, pp.38–51.

21. Li, Z.W., Zhou, M.C. (2006) Clarifications on the definitions of elementary siphons of Petri nets. *IEEE Transactions on Systems, Man, and Cybernetics, Part A*, vol.36, no.6, pp.1227–1229.
22. Li, Z.W., Zhao, M. (2008) On controllability of dependent siphons for deadlock prevention in generalized Petri nets. *IEEE Transactions on Systems, Man, and Cybernetics, Part A*, vol.38, no.2, pp.369–384.
23. Li, Z.W., Zhou, M.C. (2008) Control of elementary and dependent siphons of Petri nets and their application. *IEEE Transactions on Systems, Man, and Cybernetics, Part A*, vol.38, no.1, pp.133–148.
24. Park, J., Reveliotis, S.A. (2000) Algebraic synthesis of efficient deadlock avoidance policies for sequential resource allocation systems. *IEEE Transactions on Robotics and Automation*, vol.16, no.2, pp.190–195.
25. Park, J., Reveliotis, S.A. (2001) Deadlock avoidance in sequential resource allocation systems with multiple resource acquisitions and flexible routings. *IEEE Transactions on Automatic Control*, vol.46, no.10, pp.1572–1583.
26. Schrijver, A. (1998) *Theory of Linear and Integer Programming*. New York: John Wiley & Sons.
27. Tricas, F., Martinez, J. (1995) An extension of the liveness theory for concurrent sequential processes competing for shared resources. In *Proc. IEEE Int. Conf. on Systems, Man, and Cybernetics*, pp.3035–3040.
28. Tricas, F., García-Vallès, F., Colom, J.M., Ezpeleta, J. (2000) An iterative method for deadlock prevention in FMSs. In *Proc. 5th Workshop on Discrete Event Systems*, R. Boel and G. Stremersch (Eds.), pp.139–148.
29. Wang, A.R., Li, Z.W., Jia, J.Y., Zhou, M.C. (2009) An effective algorithm to find elementary siphons in a class of Petri nets. To appear in *IEEE Transactions on Systems, Man, and Cybernetics, Part A*.
30. Winston, W.L., Venkataramanan, M. (2002) *Introduction to Mathematical Programming*. Belmont CA: Duxbury Resource Center.
31. Xie, X.L., Jeng, M.D. (1999) ERCN-merged nets and their analysis using siphons. *IEEE Transactions on Robotics and Automation*, vol.15, no.4, pp.692–703.
32. Yamalidou, E., Moody, J.O., Antsaklis, P.J. (1996) Feedback control of Petri nets based on place invariants. *Automatica*, vol.32, no.1, pp.15–28.
33. Zouari, B., Barkaoui, K. (2003) Parameterized supervisor synthesis for a modular class of discrete event systems. In *Proc. IEEE Int. Conf. on Systems, Man, and Cybernetics*, pp.1874–1879.

Chapter 4
Monitor Implementation of GMECs

Abstract This chapter first presents a method to handle a class of important supervisory control specifications in discrete-event systems, generalized mutual exclusion constraints, which is considered to be a natural extension to elementary siphons. A set of generalized mutual exclusion constraints is divided into elementary and dependent ones. The latter can be implicitly enforced by explicitly enforcing their elementary constraints via properly setting the control depth variables of the former. It is shown that this method usually leads to a structurally simple Petri net supervisor given a set of constraints. Then, it is applied to the design of a liveness-enforcing (Petri net) supervisor for a generalized Petri net that can model a large class of automated manufacturing systems. Finally, an elementary constraint identification algorithm is discussed.

4.1 Introduction

This chapter considers the problem of forbidden-state specifications that can be represented by generalized mutual exclusion constraints (GMECs). A GMEC is defined as a condition that limits a weighted sum of tokens contained in a subset of places [9, 10, 30, 37] and includes both sequential and parallel mutual exclusions [46,47]. Many constraints that deal with exclusions between states and events can be transformed into GMECs [27].

As one of the most common types of control specifications, GMECs have been used in the optimal control of chemical processes [44], coordination of AGVs [30], specifications of manufacturing constraints [37], batch processing [42], and supervisory control of railway networks [11]. They are also important for the representation of deadlock prevention and liveness specifications [14,18,40,41].

The enforcement of a GMEC is usually done by adding to a plant model a controller that takes the form of a single place called a *monitor* with arcs going to and coming from the plant transitions [1–5, 9, 30, 37, 38, 45]. The monitor synthesis is shown to be maximally permissive and very efficient from the computational point of view if all transitions in a plant model are assumed to be controllable, i.e., each transition can be prevented from firing by an external control agent. In the case that a conjunction of GMECs has to be imposed, a monitor for each GMEC is needed [9]. If a GMEC is imposed to a plant in presence of uncontrollable transitions [3], the GMEC has to be transformed into another constraint. The monitor derived from the new constraint may still yield a maximally permissive supervisor. Under some conditions, the maximally permissive control law is a disjunction of GMECs [31–33]. The control specification stated as a disjunction of GMECs is considered by Basile et al. in [3]. The more complicated and hybrid control specifications consisting of GMECs and Parikh vectors are considered in [27, 37, 38] when there exist uncontrollable and unobservable transitions in a plant.

The complexity of enforcing a GMEC to a plant with uncontrollable transitions is enhanced. To enforce it to such a plant, we may need to prevent it from reaching a superset of the forbidden markings. The superset contains all those markings from which a forbidden state may be reached by firing a sequence of uncontrollable transitions.

Although GMECs have been extensively investigated in the literature, insufficient attention has been focused on a systematic approach that can minimize the number of additional monitors for a given set of GMECs. It is usually taken for granted that the number of monitors is propositional to the number of GMECs that are to be enforced. Motivated by the concept of elementary siphons [34, 35], the constraints in a set of GMECs are divided into elementary and dependent ones. This chapter explores the conditions under which a dependent constraint can be implicitly imposed by properly imposing its elementary constraints to the plant. This can often lead to a structurally simple Petri net supervisor. It is also useful in simplifying the structure of the liveness-enforcing (Petri net) supervisors resulting from an efficient deadlock prevention policy in which liveness requirements are represented as a set of GMECs [40].

The seminal work in this area is by Giua et al. [9, 10], where the concepts of redundancy, equivalence, and simplification of GMECs are proposed, which can be decided by solving LPPs. What is presented in this chapter aims to generalize and extend the related results in the literature. Specifically, conditions are developed under which a dependent constraint is implicitly enforced.

4.2 Generalized Mutual Exclusion Constraints

The following results are mainly from [9] and [37].

Definition 4.1. Let (N, M_0) be a net system with place set P. A GMEC in N is defined as a set of legal markings $\mathcal{M}(l,b) = \{M \in \mathbb{N}^{|P|} | l^T M \leq b\}$, where l is a non-negative P-vector and $b \in \mathbb{N}$ is called the constraint constant.

The markings in $\mathbb{N}^{|P|}$ that are not in $\mathcal{M}(l,b)$ are called forbidden markings with respect to constraint (l,b). In a GMEC (l,b) with $l = a_1 p_1 + a_2 p_2 + \cdots + a_n p_n$, it is denoted by $(l,b) \equiv a_1M(p_1)+a_2M(p_2)+\cdots+a_nM(p_n) \leq b$, where $\forall i \in \mathbb{N}_n$, a_i is a positive integer.

Example 4.1. (l,b) is a GMEC in Petri net (N,M_0) with $l=(2\ 0\ 3\ 1\ 0)^T$ and $b=4$. The GMEC can be denoted by $2M(p_1)+3M(p_3)+M(p_4) \leq 4$.

Definition 4.2. A set of GMECs (L,B) with $L=[l_1|l_2|\cdots|l_m]$ and $B=(b_1,b_2,\ldots,b_m)$ defines a set of legal markings $\mathcal{M}(L,B)=\{M \in \mathbb{N}^{|P|} | L^TM \leq B\} = \cap_{i=1}^m \mathcal{M}(l_i,b_i)$.

Definition 4.3. A GMEC (l,b) is redundant with respect to a set of marking $\mathcal{M} \subseteq \mathbb{N}^{|P|}$ if $\mathcal{M} \subseteq \mathcal{M}(l,b)$. It is redundant with respect to a net (N,M_0) if $R(N,M_0) \subseteq \mathcal{M}(l,b)$. A set of GMEC is redundant with respect to (N,M_0) if each constraint is redundant with respect to it.

Proposition 4.1. *A GMEC (l,b) is redundant with respect to (N,M_0) if the following LPP has an optimal solution $x^* < b+1$:*

$$x = max\ l^TM$$

s.t.

$$M = M_0 + [N]Y$$
$$M,Y \geq 0.$$

Definition 4.4. Two sets of GMECs (L_1,B_1) and (L_2,B_2) are equivalent with respect to (N,M_0) if $R(N,M_0) \cap \mathcal{M}(L_1,B_1) = R(N,M_0) \cap \mathcal{M}(L_2,B_2)$.

The equivalence of two sets of GMECs can be verified by the same approach employed to check redundancy. It hence follows that (L_1,B_1) and (L_2,B_2) are equivalent with respect to (N,M_0) if (L_1,B_1) is redundant with respect to $R(N,M_0) \cap \mathcal{M}(L_2,B_2)$ and (L_2,B_2) is redundant with respect to $R(N,M_0) \cap \mathcal{M}(L_1,B_1)$.

Example 4.2. Consider the Petri net (N,M_0) and its reachability graph as shown in Fig. 4.1a and Fig. 4.1b, respectively. Let (l_1,b_1) and (l_2,b_2) be two GMECs with $(l_1,b_1) \equiv M(p_2) \leq 2$ and $(l_2,b_2) \equiv M(p_2)+M(p_3) \leq 2$ and $\mathcal{M} = \{p_1+2p_4+2p_5, p_2+p_4+2p_5, p_3+2p_4+p_5\}$ be a set of markings. It can be verified that both (l_1,b_1) and (l_2,b_2) are redundant with respect to $\mathcal{M}$. Furthermore, they are also redundant with respect to the net (N,M_0) in Fig. 4.1(a).

4.3 Elementary and Dependent Constraints

In this section, elementary and dependent constraints in a set of GMECs are defined according to their linear dependency of the characteristic T-vectors.

Definition 4.5. l_i is called the characteristic P-vector of constraint (l_i,b_i) in $L^TM \leq B$.

Definition 4.6. $\eta_i = [N]^T l_i$ is called the characteristic T-vector of (l_i,b_i) in $L^TM \leq B$.

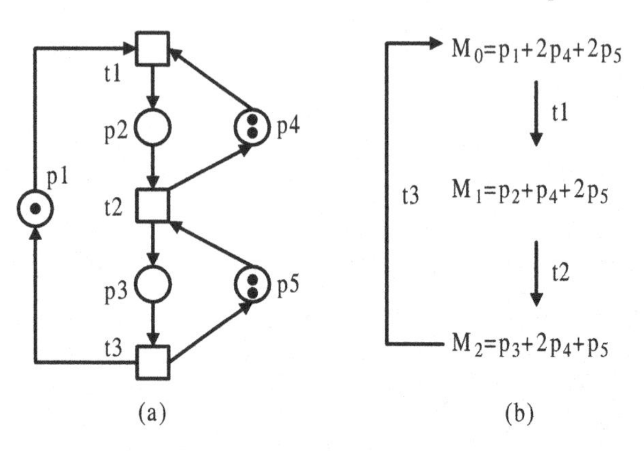

Fig. 4.1 (a) A Petri net (N, M_0) and (b) its reachability graph $R(N, M_0)$

Example 4.3. $(l_1, b_1) \equiv M(p_1) + M(p_7) \leq 6$, $(l_2, b_2) \equiv M(p_2) + M(p_6) \leq 6$, $(l_3, b_3) \equiv M(p_3) + M(p_5) + M(p_6) \leq 7$, $(l_4, b_4) \equiv M(p_2) + M(p_4) + M(p_7) \leq 10$, and $(l_5, b_5) \equiv M(p_3) + M(p_5) + M(p_6) + M(p_7) \leq 14$ are a set of GMECs that are enforced to the plant shown in Fig. 4.2. We have $\eta_1 = -t_1 + t_4$, $\eta_2 = t_2 - t_3$, $\eta_3 = t_1 - t_2 + t_4$, $\eta_4 = -2t_1 + t_2 - 3t_3 + t_4$, and $\eta_5 = -t_3 + 2t_4$.

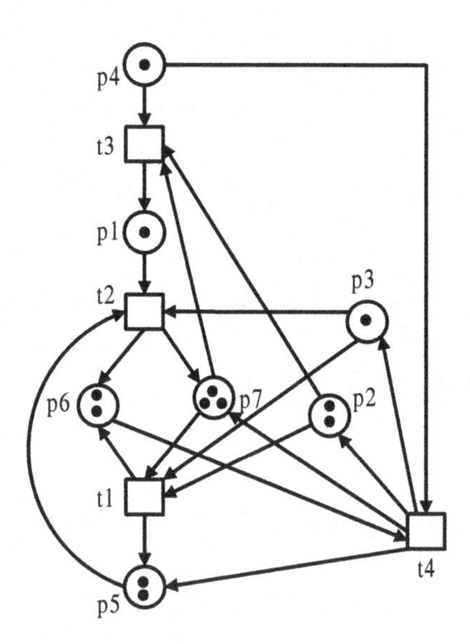

Fig. 4.2 A Petri net

Definition 4.7. Let (L, B) be a set of GMECs with n constraints. $[\eta] = [\eta_1|\eta_2|\cdots|\eta_n]^T$ is called the characteristic T-vector matrix of the set of GMECs.

Example 4.4. For the set of GMECs enforced to the plant shown in Fig. 4.2 in Example 4.3, its characteristic T-vector matrix is

$$[\eta] = \begin{pmatrix} -1 & 0 & 0 & 1 \\ 0 & 1 & -1 & 0 \\ 1 & -1 & 0 & 1 \\ -2 & 1 & -3 & 1 \\ 0 & 0 & -1 & 2 \end{pmatrix}.$$

Definition 4.8. Let η_α, η_β, ..., and η_γ ($\{\alpha,\beta,\ldots,\gamma\} \subseteq \mathbb{N}_n$) be a linearly independent maximal set of matrix $[\eta]$. Then $L_E = \{(l_\alpha,b_\alpha),(l_\beta,b_\beta),\ldots,(l_\gamma,b_\gamma)\}$ is called a set of elementary constraints in (L,B).

Definition 4.9. $(l,b) \notin L_E$ is called a strongly dependent constraint if $\eta = \sum_{(l_i,b_i)\in L_E} a_i\eta_i$, where $a_i \geq 0$.

Definition 4.10. $(l,b) \notin L_E$ is called a weakly dependent constraint if there exist non-empty sets L^1, $L^2 \subset L_E$ such that $L^1 \cap L^2 = \emptyset$ and $\eta = \sum_{(l_i,b_i)\in L^1} a_i\eta_i - \sum_{(l_j,b_j)\in L^2} a_j\eta_j$, where $a_i, a_j > 0$.

If (l,b) is a strongly dependent constraint with respect to $l_1 - l_n$, i.e., $\eta = \sum_{i=1}^{n} a_i\eta_i$, $(l_1,b_1) - (l_n,b_n)$ are called the elementary constraints of (l,b). Similarly, $(l',b') \in L^1 \cup L^2$ is called an elementary constraint of (l,b) if (l,b) is weakly dependent.

Let (L,B) be a set of GMECs that are enforced to a Petri net (N,M_0) with $N = (P,T,F,W)$. Concerning the number of elementary constraints in (L,B), we have the following important results.

Theorem 4.1. *$|L_E|$ is equal to rank$([\eta])$, where rank$([\eta])$ is the rank of $[\eta]$.*

Proof. It follows immediately from the definition of elementary constraints. □

Theorem 4.2. $|L_E| \leq min\{|P|,|T|\}$.

Proof. Since $[\eta] = L^T[N]$, we have $rank([\eta]) \leq min\{rank(L), rank([N])\}$. It implies that $rank([\eta]) \leq rank([N])$. Hence $|L_E| \leq min\{|P|,|T|\}$ is true. □

This result indicates that the number of elementary constraints in a set of GMECs that are enforced to a Petri net is bounded by the smaller of its place and transition counts.

Example 4.5. In Example 4.3, the five GMECs enforced to the Petri net shown in Fig. 4.2 lead to $\eta_5 = \eta_1 + \eta_2 + \eta_3$. It is easy to verify $rank([\eta]) = 4$, indicating that there are four elementary and one dependent constraints. If $L_E = \{(l_1,b_1)$, (l_2,b_2), (l_3,b_3), $(l_4,b_4)\}$ is selected as elementary constraints, (l_5,b_5) is strongly dependent. If $L_E = \{(l_1,b_1),(l_2,b_2)$, (l_4,b_4), $(l_5,b_5)\}$ is selected, (l_3,b_3) is weakly dependent.

4.4 Implicit Enforcement of Dependent Constraints

This section focuses on the conditions under which a dependent constraint is implicitly enforced by properly enforcing its elementary constraints. First a method from [9] that enforces a GMEC by adding a monitor is reviewed.

Let (l,b) be a GMEC enforced to a plant (N,M_0) with $N=(P,T,F,W)$ such that M_0 satisfies the constraint (l,b), i.e., $l^T M_0 \leq b$.

Definition 4.11. Let (N^S,M_0^S) be the resultant net with the addition of monitor V such that $[N^S](V,\cdot) = -l^T[N]$, $\forall p \in P$, $M_0^S(p) = M_0(p)$, and $M_0^S(V) = b - l^T M_0$.

Proposition 4.2. *Monitor V implements constraint (l,b), i.e., $\forall M \in R(N^S,\ M_0^S)$, $l^T M \leq b$, and minimally restricts the behavior of (N^S,M_0^S) in the sense that it prevents only transition firings that yield forbidden markings.*

Example 4.6. Suppose that V_1 is the monitor that enforces $(l_1,b_1) \equiv M(p_1) + M(p_7) \leq 6$ to the Petri net shown in Fig. 4.2. We have

$$[N](V_1,\cdot) = -l_1^T[N] = -\begin{pmatrix} 1\,0\,0\,0\,0\,0\,1 \end{pmatrix} \begin{pmatrix} 0 & -1 & 1 & 0 \\ -1 & 0 & -1 & 1 \\ -1 & -1 & 0 & 1 \\ 0 & 0 & -1 & -1 \\ 1 & -1 & 0 & 1 \\ 1 & 1 & 0 & -1 \\ -1 & 1 & -1 & 1 \end{pmatrix} = \begin{pmatrix} 1\,0\,0\,-1 \end{pmatrix}$$

and $M_0^S(V_1) = b_1 - l_1^T M_0 = 6 - 4 = 2$, as shown in Fig. 4.3. Similarly, we can derive the following monitors $V_2 - V_5$:

$[N^S](V_2,\cdot) = -t_2 + t_3$,
$[N^S](V_3,\cdot) = -t_1 + t_2 - t_4$,
$[N^S](V_4,\cdot) = 2t_1 - t_2 + 3t_3 - t_4$,
$[N^S](V_5,\cdot) = t_3 - 2t_4$,
$M_0^S(V_2) = b_2 - l_2^T M_0 = 2$,
$M_0^S(V_3) = b_3 - l_3^T M_0 = 2$,
$M_0^S(V_4) = b_4 - l_4^T M_0 = 4$,
$M_0^S(V_5) = b_5 - l_5^T M_0 = 6$.

According to Proposition 4.2, they implement all the constraints and minimally restrict the net behavior.

Next a parameterized constraint enforcement approach is proposed, which plays an important role in the development of the condition under which a dependent constraint is implicitly enforced.

Proposition 4.3. *Given a constraint (l,b), the incidence vector of its monitor $[N^S](V,\cdot)$ is defined in Definition 4.11. $\forall M \in R(N^S,M_0^S)$, $l^T M \leq b$ if $M_0^S(V) = b - l^T M_0 - \xi$, where $0 \leq \xi \leq b - l^T M_0$.*

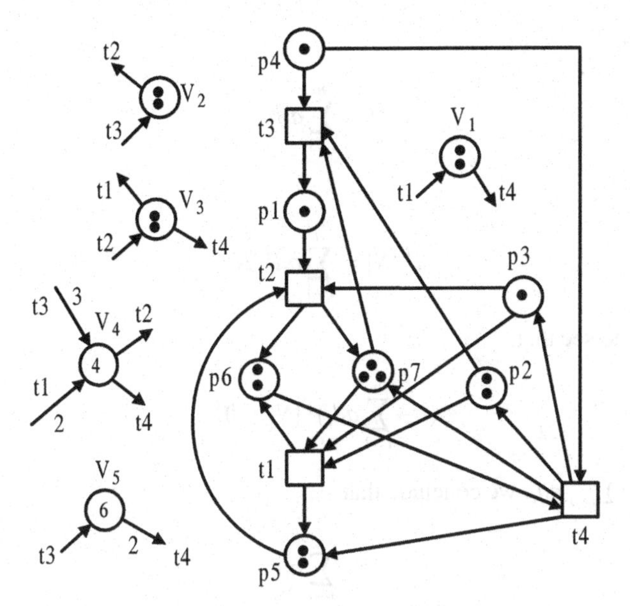

Fig. 4.3 A controlled system (N^S, M_0^S)

Proof. Since $0 \leq \xi \leq b - l^T M_0$, $\forall M \in R(N, M_0)$, $l^T M \leq b - \xi \leq b$. That is to say, any reachable marking in (N^S, M_0^S) satisfies the constraint (l, b). □

Similarly to the control depth variable of a siphon in [34], ξ is called the control depth variable of a constraint (l, b). For $(l_1, b_1) \equiv M(p_1) + M(p_7) \leq 6$ in Fig. 4.2, $\xi_1 = 1$ means that $M_0^S(V_1) = 1$.

Let (l, b) be a GMEC to (N, M_0). We define

$$M_{max}^l = max\{l^T M | M \in R(N, M_0)\} \tag{4.1}$$

and

$$M_{min}^l = min\{l^T M | M \in R(N, M_0)\}. \tag{4.2}$$

Example 4.7. The Petri net shown in Fig. 4.2 has 11 reachable states. For (l_1, b_1), (l_2, b_2), and (l_3, b_3), it is easy to find that $M_{max}^{l_1} = 5$, $M_{min}^{l_1} = 3$, $M_{max}^{l_2} = 5$, $M_{min}^{l_2} = 3$, $M_{max}^{l_3} = 8$, and $M_{min}^{l_3} = 4$.

Theorem 4.3. *Let (l, b) be a strongly dependent constraint with $\eta = \sum_{i=1}^{n} a_i \eta_i$ and $l \neq \sum_{i=1}^{n} a_i l_i$. (l, b) is redundant with respect to (N, M_0) if*

$$l^T M_0 - \sum_{i=1}^{n} a_i l_i^T M_0 + \sum_{i=1}^{n} a_i M_{max}^{l_i} \leq b.$$

Proof. Since

$$\eta = \sum_{i=1}^{n} a_i \eta_i,$$

we have

$$l^T[N] = \sum_{i=1}^{n} a_i l_i^T[N].$$

It is easy to see that

$$(l - \sum_{i=1}^{n} a_i l_i)^T [N] = \mathbf{0}^T.$$

Since $l \neq \sum_{i=1}^{n} a_i l_i$, we conclude that

$$l - \sum_{i=1}^{n} a_i l_i$$

is a P-invariant of N.
This implies that $\forall M \in R(N, M_0)$,

$$(l - \sum_{i=1}^{n} a_i l_i)^T M = (l - \sum_{i=1}^{n} a_i l_i)^T M_0. \tag{4.3}$$

From (4.3), we have

$$l^T M = l^T M_0 - \sum_{i=1}^{n} a_i l_i^T M_0 + \sum_{i=1}^{n} a_i l_i^T M \tag{4.4}$$

and

$$l^T M \leq l^T M_0 - \sum_{i=1}^{n} a_i l_i^T M_0 + \sum_{i=1}^{n} a_i M_{max}^{l_i}.$$

Clearly, $l^T M \leq b$ is true if

$$l^T M_0 - \sum_{i=1}^{n} a_i l_i^T M_0 + \sum_{i=1}^{n} a_i M_{max}^{l_i} \leq b. \tag{4.5}$$

□

Example 4.8. As stated previously, $(l_5, b_5) \equiv M(p_3) + M(p_5) + M(p_6) + M(p_7) \leq 14$ in Fig. 4.2 is a strongly dependent constraint with respect to (l_1, b_1), (l_2, b_2), and (l_3, b_3) with $\eta_5 = \eta_1 + \eta_2 + \eta_3$. We have
$l_5^T M_0 = M_0(p_3) + M_0(p_5) + M_0(p_6) + M_0(p_7) = 8$,
$l_1^T M_0 = M_0(p_1) + M_0(p_7) = 4$,
$l_2^T M_0 = M_0(p_2) + M_0(p_6) = 4$,

$l_3^T M_0 = M_0(p_3) + M_0(p_5) + M_0(p_6) = 5$,
$M_{max}^{l_1} = 5$,
$M_{max}^{l_2} = 5$,
$M_{max}^{l_3} = 8$.

Considering $b_5 = 14$, we have $l_5^T M_0 - \sum_{i=1}^{3} a_i l_i^T M_0 + \sum_{i=1}^{3} a_i M_{max}^{l_i} < b_5$. As a result, $(l_5, b_5) \equiv M(p_3) + M(p_5) + M(p_6) + M(p_7) \leq 14$ is redundant with respect to the Petri net in Fig. 4.2.

Theorem 4.4. *Let (l,b) be a strongly dependent constraint with $\eta = \sum_{i=1}^{n} a_i \eta_i$ and $l = \sum_{i=1}^{n} a_i l_i$ in a Petri net (N, M_0). (l,b) is redundant with respect to (N, M_0) if*

$$\sum_{i=1}^{n} a_i b_i \leq b.$$

Proof. $\forall M \in R(N, M_0)$, we have $l^T M = \sum_{i=1}^{n} a_i l_i^T M \leq \sum_{i=1}^{n} a_i b_i \leq b$. This result is true. □

Example 4.9. Suppose that there is a set of GMECs $(l_1, b_1) \equiv M(p_2) \leq 2$, $(l_2, b_2) \equiv M(p_3) \leq 1$, and $(l_3, b_3) \equiv 2M(p_2) + M(p_3) \leq 10$ for the Petri net shown in Fig. 4.4(a). Their corresponding monitors are shown in Fig. 4.4(b). The fact $l_3 = 2l_1 + l_2$ can be verified. Clearly, we have $2b_1 + b_2 < b_3$ and thus conclude that (l_3, b_3) is redundant.

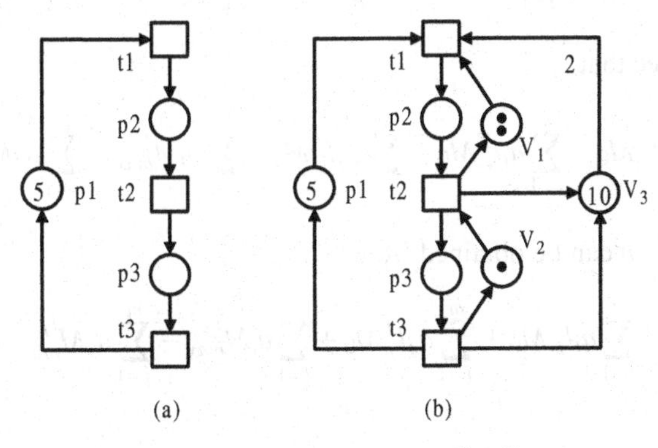

Fig. 4.4 (a) a plant (N, M_0) and (b) the controlled system (N^S, M_0^S)

Theorem 4.5. *Let (l,b) be a weakly dependent GMEC with $\eta = \sum_{i=1}^{n} a_i \eta_i - \sum_{j=1}^{m} a_j \eta_j$ and $l \neq \sum_{i=1}^{n} a_i l_i - \sum_{j=1}^{m} a_j l_j$ in a plant (N, M_0). It is redundant with respect to (N, M_0) if*

$$l^T M_0 - \sum_{i=1}^{n} a_i l_i^T M_0 + \sum_{j=1}^{m} a_j l_j^T M_0 + \sum_{i=1}^{n} a_i M_{max}^{l_i} - \sum_{j=1}^{m} a_j M_{min}^{l_j} \leq b.$$

Proof. Since

$$\eta = \sum_{i=1}^{n} a_i \eta_i - \sum_{j=1}^{m} a_j \eta_j \tag{4.6}$$

we have

$$(l - \sum_{i=1}^{n} a_i l_i + \sum_{j=1}^{m} a_j l_j)^T [N] = \mathbf{0}^T, \tag{4.7}$$

indicating that

$$l - \sum_{i=1}^{n} a_i l_i + \sum_{j=1}^{m} a_j l_j$$

is a P-invariant of N.
As a result, $\forall M \in R(N, M_0)$,

$$l^T M_0 - \sum_{i=1}^{n} a_i l_i^T M_0 + \sum_{j=1}^{m} a_j l_j^T M_0 = l^T M - \sum_{i=1}^{n} a_i l_i^T M + \sum_{j=1}^{m} a_j l_j^T M,$$

i.e.,

$$l^T M = l^T M_0 - \sum_{i=1}^{n} a_i l_i^T M_0 + \sum_{j=1}^{m} a_j l_j^T M_0 + \sum_{i=1}^{n} a_i l_i^T M - \sum_{j=1}^{m} a_j l_j^T M.$$

It is easy to see that

$$l^T M \leq l^T M_0 - \sum_{i=1}^{n} a_i l_i^T M_0 + \sum_{j=1}^{m} a_j l_j^T M_0 + \sum_{i=1}^{n} a_i M_{max}^{l_i} - \sum_{j=1}^{m} a_j M_{min}^{l_j}.$$

Hence, $l^T M \leq b$ can be obtained from

$$l^T M_0 - \sum_{i=1}^{n} a_i l_i^T M_0 + \sum_{j=1}^{m} a_j l_j^T M_0 + \sum_{i=1}^{n} a_i M_{max}^{l_i} - \sum_{j=1}^{m} a_j M_{min}^{l_j} \leq b.$$

□

Theorem 4.6. *Let (l, b) be a weakly dependent GMEC with $\eta = \sum_{i=1}^{n} a_i \eta_i - \sum_{j=1}^{m} a_j \eta_j$ and $l = \sum_{i=1}^{n} a_i l_i - \sum_{j=1}^{m} a_j l_j$. It is redundant with respect to (N, M_0) if*

$$\sum_{i=1}^{n} a_i b_i - \sum_{j=1}^{m} a_j b_j \leq b.$$

Proof. Similar to the proof of Theorem 4.4. □

Note that in [9], the redundancy of a GMEC such as $(l_3, b_3) \equiv 2M(p_2) + M(p_3) \leq 10$ in Example 4.9 can be verified by solving an LPP, as stated in Proposition 4.1. It is clear that the computational complexity of Theorem 4.3 to decide the redundancy of a dependent constraint is worse than Proposition 4.1 since the complete state enumeration is needed in Theorem 4.3. However, it is shown next that the redundancy of all dependent GMECs can be determined by solving $2|L_E|$ LPPs only, where $|L_E| \leq min\{|P|, |T|\}$.

Let (l, b) be a GMEC in a Petri net (N, M_0). We define

$$M^l_{MAX} = max\{l^T M | M = M_0 + [N]Y, M \geq 0, Y \geq 0\} \tag{4.8}$$

and

$$M^l_{MIN} = min\{l^T M | M = M_0 + [N]Y, M \geq 0, Y \geq 0\}. \tag{4.9}$$

Clearly, we have

$$M^l_{MAX} \geq M^l_{max} \tag{4.10}$$

and

$$M^l_{MIN} \leq M^l_{min} \tag{4.11}$$

It is trivial that (4.8) and (4.9) can be found in polynomial time by solving LPPs. In Fig. 4.2, it is easy to find that $M^{l_1}_{MAX} = 5$, $M^{l_2}_{MAX} = 5$, $M^{l_3}_{MAX} = 8$, $M^{l_1}_{MIN} = 3$, $M^{l_2}_{MIN} = 3$, and $M^{l_3}_{MIN} = 4$.

Corollary 4.1. *Let (l, b) be a strongly dependent constraint with $\eta = \sum_{i=1}^n a_i \eta_i$ and $l \neq \sum_{i=1}^n a_i l_i$ in (N, M_0). (l, b) is redundant with respect to (N, M_0) if*

$$l^T M_0 - \sum_{i=1}^{n} a_i l_i^T M_0 + \sum_{i=1}^{n} a_i M^{l_i}_{MAX} \leq b.$$

Proof. It follows immediately from Theorem 4.3 and (4.10). □

Corollary 4.2. *Let (l, b) be a weakly dependent GMEC in Petri net (N, M_0) with $\eta = \sum_{i=1}^n a_i \eta_i - \sum_{j=1}^m a_j \eta_j$ and $l \neq \sum_{i=1}^n a_i l_i - \sum_{j=1}^m a_j l_j$. It is redundant with respect to (N, M_0) if*

$$l^T M_0 - \sum_{i=1}^{n} a_i l_i^T M_0 + \sum_{j=1}^{m} a_j l_j^T M_0 + \sum_{i=1}^{n} a_i M^{l_i}_{MAX} - \sum_{j=1}^{m} a_j M^{l_j}_{MIN} \leq b.$$

Corollaries 4.1 and 4.2 indicate that the redundancy of (weakly and strongly) dependent constraints can be decided provided that $M^{l_i}_{MAX}$ and $M^{l_i}_{MIN}$ are known, where $i \in \mathbb{N}_{|L_E|}$. That is to say, due to the two corollaries, the redundancy of dependent constraints can be decided by solving $2|L_E|$ LPPs no matter how many dependent constraints in a set of GMECs. However, as stated in Proposition 4.1, the verification of the redundancy of each GMEC needs to solve an LPP.

A parameterized approach is next presented for a dependent constraint (l,b) with $\eta = \sum_{i=1}^{n} a_i\eta_i$ that can be implicitly enforced to a plant (N,M_0) by properly setting the control depth variables of its elementary constraints. The implicit enforcement of (l,b) implies that it is redundant with respect to the augmented net with additional monitors (N^S, M_0^S).

Theorem 4.7. *Let (l,b) be a strongly dependent constraint of Petri net (N,M_0) with $\eta = \sum_{i=1}^{n} a_i\eta_i$ and $l \neq \sum_{i=1}^{n} a_i l_i$. $\forall i \in \mathbb{N}_n$, (l_i,b_i) is enforced by adding a monitor V_i with its control depth variable ξ_i, as stated in Proposition 4.3. Then, (l,b) is implicitly enforced to (N,M_0) if*

$$\sum_{i=1}^{n} a_i\xi_i \geq (l^T M_0 - b) + \sum_{i=1}^{n} a_i(b_i - l_i^T M_0).$$

Proof. According to Theorem 4.3, (l,b) is implicitly enforced if

$$l^T M_0 - \sum_{i=1}^{n} a_i l_i^T M_0 + \sum_{i=1}^{n} a_i M_{max}^{l_i} \leq b.$$

By Proposition 4.3, $\forall i \in \mathbb{N}_n$, if a monitor V_i is added such that (l_i,b_i) is enforced with $M_0^S(V_i) = b_i - l_i^T M_0 - \xi_i$, we have $M_{max}^{l_i} = b_i - \xi_i$. As a result, (l,b) is implicitly enforced if

$$l^T M_0 - \sum_{i=1}^{n} a_i l_i^T M_0 + \sum_{i=1}^{n} a_i(b_i - \xi_i) \leq b,$$

i.e.,

$$\sum_{i=1}^{n} a_i\xi_i \geq (l^T M_0 - b) + \sum_{i=1}^{n} a_i(b_i - l_i^T M_0).$$

□

Example 4.10. Suppose that $(l_5,b_5) \equiv M(p_3)+M(p_5)+M(p_6)+M(p_7) \leq 9$ is a GMEC in Fig. 4.2. As is known, $\eta_5 = \eta_1+\eta_2+\eta_3$. It is not redundant with respect to its elementary constraints (l_1,b_1), (l_2,b_2), and (l_3,b_3). However, it can be made redundant if setting the control depth variables by $\sum_{i=1}^{3} \xi_i \geq (l_5^T M_0 - b_5) + (b_1 - l_1^T M_0) + (b_2 - l_2^T M_0) + (b_3 - l_3^T M_0)$, where $b_5 = 9$, $l_5^T M_0 = 8$, $b_1 = 6$, $b_2 = 6$, $b_3 = 7$, $l_1^T M_0 = 4$, $l_2^T M_0 = 4$, and $l_3^T M_0 = 5$. That is to say, $(l_5,b_5) \equiv M(p_3)+M(p_5)+M(p_6)+M(p_7) \leq 9$ is redundant if $\sum_{i=1}^{3} \xi_i \geq (8-9)+(6-4)+(6-4)+(7-5) = 5$. Specifically, let $\xi_1 = \xi_2 = 2$ and $\xi_3 = 1$. $(l_5,b_5) \equiv M(p_3)+M(p_5)+M(p_6)+M(p_7) \leq 9$ is redundant if (l_1,b_1), (l_2,b_2), and (l_3,b_3) are enforced by monitors V_1,

V_2, and V_3 with $M_0^S(V_1) = b_1 - \xi_1 - l_1^T M_0 = 0$, $M_0^S(V_2) = b_2 - \xi_2 - l_2^T M_0 = 0$, and $M_0^S(V_3) = b_3 - \xi_3 - l_3^T M_0 = 1$.

Theorem 4.8. *Let (l,b) be a strongly dependent constraint of Petri net (N,M_0) with $\eta = \sum_{i=1}^n a_i \eta_i$ and $l = \sum_{i=1}^n a_i l_i$, where $\eta_i = [N]^T l_i$. $\forall i \in \mathbb{N}_n$, (l_i, b_i) is enforced by adding a monitor V_i with its control depth variable ξ_i, as stated in Proposition 4.3. (l,b) is implicitly enforced if*

$$\sum_{i=1}^{n} a_i (b_i - \xi_i) \leq b.$$

Proof. It follows from Theorem 4.4 and Proposition 4.3. □

Theorem 4.9. *Let (l,b) be a weakly dependent constraint in a Petri net (N,M_0) with $\eta = \sum_{i=1}^n a_i \eta_i - \sum_{j=1}^m a_j \eta_j$ and $l \neq \sum_{i=1}^n a_i l_i - \sum_{j=1}^m a_j l_j$. $\forall i \in \mathbb{N}_n$, (l_i, b_i) is enforced by adding a monitor V_i with control depth variable ξ_i according to Proposition 4.3. (l,b) is implicitly enforced to (N,M_0) if*

$$\sum_{i=1}^{n} a_i \xi_i \geq (l^T M_0 - b) + \sum_{i=1}^{n} a_i (b_i - l_i^T M_0) + \sum_{j=1}^{m} a_j l_j^T M_0.$$

Proof. Due to Theorem 4.5, (l,b) is implicitly enforced if

$$l^T M_0 - \sum_{i=1}^{n} a_i l_i^T M_0 + \sum_{j=1}^{m} a_j l_j^T M_0 + \sum_{i=1}^{n} a_i M_{max}^{l_i} - \sum_{j=1}^{m} a_j M_{min}^{l_j} \leq b. \tag{4.12}$$

From Proposition 4.3, we have $M_{max}^{l_i} = b_i - \xi_i$ and $M_{min}^{l_j} \geq 0$. As a result,

$$l^T M_0 - \sum_{i=1}^{n} a_i l_i^T M_0 + \sum_{j=1}^{m} a_j l_j^T M_0 + \sum_{i=1}^{n} a_i (b_i - \xi_i) \leq b \tag{4.13}$$

implies the truth of (4.12).

From (4.13), we have

$$\sum_{i=1}^{n} a_i \xi_i \geq (l^T M_0 - b) + \sum_{i=1}^{n} a_i (b_i - l_i^T M_0) + \sum_{j=1}^{m} a_j l_j^T M_0.$$

□

Theorem 4.10. *Let (l,b) be a weakly dependent constraint to a Petri net (N,M_0) with $\eta = \sum_{i=1}^n a_i \eta_i - \sum_{j=1}^m a_j \eta_j$ and $l = \sum_{i=1}^n a_i l_i - \sum_{j=1}^m a_j l_j$. $\forall i \in \mathbb{N}_n$, (l_i, b_i) is enforced by adding a monitor V_i with control depth variable ξ_i according to Proposition 4.3. It is implicitly enforced to (N,M_0) if*

$$\sum_{i=1}^{n} a_i (b_i - \xi_i) - \sum_{j=1}^{m} a_j b_j \leq b.$$

4.5 Application to Deadlock Prevention

This section presents the application of the methods presented in this chapter to a deadlock prevention policy developed by Park and Reveliotis in [40]. It is shown that these methods may reduce the structural complexity of the liveness-enforcing supervisor for a class of Petri nets, S^3PGR^2. It stands for *system of simple sequential processes with general resource requirements*. As stated in Chap. 1, an S^3PGR^2 is equivalent to an S^4R. For simplicity, we use S^4R to denote the class of Petri nets in the rest of this book.

Definition 4.12. A well-marked S^4R net is a marked Petri net $N = (P,T,F,W)$ with initial marking M_0 such that:

1. $P = P_A \cup P^0 \cup P_R$, where $P_A = \cup_{j=1}^{n} P_{A_j}$ is called the set of operation places such that $P_{A_i} \cap P_{A_j} = \emptyset$, $\forall i \neq j$, $P^0 = \cup_{i=1}^{n}\{p_i^0\}$ is called the set of idle places with $P^0 \cap P_A = \emptyset$, and $P_R = \{r_1, r_2, \cdots, r_m\}$ is called the set of resource places such that $(P^0 \cup P_A) \cap P_R = \emptyset$.
2. $T = \cup_{j=1}^{n} T_j$, and $\forall i \neq j$, $T_i \cap T_j = \emptyset$.
3. $W = W_A \cup W_R$, where $W_A : ((P_A \cup P^0) \times T) \cup (T \times (P_A \cup P^0)) \rightarrow \{0,1\}$ such that $\forall j \neq i$, $((P_{A_j} \cup \{p_j^0\}) \times T_i) \cup (T_i \times (P_{A_j} \cup \{p_j^0\})) \rightarrow \{0\}$, and $W_R : (P_R \times T) \cup (T \times P_R) \rightarrow \mathbb{N}$.
4. $\forall j \in \mathbb{N}_n$, the subnet N_j derived from $P_{A_j} \cup \{p_j^0\} \cup T_j$ is a strongly connected state machine such that every circuit contains p_j^0.
5. $\forall r \in P_R$, there exists a unique P-semiflow I_r such that $||I_r|| \cap P_R = \{r\}$, $||I_r|| \cap P^0 = \emptyset$, $||I_r|| \cap P_A \neq \emptyset$, and $I_r(r) = 1$. Furthermore, $P_A = (\cup_{r \in P_R} ||I_r||) \setminus P_R$.
6. N is pure and strongly connected.
7. $\forall p \in P_A$, $M_0(p) = 0$; $\forall r \in P_R$, $M_0(r) \geq max_{p \in ||I_r||} I_r(p)$; and $\forall p_j^0 \in P^0$, $M_0(p_j^0) \geq 1$.

From its definition, an S^4R is equivalent to an S^4PR. The following result about the rank of the incidence matrix of an S^4R is from [43].

Theorem 4.11. *Let* $N = (P_A \cup P^0 \cup P_R, T, F, W)$ *be an* S^4R*. Then,* $rank([N]) = |P_A|$.

For an S^4R, Park and Reveliotis [40] develop a deadlock prevention policy that is of polynomial complexity. The liveness requirements in an S^4R are represented by a set of GMECs that can be implemented by a set of monitors. The policy considers a system that is formally defined by a set of resources $R = \{r_i | i \in \mathbb{N}_m\}$ and a set of jobs $J = \{J_j | j \in \mathbb{N}_n\}$. Each resource type r_i has a capacity $C_i \in \mathbb{N}^+$. Each job type J_j is defined by a set of operations $\{p_{jk} | k \in \mathbb{N}_{\lambda_j}, \lambda_j \in \mathbb{N}^+\}$, which is partially ordered through a set of precedence constraints. Each job operation p_{jk} is associated with a conjunctive resource allocation requirement, formally expressed by an m-dimensional vector $a_{p_{jk}}$, with $a_{p_{jk}}[i]$, $i \in \mathbb{N}_m$, indicating how many units of resource r_i are required to support the operation execution. Such a system can be modeled by a class of Petri nets, namely S^4R, that is more general than S^3PR [8]. Let (N, M_0)

denote an S^{4}R, where $N = (P^0 \cup P_A \cup P_R, T, F, W)$ and P^0 (P_A; P_R) is the set of idle (operation; resource) places.

Let $P^0 = \{p_{10}, p_{11}\}$, $P_A = \{p_1, p_2, p_3, p_4, p_5, p_6\}$, and $P_R = \{r_1, r_2, r_3\}$. The Petri net shown in Fig. 4.5 is an S^{4}R that is the model of an FMS that can produce two job types J_1 and J_2 with $J_1 = \{p_1, p_2, p_3\}$ and $J_2 = \{p_4, p_5, p_6\}$. There are three resources r_1, r_2, and r_3 in the system. Furthermore, we have $C_1 = 3$, $C_2 = 2$, and $C_3 = 1$. An operation place that has no subsequent processing stage is called a terminal operation. For example, p_3 and p_6 are terminal operation places.

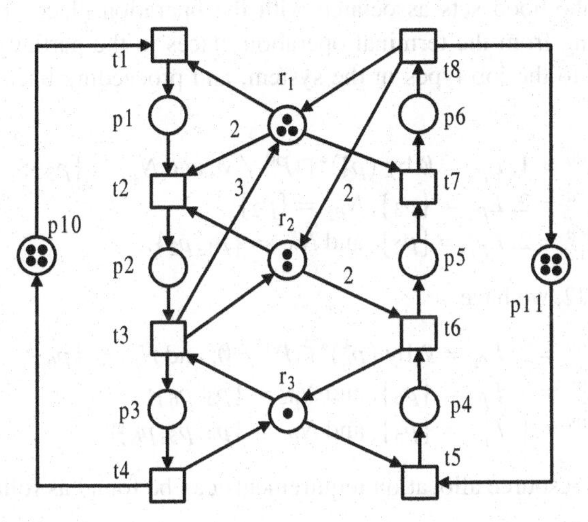

Fig. 4.5 An S^{4}R (N, M_0)

Note that each resource place r corresponds to a minimal P-semiflow I_r. As a result, we have three minimal P-semiflows that are associated with resources. They are

$I_{r_1} = r_1 + p_1 + 3p_2 + p_6$,
$I_{r_2} = r_2 + p_2 + 2p_5 + 2p_6$,
$I_{r_3} = r_3 + p_3 + p_4$.

A minimal P-semiflow I_r associated with the resource place r shows the operation places whose execution requires resource r and how many units of resource r are needed to support the operation execution. Hence, the conjunctive resource requirements of operations can be easily obtained as follows: $a_{p_1} = (1,0,0)^T$, $a_{p_2} = (3,1,0)^T$, $a_{p_3} = (0,0,1)^T$, $a_{p_4} = (0,0,1)^T$, $a_{p_5} = (0,2,0)^T$, and $a_{p_6} = (1,2,0)^T$.

Let $o_i \equiv O(r_i)$, $O: R \rightarrow \mathbb{N}_m$ be any partial order imposed on the resource set R. Given $p \in P_A$, let $\rho_p^{max} = max\{o_i | a_p[i] > 0, i \in \mathbb{N}_m\}$ and $\rho_p^{min} = min\{o_i | a_p[i] > 0, i \in \mathbb{N}_m\}$. Also, let $L_p = \{q | q \in (p^\bullet)^\bullet \cap P_A \wedge \rho_q^{max} = min_{v \in (p^\bullet)^\bullet \cap P_A} \rho_v^{max}\}$. By convention, $L_p = \emptyset$ if $(p^\bullet)^\bullet \cap P^0 \neq \emptyset$. Then:

1. The neighborhood set N_p of $p \in P_A$ is defined by $N_p = \{p\} \cup \{q | q \in \cup_{v \in L_p} N_v \wedge \rho_p^{min} \leq \rho_q^{max}\}$.

2. The adjusted resource allocation requirement $\hat{a}_p$ for $p \in P_A$ under partial order $O()$ (resource ordering) is given by $\hat{a}_p[i] = max\{a_q[i] | q \in N_p\}$ if $o_i \geq \rho_p^{min}$; otherwise $\hat{a}_p[i] = 0, \forall i \in \mathbb{N}_m$.
3. The policy-imposed constraint on the system operation is expressed by the requirement that no resource is over-allocated with respect to the adjusted operation requirements specified by $\hat{a}_p[i]$.

Consider the net shown in Fig. 4.5 under the resource order $o_1 = 2$, $o_2 = 3$, and $o_3 = 1$. Note that different resource orderings may lead to different sets of GMECs [39]. The neighborhood sets associated with the operation places $p \in P_A$ can be computed starting from the terminal operation places in the partially ordered sets that correspond to the job types in the system, and proceeding backward. For job type J1, we have

- $\rho_{p_3}^{max} = 1$, $\rho_{p_3}^{min} = 1$, $L_{p_3} = \emptyset$ by $(p_3^\bullet)^\bullet \cap P^0 \neq \emptyset$, and $N_{p_3} = \{p_3\}$;
- $\rho_{p_2}^{max} = 3$, $\rho_{p_2}^{min} = 2$, $L_{p_2} = \{p_3\}$, $N_{p_2} = \{p_2\}$;
- $\rho_{p_1}^{max} = 2$, $\rho_{p_1}^{min} = 2$, $L_{p_1} = \{p_2\}$, and $N_{p_1} = \{p_1, p_2\}$.

For job type J2, we have

- $\rho_{p_6}^{max} = 3$, $\rho_{p_6}^{min} = 2$, $L_{p_6} = \emptyset$ by $(p_6^\bullet)^\bullet \cap P^0 \neq \emptyset$, and $N_{p_6} = \{p_6\}$;
- $\rho_{p_5}^{max} = 3$, $\rho_{p_5}^{min} = 3$, $L_{p_5} = \{p_6\}$, and $N_{p_5} = \{p_5, p_6\}$;
- $\rho_{p_4}^{max} = 1$, $\rho_{p_4}^{min} = 1$, $L_{p_4} = \{p_5\}$, and $N_{p_4} = \{p_4, p_5, p_6\}$.

The adjusted resource allocation requirements can be found as follows:

$$\hat{a}_{p_1} = \begin{pmatrix} 3 \\ 1 \\ 0 \end{pmatrix}, \ \hat{a}_{p_2} = \begin{pmatrix} 3 \\ 1 \\ 0 \end{pmatrix}, \ \hat{a}_{p_3} = \begin{pmatrix} 0 \\ 0 \\ 1 \end{pmatrix}, \ \hat{a}_{p_4} = \begin{pmatrix} 1 \\ 2 \\ 1 \end{pmatrix}, \ \hat{a}_{p_5} = \begin{pmatrix} 0 \\ 2 \\ 0 \end{pmatrix}, \ \hat{a}_{p_6} = \begin{pmatrix} 1 \\ 2 \\ 0 \end{pmatrix}.$$

By imposing on the operation places a set of linear inequality constraints that are implemented by monitors, the supervised system has no deadlock states, implying that liveness is ensured. The set of inequality constraints takes the form of

$$\hat{A}_p \cdot M_P \leq f_p \tag{4.14}$$

where the column vector in $\hat{A}_p$ corresponding to an operation place p is $\hat{a}_p$, vector M_P is the restriction of marking M to operation places, and f_p is the capacity vector of resources, i.e., $f_p(i) = C_i$, $i \in \mathbb{N}_{|R|}$.

For the net shown in Fig. 4.5, we have

$$\begin{pmatrix} 3\,3\,0\,1\,0\,1 \\ 1\,1\,0\,2\,2\,2 \\ 0\,0\,1\,1\,0\,0 \end{pmatrix} \cdot \begin{pmatrix} M(p_1) \\ M(p_2) \\ M(p_3) \\ M(p_4) \\ M(p_5) \\ M(p_6) \end{pmatrix} \leq \begin{pmatrix} 3 \\ 2 \\ 1 \end{pmatrix}.$$

By extending $\hat{A}_p$ and M_P, the liveness control requirements stated in (4.14) can be described as

$$\hat{A} \cdot M \leq f_p \tag{4.15}$$

where $\hat{A}$ is derived from $\hat{A}_p$ by adding zero column vectors that correspond to idle and resource places and M is any marking in $R(N, M_0)$. It is easy to see that (4.15) is a typical set of GMECs $(\hat{A}, f_p)$.

For the net in Fig. 4.5, its liveness requirement represented by $\hat{A} \cdot M \leq f_p$ takes the following specific form, where we assume that the places are ordered in the incidence matrix according to the sequence $\langle p_1, p_2, p_3, p_4, p_5, p_6, p_{10}, p_{11}, r_1, r_2, r_3\rangle$. Figure 4.6 shows the controlled system of the S^{4}R in Fig. 4.5.

$$\begin{pmatrix} 3\,3\,0\,1\,0\,1\,0\,0\,0\,0\,0 \\ 1\,1\,0\,2\,2\,2\,0\,0\,0\,0\,0 \\ 0\,0\,1\,1\,0\,0\,0\,0\,0\,0\,0 \end{pmatrix} \cdot \begin{pmatrix} M(p_1) \\ M(p_2) \\ M(p_3) \\ M(p_4) \\ M(p_5) \\ M(p_6) \\ M(p_{10}) \\ M(p_{11}) \\ M(r_1) \\ M(r_2) \\ M(r_3) \end{pmatrix} \leq \begin{pmatrix} 3 \\ 2 \\ 1 \end{pmatrix}.$$

If we can distinguish redundant constraints imposed on the plant, a structurally simple liveness-enforcing supervisor may be found. To demonstrate this, a flexible manufacturing example is investigated. Its liveness-enforcing supervisor is first computed by the deadlock prevention policy in [40]. Then, we show how a simpler supervisor is synthesized.

Example 4.11. Figure 4.7 is the Petri net model of an FMS. It consists of nine resource places R1–R9, two idle places p_1 and p_6, and seven operation places p_2–p_5 and p_7–p_9. The Petri net is an S^{4}R that contains two processes with operation sets $P_1 = \{p_2, p_3, p_4, p_5\}$ and $P_2 = \{p_7, p_8, p_9\}$. The resource capacity vector is $f_p = (1\ 1\ 1\ 3\ 4\ 1\ 3\ 3\ 2)^T$. The conjunctive resource requirements of operations are

$$a_{p_2} = \begin{pmatrix} 1 \\ 0 \\ 0 \\ 0 \\ 1 \\ 0 \\ 0 \\ 0 \\ 0 \end{pmatrix}, \quad a_{p_3} = \begin{pmatrix} 0 \\ 1 \\ 0 \\ 0 \\ 1 \\ 0 \\ 1 \\ 0 \\ 0 \end{pmatrix}, \quad a_{p_4} = \begin{pmatrix} 0 \\ 0 \\ 0 \\ 1 \\ 0 \\ 1 \\ 1 \\ 0 \\ 0 \end{pmatrix}, \quad a_{p_5} = \begin{pmatrix} 0 \\ 0 \\ 1 \\ 0 \\ 0 \\ 0 \\ 1 \\ 0 \\ 0 \end{pmatrix},$$

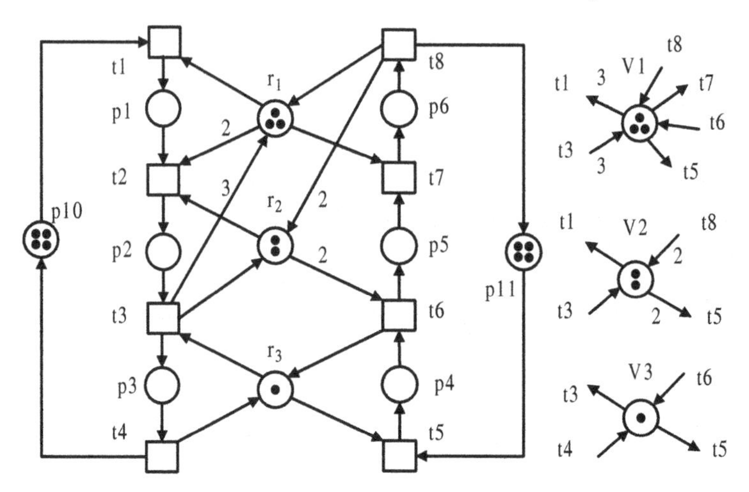

Fig. 4.6 The controlled system of the S^{4}R in Fig. 4.5

$$a_{p_7} = \begin{pmatrix} 0 \\ 0 \\ 1 \\ 1 \\ 0 \\ 0 \\ 0 \\ 0 \\ 1 \end{pmatrix}, \quad a_{p_8} = \begin{pmatrix} 0 \\ 1 \\ 0 \\ 0 \\ 0 \\ 0 \\ 0 \\ 1 \\ 1 \end{pmatrix}, \quad a_{p_9} = \begin{pmatrix} 1 \\ 0 \\ 0 \\ 0 \\ 0 \\ 0 \\ 0 \\ 1 \\ 0 \end{pmatrix}.$$

Under the resource order $o_1 = 1$, $o_2 = 1$, $o_3 = 1$, $o_4 = 3$, $o_5 = 4$, $o_6 = 1$, $o_7 = 3$, $o_8 = 3$, and $o_9 = 3$, we have

$$\hat{a}_{p_2} = \begin{pmatrix} 1 \\ 1 \\ 1 \\ 1 \\ 1 \\ 1 \\ 1 \\ 0 \\ 0 \end{pmatrix}, \quad \hat{a}_{p_3} = \begin{pmatrix} 0 \\ 1 \\ 1 \\ 1 \\ 1 \\ 1 \\ 1 \\ 0 \\ 0 \end{pmatrix}, \quad \hat{a}_{p_4} = \begin{pmatrix} 0 \\ 0 \\ 0 \\ 1 \\ 0 \\ 1 \\ 1 \\ 0 \\ 0 \end{pmatrix}, \quad \hat{a}_{p_5} = \begin{pmatrix} 0 \\ 0 \\ 1 \\ 0 \\ 0 \\ 0 \\ 1 \\ 0 \\ 0 \end{pmatrix},$$

$$\hat{a}_{p_7} = \begin{pmatrix} 1 \\ 1 \\ 1 \\ 1 \\ 0 \\ 0 \\ 0 \\ 1 \\ 1 \end{pmatrix}, \quad \hat{a}_{p_8} = \begin{pmatrix} 1 \\ 1 \\ 0 \\ 0 \\ 0 \\ 0 \\ 0 \\ 1 \\ 1 \end{pmatrix}, \quad \hat{a}_{p_9} = \begin{pmatrix} 1 \\ 0 \\ 0 \\ 0 \\ 0 \\ 0 \\ 0 \\ 1 \\ 0 \end{pmatrix}.$$

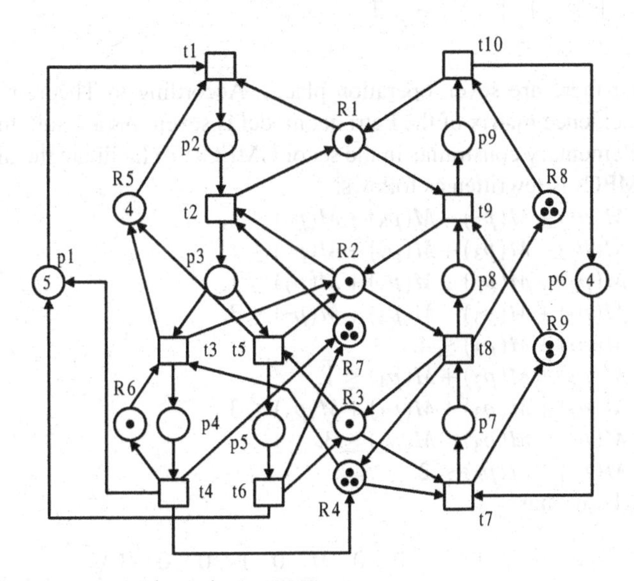

Fig. 4.7 The Petri net model (N, M_0) of an FMS

From $\hat{A}_p = [\hat{a}_{p_2}|\hat{a}_{p_3}|\hat{a}_{p_4}|\hat{a}_{p_5}|\hat{a}_{p_7}|\hat{a}_{p_8}|\hat{a}_{p_9}]$, the liveness control specifications are represented by a set of GMECs, i.e., $\hat{A} \cdot M \leq f_p$, or,

$$\begin{pmatrix} 0\,1\,0\,0\,0\,0\,1\,1\,1\,0\,0\,0\,0\,0\,0\,0\,0\,0 \\ 0\,1\,1\,0\,0\,0\,1\,1\,0\,0\,0\,0\,0\,0\,0\,0\,0\,0 \\ 0\,1\,1\,0\,1\,0\,1\,0\,0\,0\,0\,0\,0\,0\,0\,0\,0\,0 \\ 0\,1\,1\,1\,0\,0\,1\,0\,0\,0\,0\,0\,0\,0\,0\,0\,0\,0 \\ 0\,1\,1\,0\,0\,0\,0\,0\,0\,0\,0\,0\,0\,0\,0\,0\,0\,0 \\ 0\,1\,1\,1\,0\,0\,0\,0\,0\,0\,0\,0\,0\,0\,0\,0\,0\,0 \\ 0\,1\,1\,1\,1\,0\,0\,0\,0\,0\,0\,0\,0\,0\,0\,0\,0\,0 \\ 0\,0\,0\,0\,0\,0\,1\,1\,1\,0\,0\,0\,0\,0\,0\,0\,0\,0 \\ 0\,0\,0\,0\,0\,0\,1\,1\,0\,0\,0\,0\,0\,0\,0\,0\,0\,0 \end{pmatrix} \cdot M \leq \begin{pmatrix} 1 \\ 1 \\ 1 \\ 3 \\ 4 \\ 1 \\ 3 \\ 3 \\ 2 \end{pmatrix}.$$

The monitors that implement the above set of GMECs are shown in Table 4.1. For each resource, a monitor is added, i.e., there are a total of nine monitors due to the deadlock prevention policy in [40] for the Petri net model.

Table 4.1 Monitors added using the policy in [40]

V_S	$M_0^S(\cdot)$	Preset	Postset	V_S	$M_0^S(\cdot)$	Preset	Postset
V_1	1	t_2, t_{10}	t_1, t_7	V_2	1	t_3, t_5, t_9	t_1, t_7
V_3	1	t_3, t_6, t_8	t_1, t_7	V_4	3	t_4, t_5, t_8	t_1, t_7
V_5	4	t_3, t_5	t_1	V_6	1	t_4, t_5	t_1
V_7	3	t_4, t_6	t_1	V_8	3	t_{10}	t_7
V_9	2	t_9	t_7				

Notice that there are seven operation places. According to Theorem 4.11, the rank of the incidence matrix of the Petri net model is seven. As a result, there are at most seven elementary constraints in the set of GMECs. To facilitate the discussion, the set of GMECs is rewritten as follows:

$(l_1, b_1) \equiv M(p_2) + M(p_7) + M(p_8) + M(p_9) \leq 1,$
$(l_2, b_2) \equiv M(p_2) + M(p_3) + M(p_7) + M(p_8) \leq 1,$
$(l_3, b_3) \equiv M(p_2) + M(p_3) + M(p_5) + M(p_7) \leq 1,$
$(l_4, b_4) \equiv M(p_2) + M(p_3) + M(p_4) + M(p_7) \leq 3,$
$(l_5, b_5) \equiv M(p_2) + M(p_3) \leq 4,$
$(l_6, b_6) \equiv M(p_2) + M(p_3) + M(p_4) \leq 1,$
$(l_7, b_7) \equiv M(p_2) + M(p_3) + M(p_4) + M(p_5) \leq 3,$
$(l_8, b_8) \equiv M(p_7) + M(p_8) + M(p_9) \leq 3,$
$(l_9, b_9) \equiv M(p_7) + M(p_8) \leq 2.$

Accordingly, we have

$$[\eta] = \begin{pmatrix} 1 & -1 & 0 & 0 & 0 & 0 & 1 & 0 & 0 & -1 \\ 1 & 0 & -1 & 0 & -1 & 0 & 1 & 0 & -1 & 0 \\ 1 & 0 & -1 & 0 & 0 & -1 & 1 & -1 & 0 & 0 \\ 1 & 0 & 0 & -1 & -1 & 0 & 1 & -1 & 0 & 0 \\ 1 & 0 & -1 & 0 & -1 & 0 & 0 & 0 & 0 & 0 \\ 1 & 0 & 0 & -1 & -1 & 0 & 0 & 0 & 0 & 0 \\ 1 & 0 & 0 & -1 & 0 & -1 & 0 & 0 & 0 & 0 \\ 0 & 0 & 0 & 0 & 0 & 0 & 1 & 0 & 0 & -1 \\ 0 & 0 & 0 & 0 & 0 & 0 & 1 & 0 & -1 & 0 \end{pmatrix}.$$

The rank of $[\eta]$ is seven, implying that there are seven elementary and two dependent constraints. Let $L_E = \{(l_1, b_1), (l_2, b_2), (l_3, b_3), (l_4, b_4), (l_6, b_6), (l_8, b_8), (l_9, b_9)\}$. We have $\eta_5 = \eta_2 - \eta_9$, and $\eta_7 = -\eta_2 + \eta_3 - \eta_4 + 2\eta_6 + \eta_9$. Next we check the redundancy of the dependent constraints.

By solving LPP, we have

$M_{MAX}^{l_1} = 3,$
$M_{MAX}^{l_2} = 3,$

$M^{l_3}_{MAX} = 3$,
$M^{l_4}_{MAX} = 4$,
$M^{l_6}_{MAX} = 3$,
$M^{l_8}_{MAX} = 3$,
$M^{l_9}_{MAX} = 2$,
$\forall i \in \{1,2,\dots,9\}, M^{l_i}_{MIN} = 0$.

For (l_5, b_5) with $\eta_5 = \eta_2 - \eta_9$, by Corollary 4.2 we need to verify that

$$l_5^T M_0 - l_2^T M_0 + l_9^T M_0 + M^{l_2}_{MAX} - M^{l_9}_{MIN} \le b_5.$$

Since $l_5^T M_0 = l_2^T M_0 = l_9^T M_0 = M^{l_9}_{MIN} = 0$, $M^{l_2}_{MAX} = 3$, and $b_5 = 4$, we conclude that (l_5, b_5) is redundant and V_5 can be removed from the supervisor of the FMS.

For (l_7, b_7) with $\eta_7 = -\eta_2 + \eta_3 - \eta_4 + 2\eta_6 + \eta_9$, by Corollary 4.2 we need to verify that

$$l_7^T M_0 + l_2^T M_0 - l_3^T M_0 + l_4^T M_0 - 2l_6^T M_0 - l_9^T M_0 + M^{l_3}_{MAX} + 2M^{l_6}_{MAX} + M^{l_9}_{MAX} - M^{l_2}_{MIN} - M^{l_4}_{MIN} \le b_7.$$

Since $l_7^T M_0 = l_2^T M_0 = l_3^T M_0 = l_4^T M_0 = l_6^T M_0 = l_9^T M_0 = M^{l_2}_{MIN} = M^{l_4}_{MIN} = 0$, $M^{l_3}_{MAX} = M^{l_6}_{MAX} = 3$, $M^{l_9}_{MAX} = 2$, and $b_7 = 3$, we conclude that (l_7, b_7) is not redundant and V_7 cannot be removed.

In summary, a simplified supervisor (N^S, M_0^S) with monitors V_1, V_2, V_3, V_4, V_6, V_7, V_8, and V_9 can be obtained. By Proposition 4.1, it can also be verified that (l_2, b_2), (l_4, b_4), (l_7, b_7), (l_8, b_8), and (l_9, b_9) are redundant with respect to (N^S, M_0^S).

Although the structurally simplified liveness-enforcing supervisor contains the monitors that implement the elementary constraints whose characteristic T-vectors are linearly independent, and only one monitor that implements the dependent constraint, there may still exist implicit monitors whose removal does not change the behavior of the supervised system or the monitors whose removal keeps the liveness of the supervised system but with more resultant permissive behavior. This can be demonstrated by the same manufacturing example. By solving the following LPP:

$$x_2 = max\{l_2^T M\}$$

s.t.

$$M = M_0^{S\prime} + [N^{S\prime}]Y, M, Y \ge 0$$

we have $x_2^* = 1$, where $N^{S\prime}$ is the net resulting from removing V_2 and related arcs from N^S and $M_0^{S\prime}$ is the initial marking of $N^{S\prime}$. Thus, we claim that (l_2, b_2) is redundant due to $x_2^* < b_2 + 1$.

In fact, it is easy to verify that monitors V_3 and V_6 can also be removed by using MIP-based deadlock detection method [6]. Their removal can lead to a more permissive supervisor. Note that the removal of V_3 and V_6 is legal for the purpose of deadlock controls, i.e., the resultant system remains live. However, the removal of monitors by the MIP-based method may violate the GMECs represented by (4.15).

Specifically, the supervisor with monitors V_1, V_3, and V_6 has 11 reachable states and the one only with V_1 has 24 reachable states.

4.6 Some Further Results About S^{4}R Nets

S^{4}R represents an important subclass of Petri nets that can model a large class of FMSs. In this section, the redundancy decision conditions of a dependent constraint in an S^{4}R stated by Corollaries 4.1 and 4.2 and their related results can be simplified by the fact that its operation places are unmarked at the initial marking. In what follows, (N, M_0) is used to represent an S^{4}R where there is no confusion.

Corollary 4.3. *Let (l,b) be a strongly dependent constraint with $\eta = \sum_{i=1}^{n} a_i\eta_i$ and $l \neq \sum_{i=1}^{n} a_i l_i$ in (N, M_0). (l,b) is redundant if*

$$\sum_{i=1}^{n} a_i M_{MAX}^{l_i} \leq b.$$

Proof. By Corollary 4.1, this follows from the fact that $(\cup_{i=1}^{n} ||l_i||) \cup (||l||) \subseteq P_A$ and $\forall p \in P_A$, $M_0(p) = 0$. □

Corollary 4.4. *Let (l,b) be a weakly dependent constraint in (N, M_0) with $\eta = \sum_{i=1}^{n} a_i\eta_i - \sum_{j=1}^{m} a_j\eta_j$ and $l \neq \sum_{i=1}^{n} a_i l_i - \sum_{j=1}^{m} a_j l_j$. It is redundant if*

$$\sum_{i=1}^{n} a_i M_{MAX}^{l_i} \leq b.$$

Proof. From Corollary 4.2, it is redundant if

$$l^T M_0 - \sum_{i=1}^{n} a_i l_i^T M_0 + \sum_{j=1}^{m} a_j l_j^T M_0 + \sum_{i=1}^{n} a_i M_{MAX}^{l_i} - \sum_{j=1}^{m} a_j M_{MIN}^{l_j} \leq b.$$

According to the definition of an S^{4}R, $\forall p \in P_A$, $M_0(p) = 0$. We have $l^T M_0 = \sum_{i=1}^{n} a_i l_i^T M_0 = \sum_{j=1}^{m} a_j l_j^T M_0 = \sum_{j=1}^{m} a_j M_{MIN}^{l_j} = 0$. Hence, this result is true. □

This result indicates that in an S^{4}R the redundancy of a weakly dependent constraint has nothing to do with $\sum_{j=1}^{m} a_j\eta_j$. Consequently, if (l,b) is (weakly or strongly) dependent, its redundancy depends on (l_1,b_1), (l_2,b_2), ..., and (l_n,b_n) only. Furthermore, we have the following results.

Corollary 4.5. *Let (l,b) be a strongly dependent constraint in (N, M_0) with $\eta = \sum_{i=1}^{n} a_i\eta_i$ and $l \neq \sum_{i=1}^{n} a_i l_i$. (l,b) is redundant if*

$$\sum_{i=1}^{n} a_i M_{max}^{l_i} \leq b.$$

Proof. It is true by Theorem 4.3 since $\forall p \in ||l|| \cup (\cup_{i=1}^{n} ||l_i||)$, $M_0(p) = 0$. □

Corollary 4.6. *Let (l, b) be a weakly dependent constraint in (N, M_0) with $\eta = \sum_{i=1}^{n} a_i \eta_i - \sum_{j=1}^{m} a_j \eta_j$ and $l \neq \sum_{i=1}^{n} a_i l_i - \sum_{j=1}^{m} a_j \eta_j$. (l, b) is redundant if*

$$\sum_{i=1}^{n} a_i M_{max}^{l_i} \leq b.$$

Corollary 4.7. *Let (l, b) be a strongly dependent constraint in (N, M_0) with $\eta = \sum_{i=1}^{n} a_i \eta_i$ and $l \neq \sum_{i=1}^{n} a_i l_i$. $\forall i \in \mathbb{N}_n$, monitor V_i is added to enforce (l_i, b_i). (l, b) is implicitly enforced if*

$$\sum_{i=1}^{n} a_i b_i \leq b.$$

Proof. After V_i is added, (l_i, b_i) is hence enforced. The resultant net with these monitors $V_1 - V_n$ is denoted by (N^S, M_0^S). This indicates the truth of $b_i \geq M_{max}^{l_i}$, where $M_{max}^{l_i} = max\{l_i^T M | M \in R(N^S, M_0^S)\}$. $\sum_{i=1}^{n} a_i b_i \leq b$ implies $\sum_{i=1}^{n} a_i M_{max}^{l_i} \leq b$. □

Corollary 4.8. *Let (l, b) be a weakly dependent constraint in (N, M_0) with $\eta = \sum_{i=1}^{n} a_i \eta_i - \sum_{j=1}^{m} a_j \eta_j$ and $l \neq \sum_{i=1}^{n} a_i l_i - \sum_{j=1}^{m} a_j l_j$. $\forall i \in \mathbb{N}_n$, monitor V_i is added such that (l_i, b_i) is enforced. (l, b) is implicitly enforced if*

$$\sum_{i=1}^{n} a_i b_i \leq b.$$

Proof. It follows immediately from Corollary 4.6. □

Example 4.12. For the FMS example in Fig. 4.7, (l_5, b_5) is weakly dependent with $\eta_5 = \eta_2 - \eta_9$. Suppose that V_2 is added to enforce (l_2, b_2). Then, (l_5, b_5) is implicitly enforced since $b_2 < b_5$. Furthermore, (l_7, b_7) is weakly dependent with $\eta_7 = -\eta_2 + \eta_3 - \eta_4 + 2\eta_6 + \eta_9$. Suppose that V_3, V_6 and V_9 are added to enforce (l_3, b_3), (l_6, b_6) and (l_9, b_9), respectively. However, (l_7, b_7) is not implicitly enforced since $b_3 + 2b_6 + b_9 > b_7$.

In summary, for a set of GMECs to be enforced to (N, M_0) for deadlock control purpose by the policy in [40], the redundancy of a dependent constraint can be simply determined by the truth of an inequality. From the computational point of view, the methodology proposed in this section is much more efficient. Next a parameterized and more general result concerning the redundancy of a dependent constraint is shown.

Theorem 4.12. *Let (l,b) be a weakly or strongly dependent constraint in (N,M_0). It is redundant if*

$$\sum_{i=1}^{n} a_i\xi_i \geq \sum_{i=1}^{n} a_i b_i - b.$$

Proof. It follows from Corollaries 4.5, 4.6–4.8. □

This result indicates that a dependent constraint can be made redundant by properly setting the control depth variables of elementary constraints even if $\sum_{i=1}^{n} a_i b_i \leq b$ is not true.

Let (L,B) be a set of GMECs in (N,M_0), (l_i,b_i), $i \in \mathbb{N}_n$, be elementary constraints, and (l_j,b_j), $j \in \{n+1,n+2,\ldots,m\}$, be dependent constraints. The control depth variables of the elementary constraints can be decided by solving the following LPP once monitors $V_1 - V_n$ are explicitly added:

$$min \sum_{i=1}^{n} \xi_i$$

s.t.

$$\sum_{i=1}^{n} a_i\xi_i \geq \sum_{i=1}^{n} a_i b_i - b_j, a_i \geq 0, j = n+1, n+2, \ldots, m.$$

$$0 \leq \xi_i \leq b_i.$$

The existence of an optimal solution indicates that a set of GMECs can be enforced by using a set of monitors that are explicitly added for elementary constraints whose number is bounded by the smaller of place and transition counts.

Example 4.13. For the FMS example in Fig. 4.7, the control depth variables of the elementary constraints can be decided by solving the following LPP:

$min\ x = \xi_1 + \xi_2 + \xi_3 + \xi_4 + \xi_6 + \xi_8 + \xi_9$
s.t.
$\xi_2 \geq b_2 - b_5 = -3,$
$\xi_3 + 2\xi_6 + \xi_9 \geq b_3 + 2b_6 + b_9 - b_7 = 2,$
$0 \leq \xi_1 \leq b_1, 0 \leq \xi_2 \leq b_2, 0 \leq \xi_3 \leq b_3, 0 \leq \xi_4 \leq b_4,$
$0 \leq \xi_6 \leq b_6, 0 \leq \xi_8 \leq b_8, 0 \leq \xi_9 \leq b_9,$
$b_1 = 1, b_2 = 1, b_3 = 1, b_4 = 3, b_5 = 4, b_6 = 1, b_7 = 3, b_8 = 3,$ and $b_9 = 2.$

This problem has an optimal solution $x^* = 1$, implying that the control depth variable $\xi_1 = \xi_2 = \xi_3 = \xi_4 = \xi_8 = \xi_9 = 0$, and $\xi_6 = 1$. That is to say, all dependent constraints are redundant if monitors are added for the elementary constraints with their control depth variables being zero except ξ_6.

4.7 Identification of Elementary Constraints

The results in Sect. 4.6 indicate that the redundancy conditions of a dependent constraint (l, b) are easily satisfied when b is large and b_i ($i \in \mathbb{N}_n$) is small, where (l_1, b_1) – (l_n, b_n) are the elementary constraints of (l, b). This section develops an algorithm to identify a set of elementary constraints such that the redundancy condition of a dependent constraint is easily satisfied without increasing the control depth variables of its elementary constraints. Let (N, M_0) be a Petri net with $N = (P, T, F, W)$. Without loss of generality, we assume that any two vectors in $[\eta]$ are not identical.

Algorithm 4.1 Identification of elementary constraints

Input: $(L, B) = \{(l_i, b_i) | i \in \mathbb{N}_n\}$

Output: L_E, a set of elementary constraints

1: Find $[\eta]$ from (L, B)
2: $m := rank([\eta])$
3: By the *merge sort* algorithm, sort the sequence b_1, b_2, $\cdots$, and b_n to be b_{k_1}, b_{k_2}, $\ldots$, and b_{k_n} in an ascending order, where $\{k_1, k_2, \ldots, k_n\} = \mathbb{N}_n$
4: **for** $j = 1$ to n **do**
5: $\quad \chi_j := \eta_{k_j}$
6: **end for**
7: $L_E := \{(l_{k_1}, b_{k_1})\}$
8: $A := [\chi_1]^T$
9: $i := 2$
10: **while** $(m \neq 1)$ **do**
11: $\quad$ **for** $i = 2$ to n **do**
12: $\quad\quad A_E := [A^T | \chi_i]^T$
13: $\quad\quad L_E := L_E \cup \{(l, b) | \eta = \chi_i\}$
14: $\quad\quad$ **while** $(rank(A_E) \neq m)$ **do**
15: $\quad\quad\quad$ **if** $rank(A_E) - rank(A) = 0$ **then**
16: $\quad\quad\quad\quad L_E := L_E \setminus \{(l, b) | \eta = \chi_i\}$
17: $\quad\quad\quad\quad i := i + 1$
18: $\quad\quad\quad$ **else**
19: $\quad\quad\quad\quad A := A_E$
20: $\quad\quad\quad\quad i := i + 1$
21: $\quad\quad\quad$ **end if**
22: $\quad\quad$ **end while**
23: $\quad\quad i := n + 1$
24: $\quad$ **end for**
25: **end while**
26: Output L_E

This algorithm is similar to the one that identifies a set of elementary siphons in Sect. 3.6 of Chap. 3. The complexity of this algorithm is $O(2n|P||T|^2 + n\lg n)$.

Example 4.14. For the set of GMECs in the FMS example in Sect. 4.6, this algorithm outputs $L_E = \{(l_1, b_1), (l_2, b_2), (l_3, b_3), (l_4, b_4), (l_6, b_6), (l_8, b_8), (l_9, b_9)\}$.

4.8 Bibliographical Remarks

As an important class of control specifications in the supervisory control of DES, linear inequality constraints are first studied in [32]. They are also extensively investigated in a Petri net formalism. The work along this direction can be found in [3,4,9,12,13,30,36,38]. Recent tutorials and survey papers are presented by Iordache and Antsaklis [25,27]. A good reference is the book by Moody and Antsaklis [37]. The work in [7] develops a systematic method to minimize the set of monitors for forbidden-state problems in safe Petri nets.

Problems

4.1. Let $(l_1,b_1) \equiv M(p_2)+M(p_6) \le 4$, $(l_2,b_2) \equiv M(p_3)+M(p_5) \le 4$, $(l_3,b_3) \equiv M(p_2)+M(p_3)+M(p_5)+M(p_6) \le 6$, $(l_4,b_4) \equiv 3M(p_3)+2M(p_4) \le 8$, $(l_5,b_5) \equiv M(p_2)+M(p_4)+M(p_6) \le 5$, $(l_6,b_6) \equiv 3M(p_2)+4M(p_5) \le 12$, and $(l_7,b_7) \equiv M(p_2)+M(p_5)+M(p_6)+M(p_7) \le 10$ be a set of GMECs to the Petri net shown in Fig. 4.8.
1. Find a set of elementary constraints by the algorithm in Sect. 4.7;
2. Try to find a set of control depth variables of the elementary constraints under which dependent constraints are implicitly enforced;
3. Discuss the condition under which the dependent constraints cannot be implicitly enforced by adding monitors for elementary constraints only.

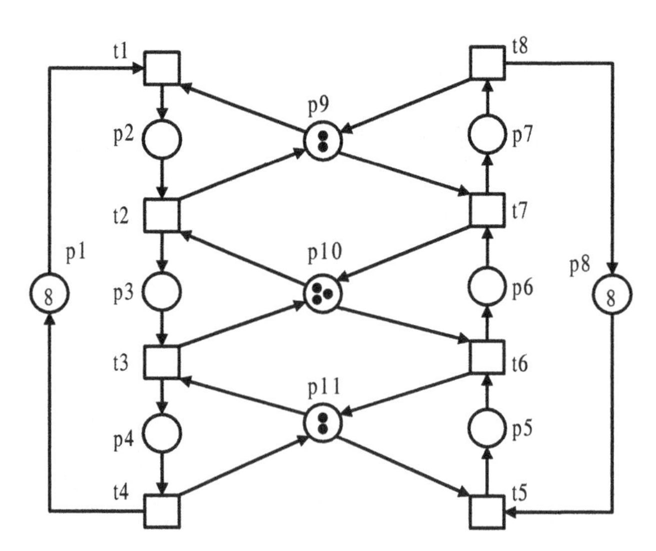

Fig. 4.8 A Petri net (N,M_0)

4.2. Extend the results in this chapter to a Petri net with uncontrollable and unobservable transitions. The reader is referred to the work by Giua, Moody, and Iordache [9,29,36–38].

4.3. Discuss the possibility of applying the results in this chapter to other classes of control specifications such as constraints involving firing vectors and time, and disjunction of inequality constraints. These control specifications are extensively investigated by Iordache et al. [14–28].

References

1. Basile, F., Chiacchio, P., Giua, A. (1998) Supervisory control of Petri nets based on suboptimal monitor places. In *Proc. IEE Int. Workshop on Discrete Event Systems*, pp.85–87.
2. Basile, F., Chiacchio, P. (2001) Optimal Petri net monitor design. In *Synthesis and Contorl of Discrete Event Systems*,B. Cailaud, X. Xie, Ph. Darondeau, and L. Lavagno (Eds.), Boston, MA: Kluwer, pp.141–154.
3. Basile, F., Carbone, C., Chiacchio, P. (2003) Petri net controllers to enforce disjunction of GMECs. In *Proc. IEEE Int. Conf. on Robotics and Automation*, pp.1440–1445.
4. Basile, F., Chiacchio, P., Giua, A. (2006) Suboptimal supervisory control of Petri nets in presence of uncontrollable transitions via monitor places. *Automatica*, vol.42, no.6, pp.995–1004.
5. Basile, F., Chiacchio, P., Giua, A. (2007) An optimization approach to Petri net monitor design. *IEEE Transactions on Automatic Control*, vol.52, no.2, pp.306–311.
6. Chu, F., Xie, X.L. (1997) Deadlock analysis of Petri nets using siphons and mathematical programming. *IEEE Transactions on Robotics and Automation*, vol.13, no.6, pp.793–804.
7. Dideban, A., Alla, H. (2007) Determination of minimal sets of control places for safe Petri nets. In *Proc. American Control Conference*, New York City, USA, pp.4975–4980.
8. Ezpeleta, J, Colom, J.M., Martinez, J. (1995) A Petri net based deadlock prevention policy for flexible manufacturing systems. *IEEE Transactions on Robotics and Automation*, vol.11, no.2, pp.173–184.
9. Giua, A., DiCesare, F., Silva, M. (1992) Generalized mutual exclusion constraints on nets with uncontrollable transitions. In *Proc. IEEE Int. Conf. on Systems, Man, and Cybernetics*, pp.974–979.
10. Giua, A., DiCesare, F., Silva, M. (1993) Petri net supervisors for generalized mutual exclusion constraints. In *Proc. 12th IFAC World Congress*, pp.267–270.
11. Giua, A. Seatzu, C. (2001) Supervisory control of railway networks with Petri nets. In *Proc. 40th IEEE Int. Conf. on Decision and Control*, pp.5004–5009.
12. Holloway, L.E., Krogh, B.H. (1990) Synthesis of feedback control logic for a class of controlled Petri nets. *IEEE Transactions on Automatic Control*, vol.35, no.5, pp.514–523.
13. Holloway, L.E., Krogh, B.H. (1992) On closed-loop liveness of discrete-event systems under maximally permissive control. *IEEE Transactions on Automatic Control*, vol.37, no.5, pp.692–697.
14. Iordache, M.V., Moody, J.O., Antsaklis, P.J. (2000) Method for the synthesis of deadlock prevention controllers in systems modeled by Petri nets. In *Proc. American Control Conference*, pp.3167–3171.
15. Iordache, M.V., Antsaklis, P.J. (2001) T-liveness enforcement in Petri nets based on structural net properties. In *Proc. IEEE Conf. on Decision and Control*, pp.4984–4989.
16. Iordache, M.V., Moody, J.O., Antsaklis, P.J. (2001) A method for the synthesis of liveness enforcing supervisors in Petri nets. In *Proc. American Control Conference*, pp.4943–4948.
17. Iordache, M.V., Antsaklis, P.J. (2002) Synthesis of supervisors enforcing general linear vector constraints in Petri nets. In *Proc. American Control Conference*, pp.154–159.

18. Iordache, M.V., Moody, J.O., Antsaklis, P.J. (2002) Synthesis of deadlock prevention supervisors using Petri nets. *IEEE Transactions on Robotics and Automation*, vol.18, no.1, pp.59–68.
19. Iordache, M.V., Antsaklis, P.J. (2003) Design of T-Liveness enforcing supervisors in Petri nets. *IEEE Transactions on Automatic Control*, vol.48, no.11, pp.1962–1974.
20. Iordache, M.V., Antsaklis, P.J. (2003) Synthesis of supervisors enforcing general linear constraints in Petri nets. *IEEE Transactions on Automatic Control*, vol.48, no.11, pp.2036–2039.
21. Iordache, M.V., Antsaklis, P.J. (2003) Decentralized control of Petri nets with constraint transformations. In *Proc. American Control Conference*, pp.314–319.
22. Iordache, M.V., Antsaklis, P.J. (2003) Admissible decentralized control of Petri nets. In *Pro. American Control Conference*, pp.332–337.
23. Iordache, M.V. (2003) Methods for the Supervisory Control of Concurrent Systems Based on Petri Net Abstractions. Doctoral Dissertation, University of Notre Dame.
24. Iordache, M.V., Antsaklis, P.J. (2004) Resilience to failures and reconfigurations in the supervision based on place invariants. In *Proc. American Control Conference*, pp.4477–4482.
25. Iordache, M.V., Antsaklis, P.J. (2004) Supervision based on place invariants: A survey. Technical Report of the ISIS Group, ISIS-2004-003, University of Notre Dame.
26. Iordache, M.V., Antsaklis, P.J. (2005) A structural approach to the enforcement of language and disjunctive constraints. In *Proc. American Control Conference*, pp.3920–3925.
27. Iordache, M.V., Antsaklis, P.J. (2006) Supervision based on place invariants: A survey. *Discrete Event Dynamic Systems: Theory and Applications*, vol.16, no.4, pp.451–492.
28. Iordache, M.V., Antsaklis, P.J. (2006) Decentralized supervision of Petri nets. *IEEE Transactions on Automatic Control*, vol.51, no.2, pp.376–381.
29. Iordache, M.V., Antsaklis, P.J. (2006) *Supervisory Control of Concurrent Systems: A Petri Net Structural Approach*. Berlin: Springer.
30. Krogh, B.H., Holloway, L.E. (1991) Synthesis of feedback control logic for discrete manufacturing systems. *Automatica*, vol.27, no.4, pp.641–645.
31. Li, Y., Wonham, W.M. (1993) Control of vector discrete-event systems. I. The base model. *IEEE Transactions on Automatic Control*, vol.38, no.8, pp.1214–1227.
32. Li, Y., Wonham, W.M. (1994) Control of vector discrete-event systems. II. Controller synthesis. *IEEE Transactions on Automatic Control*, vol.39, no.3, pp.512–531.
33. Li, Y., Wonham, W.M. (1995) Concurrent vector discrete-event systems. *IEEE Transactions on Automatic Control*, vol.40. no.4, pp.628–638.
34. Li, Z.W., Zhou, M.C. (2004) Elementary siphons of Petri nets and their application to deadlock prevention in flexible manufacturing systems. *IEEE Transactions on Systems, Man, and Cybernetics, Part A*, vol.34, no.1, pp.38–51.
35. Li, Z.W., Zhou, M.C. (2006) Clarifications on the definitions of elementary siphons of Petri nets. *IEEE Transactions on Systems, Man, and Cybernetics, Part A*, vol.36, no.6, pp.1227–1229.
36. Moody, J.O., Lemmon, M.D., Antsaklis, P.J. (1996) Supervisory control of Petri nets with uncontrollable and unobservable transitions. In *Proc. 35th IEEE Conf. on Control and Decision*, pp.4433–4438.
37. Moody, J.O., Antsaklis, P.J. (1998) *Supervisory Control of Discrete Event Systems Using Petri Nets*. Boston, MA: Kluwer.
38. Moody, J.O., Antsaklis, P.J. (2000) Petri net supervisors for DES with uncontrollable and unobservable transitions. *IEEE Transactions on Automatic Control*, vol.45, no.3, pp.462–476.
39. Park, J. (2000) Structural analysis and control of resource allocation systems using Petri nets. PhD thesis, Georgia Institute of Technology, Atlanta, GA.
40. Park, J., Reveliotis, S.A. (2001) Deadlock avoidance in sequential resource allocation systems with multiple resource acquisitions and flexible routings. *IEEE Transactions on Automatic Control*, vol.46, no.10, pp.1572–1583.
41. Reveliotis, S.A. (2007) Implicit siphon control and its role in the liveness-enforcing supervision of sequential resource allocation systems. *IEEE Transactions on Systems, Man, and Cybernetics, Part A*, vol.37, no.3, pp.319–328.
42. Tittus, M., Egardt, B. (1999) Hierarchical supervisory control for batch processes. *IEEE Transactions on Control System Technology*, vol.7, no.5, pp.542–554.

43. Tricas, F. (2003) Deadlock Analysis, Prevention and Avoidance in Sequential Resource Allocation Systems, Ph.D Dissertation, University of Zaragoza, Spain.
44. Yamalidou, E., Kantor, J. (1991) Modeling an optimal control of discrete-event chemical processes using Petri nets. *Computers and Chemical Engineering*, vol.15, no.7, pp.503–519.
45. Yamalidou, E., Moody, J.O., Antsaklis, P.J. (1996) Feedback control of Petri nets based on place invariants. *Automatica*, vol.32, no.1, pp.15–28.
46. Zhou, M.C., DiCesare, F. (1991) Parallel and sequential exclusions for Petri net modeling for manufacturing systems. *IEEE Transactions on Robotics and Automation*, vol.7, no.4, pp.515–527.
47. Zhou, M.C., DiCesare, F. (1993) *Petri Net Synthesis for Discrete Event Control of Manufacturing Systems*. Boston, MA: Kluwer.

Chapter 5
Deadlock Control Based on Elementary Siphons

Abstract This chapter first reviews a classical deadlock prevention policy for automated manufacturing systems, which is usually considered to be the first that utilizes structural theory of Petri nets to design a liveness-enforcing (Petri net) supervisor. To reduce the computational complexity to design deadlock-free supervisors, a mixed-integer-programming based deadlock detection method is presented. Then, a number of deadlock prevention policies are introduced by using the controllability results of elementary and dependent siphons, which are applicable to ordinary and generalized Petri net models. This chapter shows that the deadlock prevention policies based on elementary siphons can reduce the computational and structural complexity and improve the behavioral permissiveness of the liveness-enforcing monitor-based supervisors. Importantly, some interesting or open problems in this area are listed at the end of this chapter.

5.1 Introduction

This chapter presents a number of deadlock prevention policies. They are based on the concepts of strict minimal and elementary siphons. They can be applied to the deadlock resolution in resource allocation systems that are modeled with ordinary and generalized Petri nets. First, we review the seminal work [16] in this area. It develops a deadlock control policy that prevents strict minimal siphons from being unmarked by adding monitors for them. Then, a number of deadlock prevention approaches by using elementary siphons are presented.

5.2 Some Application Subclasses of Petri Nets

This section defines a number of subclasses of Petri nets, which can model real-life automated flexible manufacturing systems, a typical and important class of resource allocation systems in the contemporary technical domain. The policies presented in this chapter can be applied to them. First a class of Petri nets called S^3PR is defined [16]. It stands for *system of simple sequential processes with resources.*

Definition 5.1. Let (N_1,M_1) and (N_2,M_2) be two Petri nets with $N_1=(P_1,T_1,F_1,W_1)$ and $N_2=(P_2,T_2,F_2,W_2)$, where $P_1\cap P_2=P_C\neq\emptyset$ and $T_1\cap T_2=\emptyset$. (N,M) with $N=(P,T,F,W)$ is said to be the resultant net of composing (N_1,M_1) and (N_2,M_2) via the set of shared places P_C iff (1) $P=P_1\cup P_2$, $T=T_1\cup T_2$, $F=F_1\cup F_2$, and $W(x,y)=W_i(x,y)$ if $(x,y)\in F_i$, $i=1,2$; and (2) $\forall p\in P_1\setminus P_C$, $M(p)=M_1(p)$, $\forall p\in P_2\setminus P_C$, $M(p)=M_2(p)$, and $\forall p\in P_C$, $M(p)=max\{M_1(p),M_2(p)\}$.

The two nets N_1 and N_2 satisfying $P_1\cap P_2=P_C\neq\emptyset$ and $T_1\cap T_2=\emptyset$ in Definition 5.1 are said to be composable via shared places. Their composition is denoted by $N_1\circ N_2$.

Example 5.1. Two nets (N_1,M_1) and (N_2,M_2) are shown in Fig. 5.1a and Fig. 5.1b, respectively, where $P_1=\{p_1-p_4,\ p_9-p_{11}\}$, $T_1=\{t_1-t_4\}$, $P_2=\{p_5-p_{11}\}$, and $T_2=\{t_5-t_8\}$. Since $P_1\cap P_2=\{p_9,p_{10},p_{11}\}$ and $T_1\cap T_2=\emptyset$, (N_1,M_1) and (N_2,M_2) are composable. As shown in Fig. 5.1(c), the net resulting from the composition of (N_1,M_1) and (N_2,M_2) is denoted by (N,M) where $P=P_1\cup P_2=\{p_1-p_{11}\}$, $T=T_1\cup T_2=\{t_1-t_8\}$, $M(p_1)=M_1(p_1)=10$, $M(p_2)=M_1(p_2)=0$, $M(p_3)=M_1(p_3)=0$, $M(p_4)=M_1(p_4)=0$, $M(p_5)=M_2(p_5)=10$, $M(p_6)=M_2(p_6)=0$, $M(p_7)=M_2(p_7)=0$, $M(p_8)=M_2(p_8)=0$, $M(p_9)=max\{M_1(p_9),M_2(p_9)\}=2$, $M(p_{10})=max\{M_1(p_{10}),M_2(p_{10})\}=2$, and $M(p_{11})=max\{M_1(p_{11}),M_2(p_{11})\}=3$.

Definition 5.2. A simple sequential process (S^2P) is a Petri net $N=(P_A\cup\{p^0\},T,F)$ where (1) $P_A\neq\emptyset$ is called the set of operation places; (2) $p^0\notin P_A$ is called the idle process place; (3) N is a strongly connected state machine; and 4) every circuit of N contains place p^0.

Definition 5.3. A simple sequential process with resources (S^2PR) is a Petri net $N=(\{p^0\}\cup P_A\cup P_R,T,F)$ such that:

1. The subnet generated by $X=P_A\cup\{p^0\}\cup T$ is an S^2P.
2. $P_R\neq\emptyset$ and $(P_A\cup\{p^0\})\cap P_R=\emptyset$.
3. $\forall p\in P_A$, $\forall t\in{}^\bullet p$, $\forall t'\in p^\bullet$, $\exists r_p\in P_R$, ${}^\bullet t\cap P_R=t'^\bullet\cap P_R=\{r_p\}$.
4. The following statements are verified: (a) $\forall r\in P_R$, ${}^{\bullet\bullet}r\cap P_A=r^{\bullet\bullet}\cap P_A\neq\emptyset$ and (b) $\forall r\in P_R$, ${}^\bullet r\cap r^\bullet=\emptyset$.
5. ${}^{\bullet\bullet}(p^0)\cap P_R=(p^0)^{\bullet\bullet}\cap P_R=\emptyset$.

Note that ${}^\bullet r$ represents place r's input transitions. ${}^{\bullet\bullet}r=\cup_{t\in{}^\bullet r}{}^\bullet t$ is the set of all input places of all input transitions of place r. Similarly, $r^{\bullet\bullet}=\cup_{t\in r^\bullet}t^\bullet$ represents the set of all output places of all output transitions of place r. For example, in Fig. 5.1(c), ${}^\bullet p_9=\{t_2,t_8\}$ and ${}^{\bullet\bullet}p_9={}^\bullet t_2\cup{}^\bullet t_8=\{p_2,p_{10},p_8\}$. $p_9^\bullet=\{t_1,t_7\}$ and $p_9^{\bullet\bullet}=t_1^\bullet\cup t_7^\bullet=\{p_2,p_{10},p_8\}$. Clearly, ${}^{\bullet\bullet}p_9=p_9^{\bullet\bullet}$.

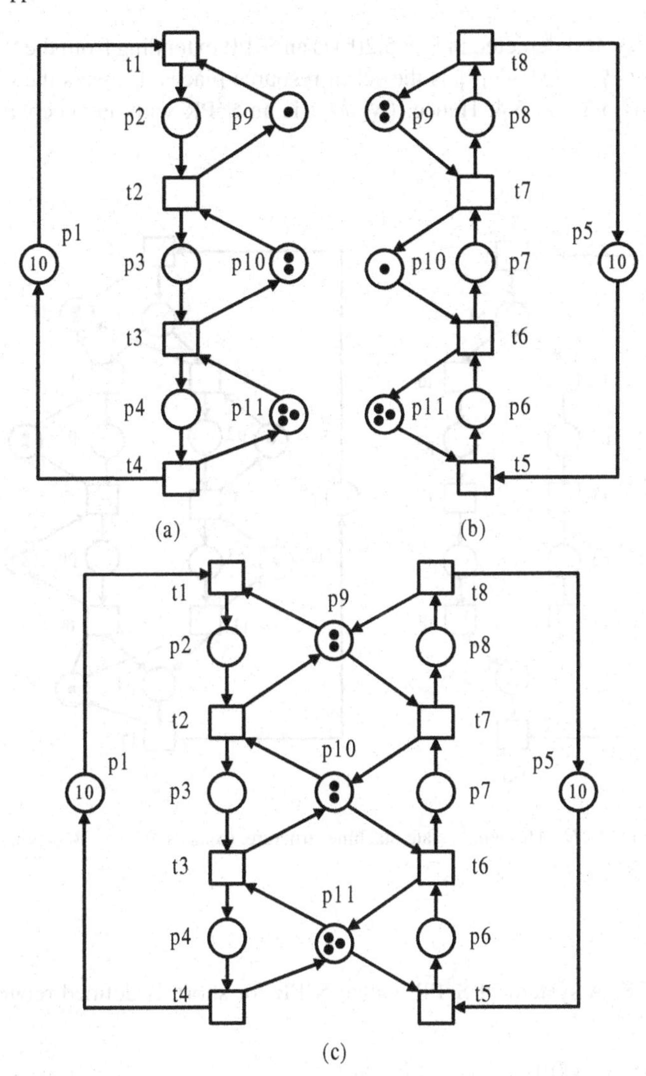

Fig. 5.1 (a) A Petri net (N_1, M_1), (b) a Petri net (N_2, M_2), and (c) the composed net (N, M)

Definition 5.4. Let $N = (P_A \cup \{p^0\} \cup P_R, T, F)$ be an S^{2}PR. An initial marking M_0 is called an acceptable initial marking for N iff (1) $M_0(p^0) \geq 1$, (2) $M_0(p) = 0$, $\forall p \in P_A$, and (3) $M_0(r) \geq 1$, $\forall r \in P_R$. An S^{2}PR with such a marking is said to be acceptably marked.

Example 5.2. The net shown in Fig. 5.2(a) is an S^{2}P, where p^0 is the idle process place and $P_A = \{p_1 - p_6\}$ is the set of operation places. It is easy to verify that the net is a strongly connected state machine.

The net (N_2, M_2) depicted in Fig. 5.2(b) is an S^{2}PR extending from the S^{2}P in Fig. 5.2(a), where $P_R = \{p_7 - p_{12}\}$ is the set of resource places. It meets the conditions in Definitions 5.3 and 5.4. Hence, (N_2, M_2) is an S^{2}PR with an acceptable initial marking.

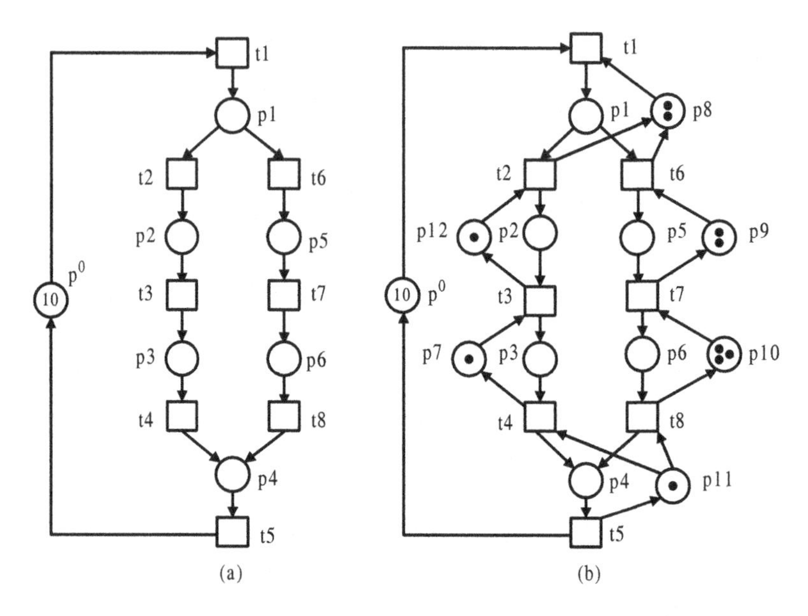

Fig. 5.2 (a) An S^{2}P (N_1, M_1) with a state machine structure, (b) an S^{2}PR (N_2, M_2) with an acceptable initial marking

Definition 5.5. A system of S^{2}PR, called S^{3}PR for short, is defined recursively as follows:

1. An S^{2}PR is an S^{3}PR.
2. Let $N_i = (P_{A_i} \cup \{p_i^0\} \cup P_{R_i}, T_i, F_i)$, $i \in \{1,2\}$, be two S^{3}PR such that $(P_{A_1} \cup \{p_1^0\}) \cap (P_{A_2} \cup \{p_2^0\}) = \emptyset$, $P_{R_1} \cap P_{R_2} = P_C \neq \emptyset$, and $T_1 \cap T_2 = \emptyset$. Then, the net $N = (P_A \cup P^0 \cup P_R, T, F)$ resulting from the composition of N_1 and N_2 via P_C defined as follows: (1) $P_A = P_{A_1} \cup P_{A_2}$, (2) $P^0 = \{p_1^0\} \cup \{p_2^0\}$, (3) $P_R = P_{R_1} \cup P_{R_2}$, (4) $T = T_1 \cup T_2$, and (5) $F = F_1 \cup F_2$ is also an S^{3}PR.

In the sequel, an S^{3}PR N composed of n S^{2}PR N_1-N_n, denoted by $N = \bigcirc_{i=1}^{n} N_i$, is defined as follows: $N = N_1$ if $n = 1$; $N = (\bigcirc_{i=1}^{n-1} N_i) \circ N_n$ if $n > 1$. $\overline{N_i}$ is used to denote the S^{2}P from which the S^{2}PR N_i is formed. Transitions in $(P^0)^\bullet$ are called source transitions that represent the entry of raw materials when a manufacturing system is modeled with an S^{3}PR.

Definition 5.6. Let N be an S^{3}PR. (N, M_0) is called an acceptably marked S^{3}PR iff one of the following statements is true:

1. (N, M_0) is an acceptably marked S^{2}PR.
2. $N = N_1 \circ N_2$, where (N_i, M_{0_i}) is an acceptably marked S^{3}PR and

   a. $\forall i \in \{1,2\}, \forall p \in P_{A_i} \cup \{p_i^0\}, M_0(p) = M_{0_i}(p)$.
   b. $\forall i \in \{1,2\}, \forall r \in P_{R_i} \setminus P_C, M_0(r) = M_{0_i}(r)$.
   c. $\forall r \in P_C, M_0(r) = max\{M_{0_1}(r), M_{0_2}(r)\}$.

Example 5.3. The net (N_1, M_1) shown in Fig. 5.1(a) is an S^{3}PR if p_1 is an idle process place, $p_2 - p_4$ are operation places, and $p_9 - p_{11}$ are resource places. Likewise, (N_2, M_2) shown in Fig. 5.1(b) is an S^{3}PR if p_5 is an idle process place, $p_6 - p_8$ are operation places, and $p_9 - p_{11}$ are resource places. Since they have common resource places, they are composable. Their composition leads to an S^{3}PR (N, M), as shown in Fig. 5.1(c). Since one can verify that it meets the conditions in Definition 5.6, the net in Fig. 5.1(c) is an acceptably marked S^{3}PR.

In what follows, when we talk about an S^{3}PR, it is assumed to be acceptably marked unless otherwise stated. For example, the net shown in Fig. 5.3 is an S^{3}PR if $P^0 = \{p_1, p_{10}\}$, $P_R = \{p_{11} - p_{15}\}$, and others are operation places. Specifically, we have $P_{A_1} = \{p_2, p_3, p_5 - p_7\}$ and $P_{A_2} = \{p_4, p_8, p_9\}$. It is easy to verify that (N, M_0) is acceptably marked.

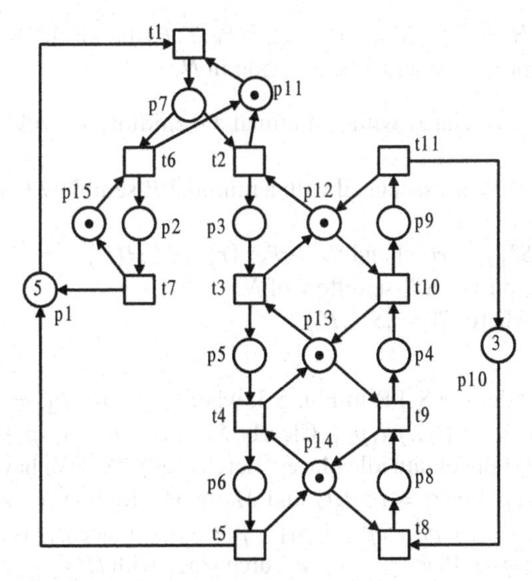

Fig. 5.3 An S^{3}PR net (N, M_0)

Let S be a strict minimal siphon in an S^{3}PR $N = (P_A \cup P^0 \cup P_R, T, F)$. Ezpeleta et al. [16] show that S does not contain idle places but consists of operation and resource places only. As a result, S can be represented by $S^A \cup S^R$, where $S^R = S \cap P_R$ and $S^A = S \setminus S^R$, i.e., $S^A = S \cap P_A$.

Definition 5.7. For $r \in P_R$, $H(r) = {}^{\bullet\bullet}r \cap P_A$, the operation places that use r, is called the set of holders of r. Let $[S] = (\cup_{r \in S^R} H(r)) \setminus S$. $[S]$ is called the complementary set of siphon S.

Suppose that $S = \{r_1, r_2, p_2, p_3\}$ is a strict minimal siphon in an S^{3}PR, where r_1 and r_2 are resource places, and p_2 and p_3 are operation places. If $H(r_1) = \{p_1, p_2\}$ and $H(r_2) = \{p_3, p_4\}$, then we have complementary set $[S] = (\{p_1, p_2\} \cup \{p_3, p_4\}) \setminus S = \{p_1, p_4\}$.

The concept of the complementary set of a siphon plays an important role in the development of the deadlock prevention policy in [16]. Intuitively, the complementary set of a siphon is a set of operation places that use the resources in it but are excluded from it. That is to say, the operation places in the complementary set compete for the limited resources with those in the siphon. When the tokens initially staying in the resource places of a siphon are completely held or "stolen" by the places in its complementary set, the siphon is emptied. As is known, if a siphon has no token, it remains free of tokens in the subsequent reachable markings. The transitions in its postset are completely disabled, leading to deadlocks.

It is shown in [16] that a strict minimal siphon S in an S^{3}PR can be emptied and $|S \cap S^R| > 1$ is true. That is to say, a strict minimal siphon in an S^{3}PR contains at least two resource places.

Property 5.1. Let $N = \bigcirc_{i=1}^{n} N_i = (P^0 \cup P_A \cup P_R, T, F)$ be an S^{3}PR consisting of n simple sequential processes and S be a siphon in N.

1. Any $p \in P_{A_i}$ is associated with a minimal P-semiflow I_p with support $||I_p|| = P_{A_i} \cup \{p_i^0\}$.
2. Any resource $r \in P_R$ is associated with a minimal P-semiflow I_r such that $||I_r|| = \{r\} \cup H(r)$.
3. $\forall p \in [S]$, $\exists r \in S^R$, $p \in H(r)$ and $\forall r' \in P_R \setminus \{r\}$, $p \notin H(r')$.
4. $[S] \cup S$ is the support of a P-semiflow of N.
5. $[S] = \cup_{i=1}^{n} [S]^i$, where $[S]^i = [S] \cap P_{A_i}$.

Example 5.4. Consider the S^{3}PR in Fig. 5.3, where $p_1^0 = p_1$, $P_{A_1} = \{p_2, p_3, p_5, p_6, p_7\}$, $p_2^0 = p_{10}$, and $P_{A_2} = \{p_4, p_8, p_9\}$. Clearly, $S_1 = \{p_5, p_9, p_{12}, p_{13}\}$ is a strict minimal siphon. It does not contain idle places. Let $S_1 = S_1^A \cup S_1^R$. We have $S_1^A = \{p_5, p_9\}$ and $S_1^R = \{p_{12}, p_{13}\}$. Let $\sigma = t_1 t_2 t_8 t_9$ and $M_0[\sigma\rangle M_1$. Siphon S_1 is emptied under marking $M_1 = 4p_1 + p_3 + p_4 + 2p_{10} + p_{11} + p_{14} + p_{15}$. Place p_{12} is a resource place with $H(p_{12}) = \{p_3, p_9\}$. Place p_{13} is a resource place with $H(p_{13}) = \{p_4, p_5\}$. Complementary set $[S_1] = H(p_{12}) \cup H(p_{13}) \setminus S_1 = \{p_3, p_9, p_5, p_4\} \setminus S_1 = \{p_3, p_4\}$. Specifically, $[S_1] = [S_1]^1 \cup [S_1]^2$, where $[S_1]^1 = \{p_3\}$ and $[S_1]^2 = \{p_4\}$. The minimal P-invariants associated with idle place p_1 and resource place p_{12} are $I_{p_1} = p_1 + p_2 + p_3 + p_5 + p_6 + p_7$ and $I_{p_{12}} = p_3 + p_9 + p_{12}$, respectively.

Note that $\forall p \in P_A \cup P^0$ ($\forall r \in P_R$), $||I_p||$ ($||I_r||$) is a minimal siphon and trap that is initially marked. Let Π be the set of strict minimal siphons in an S^{3}PR. The following result is from [16], indicating that an S^{3}PR is live iff there is no siphon that can be emptied.

Theorem 5.1. *An S^{3}PR (N, M_0) is live iff $\forall M \in R(N, M_0)$, $\forall S \in \Pi$, $M(S) > 0$.*

Next a more general class of Petri nets than S^{3}PR nets is introduced. It is first proposed in [25] and called ES^{3}PR. It stands for extended S^{3}PR.

Definition 5.8. An ES^{3}PR is an ordinary connected self-loop-free Petri net $N = \bigcirc_{i=1}^{n} N_i = (P, T, F)$ where:

1. $N_i = (P_{A_i} \cup \{p_i^0\} \cup P_{R_i}, T_i, F_i)$, $i \in \mathbb{N}_n$.
2. $P = P_A \cup P^0 \cup P_R$ is a partition such that (1) $P_A = \cup_{i=1}^{n} P_{A_i}$ is called the set of operation places, where $\forall i, j \in \mathbb{N}_n$, $i \neq j$, $P_{A_i} \neq \emptyset$ and $P_{A_i} \cap P_{A_j} = \emptyset$; (2) $P^0 = \cup_{i=1}^{n} \{p_i^0\}$ is called the set of idle places; (3) $P_R = \{r_1, r_2, \ldots, r_m | m \in \mathbb{N}^+\}$ is called the set of resource places.
3. $T = \cup_{i=1}^{n} T_i$ is called the set of transitions, where $\forall i, j \in \mathbb{N}_n$, $i \neq j$, $T_i \neq \emptyset$ and $T_i \cap T_j = \emptyset$.
4. $\forall i \in \mathbb{N}_n$, the subnet $\overline{N_i}$ generated by $P_{A_i} \cup \{p_i^0\} \cup T_i$ is a strongly connected state machine such that every circuit of the state machine contains idle place p_i^0.
5. $\forall r \in P_R$, there exists a unique minimal P-semiflow $I_r \in \mathbb{N}^{|P|}$ such that $\{r\} = ||I_r|| \cap P_R$, $P^0 \cap ||I_r|| = \emptyset$, $P_A \cap ||I_r|| \neq \emptyset$, and $\forall p \in ||I_r||$, $I_r(p) = 1$.
6. $P_A = \cup_{r \in P_R} (||I_r|| \backslash \{r\})$.

Definition 5.9. An initial marking M_0 is acceptable for an ES^{3}PR $N = (P_A \cup P^0 \cup P_R, T, F)$ iff (1) $\forall i \in \mathbb{N}_n$, $M_0(p_i^0) > 0$; (2) $\forall p \in P_A$, $M_0(p) = 0$; and (3) $\forall r \in P_R$, $M_0(r) > 0$.

From their definitions, the difference between S^{3}PR and ES^{3}PR is that the usage of the resources in the latter is more flexible than that in the former. Specifically, each operation place in an S^{3}PR needs one resource only. However, an operation place in an ES^{3}PR may need two or more resources.

Theorem 5.2. *Let (N, M_0) be an ES^{3}PR. It is live iff no siphon can become empty.*

The net shown in Fig. 5.4 is an ES^{3}PR in which p_2–p_9 are operation places that are initially unmarked, p_1 and p_{10} are idle places, and the others are resource places. It is not an S^{3}PR. It is live since no siphon can be emptied during its evolution.

As a generalization of S^{3}PR and ES^{3}PR in [25] and [50], respectively, the definition of the class of *systems of sequential systems with shared resources* (S^{4}R) is given in [1, 3]. It allows multiple arcs between a resource place and transition. The represented net models fall into the generalized Petri net class. However, an S^{4}R cannot model a manufacturing system with assembly and disassembly operations since it is composed of a set of state machines where no synchronization structure is allowed.

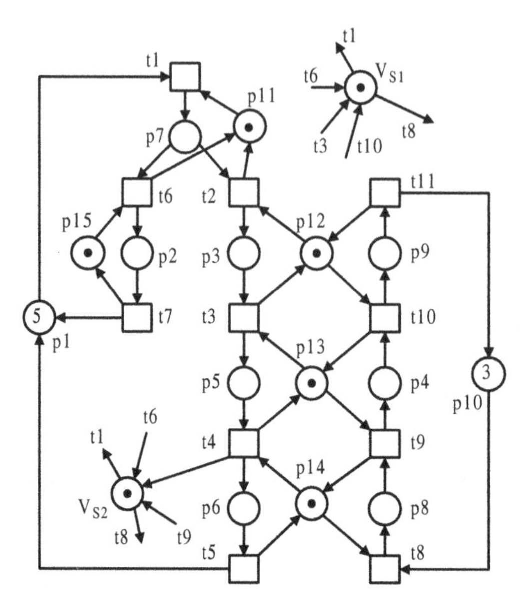

Fig. 5.4 An ES^{3}PR (N,M_0)

Definition 5.10. An S^{4}R is a generalized connected self-loop-free Petri net $N = \bigcirc_{i=1}^{n} N_i = (P,T,F,W)$ where:

1. $N_i = (P_{A_i} \cup \{p_i^0\} \cup P_{R_i}, T_i, F_i, W_i)$, $i \in \mathbb{N}_n$.
2. $P = P_A \cup P^0 \cup P_R$ is a partition such that (1) $P_A = \cup_{i=1}^{n} P_{A_i}$ is called the set of operation places, where $\forall i,j \in \mathbb{N}_n$, $i \neq j$, $P_{A_i} \neq \emptyset$ and $P_{A_i} \cap P_{A_j} = \emptyset$; (2) $P^0 = \cup_{i=1}^{n} \{p_i^0\}$ is called the set of idle places; (3) $P_R = \cup_{i=1}^{n} P_{R_i} = \{r_1, r_2, \cdots, r_m | m \in \mathbb{N}^+\}$ is called the set of resource places.
3. $T = \cup_{i=1}^{n} T_i$ is called the set of transitions, where $\forall i,j \in \mathbb{N}_n$, $i \neq j$, $T_i \neq \emptyset$ and $T_i \cap T_j = \emptyset$.
4. $\forall i \in \mathbb{N}_n$, the subnet $\overline{N_i}$ generated by $P_{A_i} \cup \{p_i^0\} \cup T_i$ is a strongly connected state machine such that every circuit of the state machine contains idle place p_i^0.
5. $\forall r \in P_R$, there exists a unique minimal P-semiflow $I_r \in \mathbb{N}^{|P|}$ such that $\{r\} = ||I_r|| \cap P_R$, $P^0 \cap ||I_r|| = \emptyset$, $P_A \cap ||I_r|| \neq \emptyset$, and $I_r(r) = 1$.
6. $P_A = \cup_{r \in P_R} (||I_r|| \setminus \{r\})$.

Definition 5.11. An initial marking M_0 is acceptable for an S^{4}R $N = (P_A \cup P^0 \cup P_R, T, F, W)$ iff (1) $\forall i \in \mathbb{N}_n$, $M_0(p_i^0) > 0$; (2) $\forall p \in P_A$, $M_0(p) = 0$; and (3) $\forall r \in P_R$, $M_0(r) \geq max\{I_r(p) | p \in P_A\}$.

Example 5.5. The net shown in Fig. 4.5 is an S^{4}R, where $I_{r_1} = r_1 + p_1 + 3p_2 + p_6$ is the minimal P-semiflow associated with resource r_1. An acceptable initial marking M_0 needs to satisfy $M_0(r_1) \geq 3$.

The class of Petri nets defined above is alternatively named to be S^{4}PR [51] by different researchers. The following result from [1] concerns the liveness of an S^{4}R.

Theorem 5.3. *[1] Let (N, M_0) be a marked S^{4}R net. It is live iff it satisfies the max-cs property* [1].

Theorem 5.4. *[1, 51] Let S be a strict minimal siphon in an S^{4}R $N = (P_A \cup P^0 \cup P_R, T, F, W)$. Then $S = S^R \cup S^A$ satisfies $S \cap P_R = S^R \neq \emptyset$ and $S \cap P_A = S^A \neq \emptyset$.*

Before the presentation of the complementary set of a siphon in a generalized Petri net, a brief theory of multisets is given in order to recognize well the role of the complementary sets of siphons in the development of deadlock prevention policies for generalized Petri nets that are used to model automated manufacturing systems. A multiset is a generalization of the concept of a set.

Definition 5.12. A multiset Ω, over a non-empty set A, is a mapping $\Omega : A \to \mathbb{N}^+$, which we represent as a formal sum $\sum_{a \in A} \Omega(a).a$.

In multiset Ω, non-negative integer $\Omega(a)$ is the coefficient of element $a \in A$, indicating the number of occurrences of a in Ω. It is said that $a \in A$ belongs to Ω, denoted by $a \in \Omega$, if $\Omega(a) > 0$. It does not belong to Ω, denoted by $a \notin \Omega$, if $\Omega(a) = 0$. Let Ω_{MS} denote the set of all multisets over A and Ω_1 and Ω_2 be two multisets in Ω_{MS}. The basic operations on multisets are union, intersection, addition, and difference, which are defined as follows.

Definition 5.13. $\Omega_1 \cup \Omega_2 := \sum_{a \in A} max\{\Omega_1(a), \Omega_2(a)\}.a$.

Definition 5.14. $\Omega_1 \cap \Omega_2 := \sum_{a \in A} min\{\Omega_1(a), \Omega_2(a)\}.a$.

Definition 5.15. $\Omega_1 + \Omega_2 := \sum_{a \in A} (\Omega_1(a) + \Omega_2(a)).a$.

Definition 5.16. $\Omega_1 - \Omega_2 := \sum_{a \in A} (\Omega_1(a) - (\Omega_1 \cap \Omega_2)(a)).a$.

A multiset without any element is denoted by $\emptyset$ as an empty set. A multiset becomes a set if the multiplicity of every element is one.

Example 5.6. $\Omega_1 = a + b$, $\Omega_2 = 2a + b + c$, and $\Omega_3 = 3a + 2c$ are three multisets over $A = \{a, b, c\}$. We have $\Omega_1 \cup \Omega_2 = 2a + b + c$, $\Omega_1 \cup \Omega_3 = 3a + b + 2c$, $\Omega_1 \cap \Omega_3 = a$, $\Omega_2 \cap \Omega_3 = 2a + c$, $\Omega_1 + \Omega_2 = 3a + 2b + c$, $\Omega_1 - \Omega_2 = \emptyset$, $\Omega_3 - \Omega_1 = 2a + 2c$, and $\Omega_3 - \Omega_2 = a + c$.

Definition 5.17. Let r be a resource place and S be a strict minimal siphon in an S^{4}R. The holder of resource r is defined as the difference of two multisets I_r and r: $H(r) = I_r - r$. As a multiset, $Th(S) = \sum_{r \in S^R} H(r) - \sum_{r \in S^R, p \in S^A} I_r(p).p$ is called the complementary set of siphon S.

---

[1] This result is unfortunately not correct since the max-cs property is found to be a sufficient but not necessary condition for the liveness of an S^{4}R. For details, the reader can be referred to [8], where the controllability condition of siphons in a generalized Petri net is relaxed by introducing the concept of max'-controlled siphons.

Definition 5.18. $||Th(S)|| = \{p | p \in \cup_{r \in S^R} H(r), p \notin S\}$ is called the support of complementary set $Th(S)$ of siphon S.

It is easy to see that $||Th(S)|| \subseteq P_A$ is true. Let $||Th(S)||^i = ||Th(S)|| \cap P_{A_i}$; we have $||Th(S)|| = \cup_{i=1}^{n} ||Th(S)||^i$. Let $\sum_{p \in ||Th(S)||} h_S(p)p$ denote $Th(S)$. Clearly, $h_S(p)$ indicates that siphon S loses $h_S(p)$ tokens if the number of tokens in p increases by one. $h_S(p)$ is called the risk coefficient of place p.

Example 5.7. The net shown in Fig. 5.5 is an S^{4}R if $p_1^0 = p_7$, $p_2^0 = p_{11}$, $P_{A_1} = \{p_1$–$p_6\}$, $P_{A_2} = \{p_8$–$p_{10}\}$, $P_{R_1} = \{p_{12}$–$p_{15}\}$, and $P_{R_2} = \{p_{12}$–$p_{14}\}$. It has three strict minimal siphons:

$S_1 = \{p_3, p_6, p_9, p_{13}, p_{14}\}$,
$S_2 = \{p_2, p_5, p_{10}, p_{12}, p_{13}\}$,
$S_3 = \{p_3, p_6, p_{10}, p_{12}, p_{13}, p_{14}\}$.

We have

$I_{p_{12}} = 2p_1 + p_{10} + p_{12}$,
$I_{p_{13}} = p_2 + p_5 + p_9 + p_{13}$,
$I_{p_{14}} = p_3 + p_6 + p_8 + p_{14}$,
$I_{p_{15}} = p_4 + p_{15}$,
$H(p_{12}) = 2p_1 + p_{10}$,
$H(p_{13}) = p_2 + p_5 + p_9$,
$H(p_{14}) = p_3 + p_6 + p_8$.

Furthermore, the complementary sets and their supports of siphons S_1–S_3 are:

$Th(S_1) = (H(p_{13}) + H(p_{14})) - (p_3 + p_6 + p_9) = p_2 + p_5 + p_8$,
$Th(S_2) = (H(p_{12}) + H(p_{13})) - (p_2 + p_5 + p_{10}) = 2p_1 + p_9$,
$Th(S_3) = \sum_{i=12}^{14} H(p_i) - (p_3 + p_6 + p_{10}) = 2p_1 + p_2 + p_5 + p_8 + p_9$,
$||Th(S_1)|| = \{p_2, p_5, p_8\}$,
$||Th(S_2)|| = \{p_1, p_9\}$,
$||Th(S_3)|| = \{p_1, p_2, p_5, p_8, p_9\}$.

For $Th(S_2)$, we have the risk coefficients of p_1 and p_9 are $h_{S_2}(p_1) = 2$ and $h_{S_2}(p_9) = 1$, respectively. It is easy to see that two tokens are removed from S_2 (in fact from p_{12}) if the number of tokens in p_1 increases by one.

Note that where there is no confusion, we often use $[S]$ to denote the set of places in $||Th(S)||$ as well as $Th(S)$ for convenience and clarity.

5.3 An MIP-Based Deadlock Detection Method

Let (N, M) be an ordinary net with $N = (P, T, F)$ and S be the maximal empty siphon at M, i.e., $\forall p \notin S$, $M(p) > 0$. As shown in [9], finding S in N is the solution of a mixed-integer-programming (MIP) problem. $\forall p \notin S$, let $v_p = 1$ and $\forall t \notin S^\bullet$, let $z_t = 1$.

It is easy to see that any p with $v_p = 1$ and any t with $z_t = 1$ are removed from the net. Since S is a siphon, we have that $\forall t \in p^\bullet$, $v_p = 0$ implies $z_t = 0$ and $\forall p \in t^\bullet$, $z_t = 1$ implies the truth of $v_p = 1$. This leads to

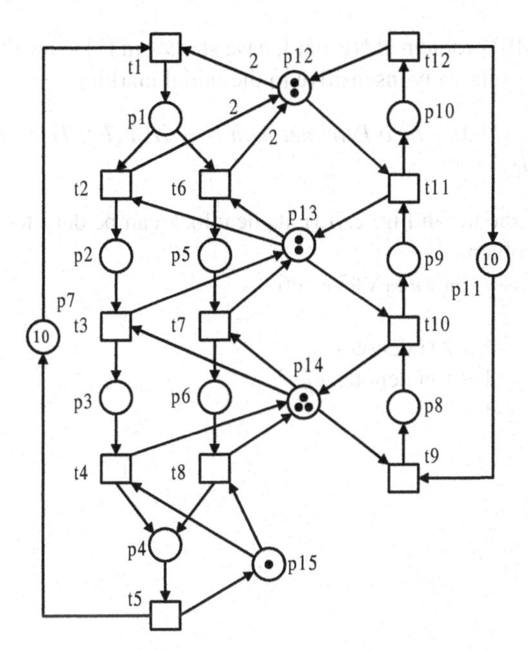

Fig. 5.5 An S^{4}R (N, M_0)

$$z_t \geq \sum_{p \in {}^\bullet t} v_p - |{}^\bullet t| + 1, \forall t \in T, \tag{5.1}$$

$$v_p \geq z_t, \forall (t,p) \in F, \tag{5.2}$$

$$v_p, z_t \in \{0,1\}. \tag{5.3}$$

For a structurally bounded net, we have

$$v_p \geq M(p)/\pi(p), \forall p \in P \tag{5.4}$$

where $\pi(p) = max\{M(p)|M = M_0 + [N]Y, M \geq 0, Y \geq 0\}$ is the structural bound of place p. As shown in [9], the maximal unmarked siphon can be determined by the following MIP problem and there exist siphons unmarked in (N, M_0) iff $G^{MIP}(M_0) < |P|$:

$$G^{MIP}(M_0) = min \sum_{p \in P} v_p$$

s.t. constraints (5.1)–(5.4) and

$$M = M_0 + [N]Y, M \geq 0, Y \geq 0.$$

Although an MIP problem is NP-hard, case studies in [9] show that its computational efficiency is relatively insensitive to the initial marking.

Theorem 5.5. *Let* (N, M_0) *be a Petri net with* $N = (P, T, F)$*. There is no emptiable siphon if* $G^{MIP}(M_0) = |P|$.

Example 5.8. For the net in Fig. 2.7(a), its deadlock can be detected by solving the following MIP problem:

```
MIN VP1+VP2+VP3+VP4+VP5+VP6
SUBJECT TO
ZT1-VP1-VP6≥-1 // ZTi denotes z_{t_i}
ZT2-VP2-VP5≥-1 // VPi denotes v_{p_i}
ZT3-VP3-VP6≥-1
ZT4-VP4≥0
VP1-ZT4≥0
VP6-ZT4≥0
VP2-ZT1≥0
VP3-ZT2≥0
VP6-ZT2≥0
VP5-ZT3≥0
VP4-ZT3≥0
2VP1-MP1≥0 // MPi denotes M(p_i)
VP2-MP2≥0
VP3-MP3≥0
VP4-MP4≥0
VP5-MP5≥0
VP6-MP6≥0
MP1+Y1-Y4=2 // Yi denotes y_i
MP2-Y1+Y2=0
MP3-Y2+Y3=0
MP4-Y3+Y4=0
MP5+Y2-Y3=1
MP6+Y1+Y3-Y2-Y4=1
END
INT VP1
INT VP2
INT VP3
INT VP4
INT VP5
INT VP6
INT ZT1
INT ZT2
INT ZT3
INT ZT4
```

Note that the above source code follows the syntax of Lindo [38], where "int x" indicates variable $x \in \{0, 1\}$. Lindo gives a feasible solution Z=2 with VP2=VP3=1 and VP1=VP4=VP5=VP6=0. It means that there is a siphon $S = \{p_1, p_4, p_5, p_6\}$ that can be emptied. When its initial marking changes to be $M_0 = p_1 + p_5 + p_6$ from $M_0 = 2p_1 + p_5 + p_6$, we have $z = 6$, the number of places, indicating that no siphon can be emptied.

5.4 A Classical Deadlock Prevention Policy

This section mainly introduces the deadlock prevention policy proposed by Ezpeleta et al. in [16]. It develops a systematic method to establish a liveness-enforcing supervisor for an S^{3}PR by adding monitors for its strict minimal siphons such that they are prevented from being unmarked. The work in [16] is usually considered to be one of the most significant contributions in deadlock control area using a Petri net formalism [21].

Before the presentation of the policy, some notations are first introduced in order that the readers can understand it well. It should be stressed that many monitor-based deadlock prevention policies in the literature are motivated by the seminal work in [16] and [29].

Let $N = (P, T, F)$ be an S^{2}P with idle process place p^0. The length of a path (circuit) in a Petri net is defined as the number of its nodes. The support of a path (circuit) is the set of its nodes.

- Let $\mathscr{C}$ be a circuit of N and x and y be two nodes of $\mathscr{C}$. Node x is said to be *previous* to y iff there exists a path in $\mathscr{C}$ from x to y, the length of which is greater than one and does not pass over the idle process place p^0. This fact is denoted by $x <_{\mathscr{C}} y$.
- Let x and y be two nodes in N. Node x is said to be *previous* to y in N iff there exists a circuit $\mathscr{C}$ such that $x <_{\mathscr{C}} y$. This fact is denoted by $x <_N y$.
- Let x and $A \subseteq P \cup T$ be a node and a set of nodes in N, respectively. Then $x <_N A$ iff there exists a node $y \in A$ such that $x <_N y$ and $A <_N x$ iff there exists a node $y \in A$ such that $y <_N x$.

Example 5.9. In the net N in Fig. 5.3, $\mathscr{C} = p_1 t_1 p_7 t_2 p_3 t_3 p_5 t_4 p_6 t_5 p_1$ is a circuit and $EP(p_7, p_6) = p_7 t_2 p_3 t_3 p_5 t_4 p_6$ is a path in $\mathscr{C}$. The support of $EP(p_7, p_6)$ is $\{p_7, t_2, p_3, t_3, p_5, t_4, p_6\}$ and the support of $\mathscr{C}$ is $\{p_1, t_1, p_7, t_2, p_3, t_3, p_5, t_4, p_6, t_5\}$. Clearly, we have $p_7 <_{\mathscr{C}} p_6$ and $p_7 <_N p_6$.

The following notations are also useful in the establishment of a deadlock prevention policy. Mathematically, given a set A, the power set (or powerset) of A, written as 2^A, is the set of all subsets of A. Note that Π is used to denote the set of strict minimal siphons in an S^{3}PR (N, M_0). First the definitions of the sets of downstream and upstream siphons of a transition are given.

Definition 5.19. Let $\Delta^+(t)$ ($\Delta^-(t)$) denote the set of downstream (upstream) siphons of a transition t and $\mathscr{P}_S$ denote the adjoint set of a siphon S in an S^{3}PR $N = \bigcirc_{i=1}^n N_i = (P^0 \cup P_A \cup P_R, T, F)$.

1. $\Delta^+ : T \to 2^\Pi$ is a mapping defined as follows: If $t \in T_i$, then $\Delta^+(t) = \{S \in \Pi | t <_{\overline{N}_i} [S]^i\}$. If $S \in \Delta^+(t)$ then the set $[S]^i$ is *reachable* from t, i.e., there exists a path in $\overline{N}_i$ leading from t to an operation place $p \in P_{A_i}$ that is not included in S but uses a resource of S, where $[S] = \cup_{i=1}^n [S]^i$, $P_A = \cup_{i=1}^n P_{A_i}$, and $[S]^i = [S] \cap P_{A_i}$.
2. $\Delta^- : T \to 2^\Pi$ is a mapping defined as follows: If $t \in T_i$, then $\Delta^-(t) = \{S \in \Pi | [S]^i <_{\overline{N}_i} t\}$.
3. $\forall i \in \mathbb{N}_n, \forall S \in \Pi$, $\mathscr{P}_S^i = [S]^i \cup \{p \in P_{A_i} | p <_{\overline{N}_i} [S]^i\}$, and $\mathscr{P}_S = \cup_{i=1}^n \mathscr{P}_S^i$.

Example 5.10. Take the net shown in Fig. 5.3 as an example. There are three strict minimal siphons $S_1 = \{p_5, p_9, p_{12}, p_{13}\}$, $S_2 = \{p_4, p_6, p_{13}, p_{14}\}$, and $S_3 = \{p_6, p_9, p_{12}, p_{13}, p_{14}\}$. Their complementary sets are $[S_1] = \{p_3, p_4\}$, $[S_2] = \{p_5, p_8\}$, and $[S_3] = \{p_3, p_4, p_5, p_8\}$, respectively. We have downstream siphons $\Delta^+(t_1) = \Delta^+(t_2) = \Delta^+(t_8) = \{S_1, S_2, S_3\}$, $\Delta^+(t_3) = \{S_2, S_3\}$, and $\Delta^+(t_4) = \Delta^+(t_{10}) = \emptyset$. Similarly, upstream siphons include $\Delta^-(t_1) = \Delta^-(t_2) = \Delta^-(t_6) = \Delta^-(t_7) = \emptyset$, $\Delta^-(t_3) = \{S_1\}$, and $\Delta^-(t_4) = \Delta^-(t_5) = \{S_1, S_2, S_3\}$.

We have adjoint sets $\mathscr{P}_{S_1} = \mathscr{P}_{S_1}^1 \cup \mathscr{P}_{S_1}^2 = (\{p_3\} \cup \{p_7\}) \cup (\{p_4\} \cup \{p_8\}) = \{p_3, p_4, p_7, p_8\}$, $\mathscr{P}_{S_2} = \mathscr{P}_{S_2}^1 \cup \mathscr{P}_{S_2}^2 = (\{p_5\} \cup \{p_7, p_3\}) \cup \{p_8\} = \{p_7, p_3, p_5, p_8\}$, and $\mathscr{P}_{S_3} = \mathscr{P}_{S_3}^1 \cup \mathscr{P}_{S_3}^2 = (\{p_3, p_5\} \cup p_7) \cup \{p_4, p_8\} = \{p_7, p_3, p_5, p_4, p_8\}$.

Definition 5.20. Let (N, M_0) be an S^{3}PR with $N = \bigcirc_{i=1}^n N_i = (P_A \cup P^0 \cup P_R, T, F)$. The net $(N_V, M_{0V}) = (P_A \cup P^0 \cup P_R \cup P_V, T, F \cup F_V, M_{0V})$ is the controlled system of (N, M_0) iff:

1. $P_V = \{V_S | S \in \Pi\}$ is a set of monitors such that there exists a bijective mapping between Π and P_V.
2. $F_V = F_V^1 \cup F_V^2 \cup F_V^3$, where
   $F_V^1 = \{(V_S, t) | S \in \Delta^+(t), t \in P^{0\bullet}\}$,
   $F_V^2 = \{(t, V_S) | t \in [S]^\bullet, S \notin \Delta^+(t)\}$,
   $F_V^3 = \cup_{i=1}^n \{(t, V_S) | t \in T_i \setminus P^{0\bullet}, S \notin \Delta^-(t), {}^\bullet t \cap P_{A_i} \subseteq \mathscr{P}_S^i, t \nless [S]^i\}$.
3. M_{0V} is defined as follows: (1) $\forall p \in P_A \cup P^0 \cup P_R$, $M_{0V}(p) = M_0(p)$ and (2) $\forall V_S \in P_V$, $M_{0V}(V_S) = M_0(S) - 1$.

Theorem 5.6. *(N_V, M_{0V}) is live [16].*

Example 5.11. For the net shown in Fig. 5.3, three monitors are needed to prevent three strict minimal siphons from being emptied. We first take $S_1 = \{p_5, p_9, p_{12}, p_{13}\}$ as an example. Since $P^0 = \{p_1, p_{10}\}$, we have $P^{0\bullet} = \{t_1, t_8\}$. As a result, $\{(V_{S_1}, t_1), (V_{S_1}, t_8)\} \subseteq F_V^1$.

Due to $[S_1] = \{p_3, p_4\}$, $[S_1]^\bullet = \{t_3, t_{10}\}$. Note that $S_1 \notin \Delta^+(t_3)$ and $S_1 \notin \Delta^+(t_{10})$. We have $\{(t_3, V_{S_1}), (t_{10}, V_{S_1})\} \subseteq F_V^2$.

Next let us find the arcs related to V_{S_1} in F_V^3. Let
$T_\alpha = (T_1 \setminus P^{0\bullet}) \cup (T_2 \setminus P^{0\bullet})$,
$T_\beta = \{t | S_1 \notin \Delta^-(t), t \in T\}$,
$T_\gamma = \{t | {}^\bullet t \cap P_{A_1} \subseteq \mathscr{P}_S^1\} \cup \{t | {}^\bullet t \cap P_{A_2} \subseteq \mathscr{P}_S^2\}$,

$T_\delta = \{t | t \nless [S_1]^1\} \cup \{t | t \nless [S_1]^2\}$.
We have
$T_\alpha = \{t_2 - t_7, t_9 - t_{11}\}$,
$T_\beta = \{t_1, t_2, t_6 - t_9\}$,
$T_\gamma = \{t_2, t_3, t_6, t_9, t_{10}\}$,
$T_\delta = \{t_3 - t_7, t_{10}, t_{11}\}$.
It is easy to see that $T_\alpha \cap T_\beta \cap T_\gamma \cap T_\delta = \{t_6\}$. Consequently, $(t_6, V_{S_1}) \in F_V^3$.
For siphons S_2 and S_3, monitors V_{S_2} and V_{S_3} can be added with $\{(V_{S_2}, t_1), (V_{S_2}, t_8), (V_{S_3}, t_1), (V_{S_3}, t_8)\} \subseteq F_V^1$, $\{(t_4, V_{S_2}), (t_9, V_{S_2}), (t_4, V_{S_3}), (t_{10}, V_{S_3})\} \subseteq F_V^2$, and $\{(t_6, V_{S_2}), (t_6, V_{S_3})\} \subseteq F_V^3$. The controlled system for (N, M_0) is shown in Fig. 5.6.

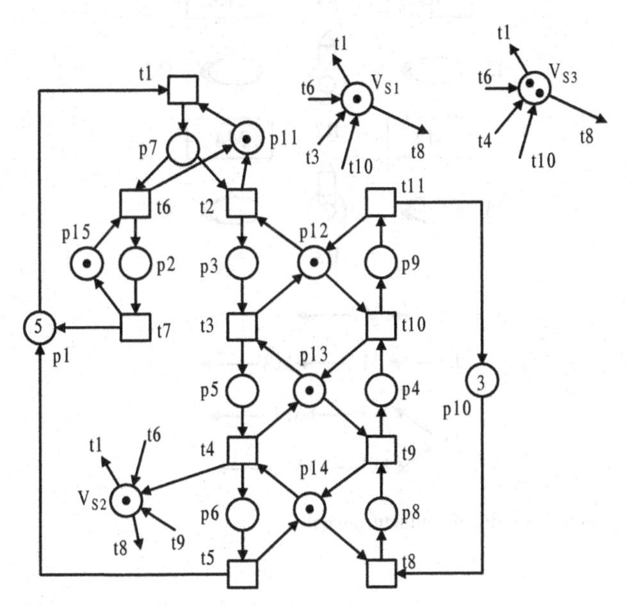

Fig. 5.6 The controlled system of (N, M_0)

According to Theorem 5.6, the net in Fig. 5.6 is live. □

Remark 5.1. For a strict minimal siphon S, this policy ensures that the maximal number of tokens held by $\mathscr{P}_S$ is not more than $M_0(S)$. Since $[S] \subseteq \mathscr{P}_S$, S cannot be emptied if a monitor V_S is added for it. For example, $S_1 = \{p_5, p_9, p_{12}, p_{13}\}$ is a strict minimal siphon with $M_0(S_1) = 2$. Its monitor guarantees that the maximal number of tokens held in $\mathscr{P}_{S_1}$ is $M_0(S_1) - 1 = 1$.

Next we present an FMS example to illustrate the deadlock prevention policy. This FMS is extensively investigated in the literature [6, 7, 16, 24, 30–33, 52, 54]. Particularly, in a recent survey paper [36], this FMS is exploited as a benchmark example to compare a variety of deadlock prevention policies in the literature.

Example 5.12. A flexible manufacturing cell as shown in Fig. 5.7 has four machine tools M1 – M4. Each machine tool can hold two parts at the same time. Also the cell contains three robots R1 – R3 and each of them can hold one part. Parts enter the cell through three loading buffers I1 – I3, and leave the cell through three unloading buffers O1 – O3. Three part types J1 – J3 are produced. The machine tools perform operations on raw parts and the robots deal with the movements of parts.

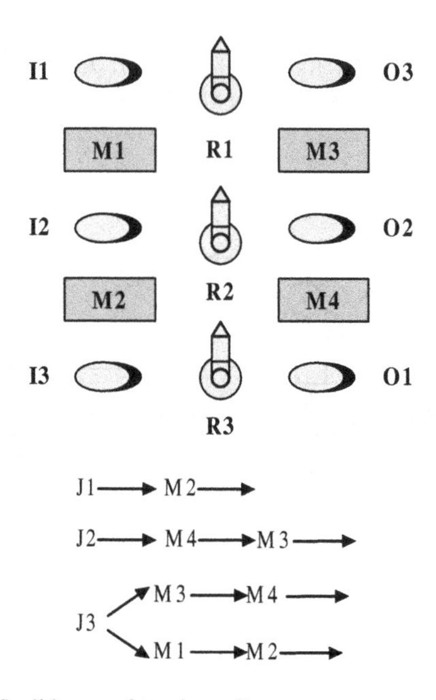

Fig. 5.7 The layout of a flexible manufacturing cell

- R1 handles part movements from I3 to M1, I3 to M3, and M3 to O2.
- R2 handles part movements from M1 to M2, M4 to M3, M3 to M4, I1 to M2, and M2 to O1.
- R3 handles part movements from I2 to M4, M2 to O3, and M4 to O3.
- M1 performs operations on J3.
- M2 performs operations on J1 and J3.
- M3 performs operations on J2 and J3.
- M4 performs operations on J2 and J3.

The production routes of the parts are as follows:

- J1: I1 $\rightarrow$ R2 $\rightarrow$ M2 $\rightarrow$ R2 $\rightarrow$ O1;
- J2: I2 $\rightarrow$ R3 $\rightarrow$ M4 $\rightarrow$ R2 $\rightarrow$ M3 $\rightarrow$ R1 $\rightarrow$ O2;
- J3: I3 $\rightarrow$ R1 $\rightarrow$ M1 $\rightarrow$ R2 $\rightarrow$ M2 $\rightarrow$ R3 $\rightarrow$ O3 or I3 $\rightarrow$ R1 $\rightarrow$ M3 $\rightarrow$ R2 $\rightarrow$ M4 $\rightarrow$ R3 $\rightarrow$ O3.

The flexible manufacturing cell contains global and local deadlocks if it is not properly supervised. Suppose that the system is in the scenario that M4 is fully occupied by machining two J3-type raw parts and R3 picks up a J2-type raw part and tries to upload M4. The processes to produce J2 and J3 remain indefinitely blocked since no further operations on them can be performed. That is to say, the system is in a local deadlock state.

While, if the system is in such a state that M4 is fully occupied by machining two J3-type raw parts, R3 has picked up a J2-type raw part from I2 and is trying to upload M4, M2 is fully occupied by machining two J1-type raw parts, and R2 has downloaded a J3-type part from M1 and is trying to upload M2, it will enter a global deadlock state. In this case, the whole system will be completely blocked.

The system can be modeled with Petri nets. Its model (N, M_0) is shown in Fig. 5.8. The physical meaning of each place is explained in Table 5.1. Let $T_1 = \{t_{11} - t_{14}\}$, $T_2 = \{t_{15} - t_{20}\}$, and $T_3 = \{t_1 - t_{10}\}$. This model belongs to S^{3}PR where p_1, p_5, and p_{14} are idle process places, $p_{20} - p_{26}$ are resource places, and the others are operation places. The occurrence of system deadlocks corresponds to the existence of unmarked siphons in the model. They can be successfully prevented by properly supervising the siphons in their net model.

There are 18 strict minimal siphons in the Petri net model of the flexible manufacturing cell. These siphons, their complementary sets, and their corresponding monitors are shown in Tables 5.2 – 5.4, respectively.

Next $S_1 = \{p_{10}, p_{18}, p_{22}, p_{26}\}$, $S_3 = \{p_2, p_4, p_8, p_{13}, p_{17}, p_{21}, p_{26}\}$, and $S_5 = \{p_2, p_4, p_8, p_{10}, p_{17}, p_{21}, p_{22}, p_{26}\}$ are taken as examples to show the way of adding monitors to prevent them from being unmarked.

According to the definition of the complementary set of a siphon, we have $[S_1] = (H(p_{22}) \cup H(p_{26})) \setminus S_1 = (\{p_{10}, p_{19}\} \cup \{p_{13}, p_{18}\}) \setminus S_1 = \{p_{13}, p_{19}\}$, $[S_3] = (\{p_2, p_4, p_8, p_{12}, p_{17}\} \cup \{p_{13}, p_{18}\}) \setminus S_3 = \{p_{12}, p_{18}\}$, and $[S_5] = (\{p_2, p_4, p_8, p_{12}, p_{17}\} \cup \{p_{10}, p_{19}\} \cup \{p_{13}, p_{18}\}) \setminus S_5 = \{p_{12}, p_{13}, p_{18}, p_{19}\}$.

Since $P^0 = \{p_1, p_5, p_{14}\}$, we have $P^{0\bullet} = \{t_1, t_{11}, t_{15}\}$. $\{S_1, S_3, S_5\} \subseteq \Delta^+(t_1)$, $\{S_1, S_3, S_5\} \subseteq \Delta^+(t_{15})$, and $\forall S \in \{S_1, S_3, S_5\}$, $S \notin \Delta^+(t_{11})$. This leads to $\{(V_{S_1}, t_1), (V_{S_1}, t_{15}), (V_{S_3}, t_1), (V_{S_3}, t_{15}), (V_{S_5}, t_1), (V_{S_5}, t_{15})\} \subseteq F_V^1$.

From $[S_1] = \{p_{13}, p_{19}\}$, $[S_1]^\bullet = \{t_{10}, t_{16}\}$. It is clear that $S_1 \notin \Delta^+(t_{10})$ and $S_1 \notin \Delta^+(t_{16})$. Therefore, $\{(t_{10}, V_{S_1}), (t_{16}, V_{S_1})\} \subseteq F_V^2$. From $[S_3] = \{p_{12}, p_{18}\}$, $[S_3]^\bullet = \{t_9, t_{17}\}$. It is clear that $S_3 \notin \Delta^+(t_9)$ and $S_3 \notin \Delta^+(t_{17})$. Therefore, $\{(t_9, V_{S_3}), (t_{17}, V_{S_3})\} \subseteq F_V^2$.

The case of S_5 is slightly different from S_1 and S_3. From $[S_5] = \{p_{12}, p_{13}, p_{18}, p_{19}\}$, we have $[S_5]^\bullet = \{t_9, t_{10}, t_{16}, t_{17}\}$. However, $S_5 \in \Delta^+(t_9)$, $S_5 \in \Delta^+(t_{16})$, $S_5 \notin \Delta^+(t_{10})$, and $S_5 \notin \Delta^+(t_{17})$. As a result, $\{(t_{10}, V_{S_5}), (t_{17}, V_{S_5})\} \subseteq F_V^2$.

In order to determine the set of additional arcs in F_V^3 for V_{S_1}, V_{S_3}, and V_{S_5}, we first find $\mathscr{P}_{S_1}$, $\mathscr{P}_{S_3}$, and $\mathscr{P}_{S_5}$.

$\mathscr{P}_{S_1} = \mathscr{P}_{S_1}^1 \cup \mathscr{P}_{S_1}^2 \cup \mathscr{P}_{S_1}^3$, where $\mathscr{P}_{S_1}^i = [S_1]^i \cup \{p \in P_{A_i} | p <_{\overline{N}_i} [S_1]^i\}$, $i = 1, 2, 3$. It is easy to see that $[S_1]^1 = \emptyset$, $[S_1]^2 = \{p_{19}\}$, $[S_1]^3 = \{p_{13}\}$, $\{p \in P_{A_1} | p <_{\overline{N}_1} [S_1]^1\} = \emptyset$, $\{p \in P_{A_2} | p <_{\overline{N}_2} [S_1]^2\} = \emptyset$, and $\{p \in P_{A_3} | p <_{\overline{N}_3} [S_1]^3\} = \{p_6, p_{11}, p_{12}\}$. Hence $\mathscr{P}_{S_1} = \{p_{13}, p_{19}, p_6, p_{11}, p_{12}\}$.

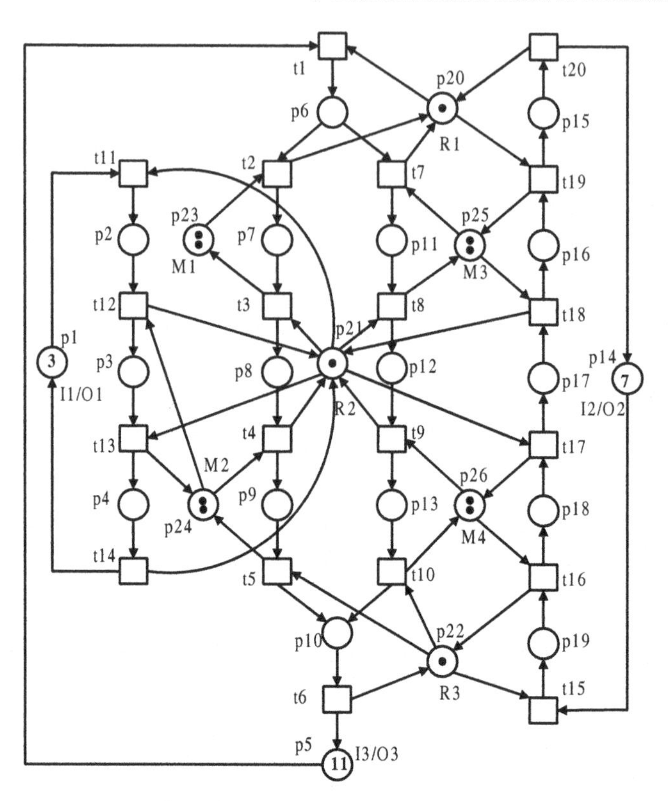

Fig. 5.8 The Petri net model (N, M_0) of a flexible manufacturing cell

$\forall t \in T_1 \setminus P^{0\bullet}$, ${}^\bullet t \cap P_{A_1} \subseteq \mathscr{P}^1_{S_1}$ is not true since ${}^\bullet t \cap P_{A_1} \neq \emptyset$ but $\mathscr{P}^1_{S_1} = \emptyset$. $\forall t \in T_2 \setminus P^{0\bullet}$, $S_1 \in \Delta^-(t)$. That is to say, there does not exist a transition $t \in (T_1 \setminus P^{0\bullet}) \cup (T_2 \setminus P^{0\bullet})$ such that $(t, V_{S_1}) \in F^3_V$. Next we check the transitions in $T_3 \setminus P^{0\bullet}$, which satisfy the conditions in F^3_V.

From $\{t | t \in T_3 \setminus P^{0\bullet}, S_1 \notin \Delta^-(t)\} = \{t_2, t_3, t_4, t_5, t_7, t_8, t_9\}$, $\{t | t \in T_3 \setminus P^{0\bullet}, {}^\bullet t \cap P_{A_3} \subseteq \mathscr{P}^3_{S_1}\} = \{t_2, t_7, t_8, t_9, t_{10}\}$, and $\{t | t \in T_3 \setminus P^{0\bullet}, t \nprec [S_1]^3\} = \{t_2, t_3, t_4, t_5, t_6, t_{10}\}$, we conclude $(t_2, V_{S_1}) \in F^3_V$.

Similarly, we have $(t_2, V_{S_3}) \in F^3_V$, $(t_2, V_{S_5}) \in F^3_V$, and $\forall t \neq t_2$, $(t, V_{S_i}) \notin F^3_V$, where $i = 1, 3, 5$.

The initial marking of these monitors is $M_{0V}(V_{S_1}) = 2$, $M_{0V}(V_{S_3}) = 2$, and $M_{0V}(V_{S_5}) = 3$. The monitors for other siphons can be accordingly determined as shown in Table 5.4. The resultant net with 18 monitors, i.e., the controlled system, is denoted by (N_V, M_{0V}), and is live with 6,287 reachable states.

Table 5.1 Meanings of the places in the net model shown in Fig. 5.8

p_1:	Raw materials in I1 available
p_2:	R2 uploads M2
p_3:	M2 machining
p_4:	R2 downloads M2 and puts finished parts in O1
p_5:	Raw materials in I3 available
p_6:	R1 uploads M1 or M3
p_7:	M1 machining
p_8:	R2 downloads M1 and uploads M2
p_9:	M2 machining
p_{10}:	R3 downloads M2 or M4 and put finished parts in O3
p_{11}:	M3 machining
p_{12}:	R2 downloads M3 and uploads M4
p_{13}:	M4 machining
p_{14}:	Raw materials in I2 available
p_{15}:	R1 downloads M3 and puts finished parts in O2
p_{16}:	M3 machining
p_{17}:	R2 downloads M4 and uploads M3
p_{18}:	M4 machining
p_{19}:	R3 uploads M4
p_{20}:	R1 available
p_{21}:	R2 available
p_{22}:	R3 available
p_{23}:	M1 available
p_{24}:	M2 available
p_{25}:	M3 available
p_{26}:	M4 available

Table 5.2 Strict minimal siphons in the model (N, M_0), where * means the corresponding siphons are dependent ones

S	Places	S	Places
S_1	$p_{10}, p_{18}, p_{22}, p_{26}$	S_2	$p_4, p_9, p_{12}, p_{17}, p_{21}, p_{24}$
S_3	$p_2, p_4, p_8, p_{13}, p_{17}, p_{21}, p_{26}$	S_4	$p_2, p_4, p_8, p_{12}, p_{16}, p_{21}, p_{25}$
S_5^*	$p_2, p_4, p_8, p_{10}, p_{17}, p_{21}, p_{22}, p_{26}$	S_6^*	$p_4, p_9, p_{12}, p_{16}, p_{21}, p_{24}, p_{25}$
S_7^*	$p_4, p_9, p_{13}, p_{17}, p_{21}, p_{24}, p_{26}$	S_8^*	$p_2, p_4, p_8, p_{13}, p_{16}, p_{21}, p_{25}, p_{26}$
S_9	$p_4, p_{10}, p_{17}, p_{21}, p_{22}, p_{24}, p_{26}$	S_{10}^*	$p_4, p_9, p_{13}, p_{16}, p_{21}, p_{24}, p_{25}, p_{26}$
S_{11}^*	$p_2, p_4, p_8, p_{10}, p_{16}, p_{21}, p_{22}, p_{25}, p_{26}$	S_{12}	$p_2, p_4, p_8, p_{12}, p_{15}, p_{20}, p_{21}, p_{23}, p_{25}$
S_{13}^*	$p_4, p_{10}, p_{16}, p_{21}, p_{22}, p_{24}, p_{25}, p_{26}$	S_{14}^*	$p_4, p_9, p_{12}, p_{15}, p_{20}, p_{21}, p_{23}, p_{24}, p_{25}$
S_{15}^*	$p_2, p_4, p_8, p_{13}, p_{15}, p_{20}, p_{21}, p_{23}, p_{25}, p_{26}$	S_{16}^*	$p_4, p_9, p_{13}, p_{15}, p_{20}, p_{21}, p_{23}-p_{26}$
S_{17}^*	$p_2, p_4, p_8, p_{10}, p_{15}, p_{20}-p_{23}, p_{25}, p_{26}$	S_{18}^*	$p_4, p_{10}, p_{15}, p_{20}-p_{26}$

5.5 An Elementary Siphon-Based Deadlock Prevention Policy

The policy proposed in [16] indicates that the number of additional monitors equals that of strict minimal siphons in an S^{3}PR. That is to say, the size of the resultant liveness-enforcing supervisor is in theory exponential with respect to the plant net size since the number of strict minimal siphons grows in a general case exponentially with respect to the net size. It implies that the supervisor is much more structurally complex than the plant model originally built.

Table 5.3 The complementary sets of the strict minimal siphons

$[S]$	Places	$[S]$	Places
$[S_1]$	p_{13}, p_{19}	$[S_2]$	p_2, p_3, p_8
$[S_3]$	p_{12}, p_{18}	$[S_4]$	p_{11}, p_{17}
$[S_5]$	$p_{12}, p_{13}, p_{18}, p_{19}$	$[S_6]$	$p_2, p_3, p_8, p_{11}, p_{17}$
$[S_7]$	$p_2, p_3, p_8, p_{12}, p_{18}$	$[S_8]$	$p_{11}, p_{12}, p_{17}, p_{18}$
$[S_9]$	$p_2, p_3, p_8, p_9, p_{12}, p_{13}, p_{18}, p_{19}$	$[S_{10}]$	$p_2, p_3, p_8, p_{11}, p_{12}, p_{17}, p_{18}$
$[S_{11}]$	$p_{11}, p_{12}, p_{13}, p_{17}, p_{18}, p_{19}$	$[S_{12}]$	$p_6, p_7, p_{11}, p_{16}, p_{17}$
$[S_{13}]$	$p_2, p_3, p_8, p_9, p_{11}, p_{12}, p_{13}, p_{17}-p_{19}$	$[S_{14}]$	$p_2, p_3, p_6, p_7, p_8, p_{11}, p_{16}, p_{17}$
$[S_{15}]$	$p_6, p_7, p_{11}, p_{12}, p_{16}, p_{17}, p_{18}$	$[S_{16}]$	$p_2, p_3, p_6, p_7, p_8, p_{11}, p_{12}, p_{16}, p_{17}, p_{18}$
$[S_{17}]$	$p_6, p_7, p_{11}, p_{12}, p_{13}, p_{16}-p_{19}$	$[S_{18}]$	$p_2, p_3, p_6, p_7, p_8, p_9, p_{11}-p_{13}, p_{16}-p_{19}$

Table 5.4 Monitors for the model (N, M_0)

Monitor	$M_{0V}(\cdot)$	Preset	Postset	Monitor	$M_{0V}(\cdot)$	Preset	Postset
V_{S_1}	2	t_2, t_{10}, t_{16}	t_1, t_{15}	V_{S_2}	2	t_4, t_7, t_{13}	t_1, t_{11}
V_{S_3}	2	t_2, t_9, t_{17}	t_1, t_{15}	V_{S_4}	2	t_2, t_8, t_{18}	t_1, t_{15}
V_{S_5}	3	t_2, t_{10}, t_{17}	t_1, t_{15}	V_{S_6}	4	t_4, t_8, t_{13}, t_{18}	t_1, t_{11}, t_{15}
V_{S_7}	4	t_4, t_9, t_{13}, t_{17}	t_1, t_{11}, t_{15}	V_{S_8}	4	t_2, t_9, t_{18}	t_1, t_{15}
V_{S_9}	5	$t_5, t_{10}, t_{13}, t_{17}$	t_1, t_{11}, t_{15}	$V_{S_{10}}$	6	t_4, t_9, t_{13}, t_{18}	t_1, t_{11}, t_{15}
$V_{S_{11}}$	5	t_2, t_{10}, t_{18}	t_1, t_{15}	$V_{S_{12}}$	5	t_3, t_8, t_{19}	t_1, t_{15}
$V_{S_{13}}$	7	$t_5, t_{10}, t_{13}, t_{18}$	t_1, t_{11}, t_{15}	$V_{S_{14}}$	7	t_4, t_8, t_{13}, t_{19}	t_1, t_{11}, t_{15}
$V_{S_{15}}$	7	t_3, t_9, t_{19}	t_1, t_{15}	$V_{S_{16}}$	9	t_4, t_9, t_{13}, t_{19}	t_1, t_{11}, t_{15}
$V_{S_{17}}$	8	t_3, t_{10}, t_{19}	t_1, t_{15}	$V_{S_{18}}$	10	$t_5, t_{10}, t_{13}, t_{19}$	t_1, t_{11}, t_{15}

Motivated by the need to reduce the structural complexity of the liveness-enforcing supervisor resulting from the policy in [16], the concept of elementary siphons is developed in [30]. This section presents an improved deadlock prevention policy for an S^{3}PR, which is based on the concept of elementary siphons. First strict minimal siphons are divided into elementary and dependent ones. Monitors are added for the elementary siphons only and the controllability of a dependent siphon is ensured by properly supervising its elementary siphons. If a dependent siphon cannot be implicitly controlled by using the monitors for elementary siphons only, a monitor is added for it.

In the deadlock prevention policy in [16], the addition of monitor V_S puts its corresponding siphon S under control. In what follows, (N_V, M_{0V}) is used to denote a controlled system for a plant S^{3}PR (N, M_0).

Lemma 5.1. *Let S and V_S be a siphon and its corresponding monitor, respectively. Then $\forall M \in R(N_V, M_{0V})$, the following invariant relation holds:*

$$M(V_S) + \sum_{i=1}^{n} M(\mathscr{P}_S^i) = M_{0V}(V_S).$$

Consider $\cup_{i=1}^n \mathscr{P}_S^i = \mathscr{P}_S$ and $\forall i \neq j$, $\mathscr{P}_S^i \cap \mathscr{P}_S^j = \emptyset$, the token invariant relation in Lemma 5.1 can be rewritten as $M(V_S) + M(\mathscr{P}_S) = M_{0V}(V_S)$. It is the token invariant relation that ensures the controllability of a strict minimal siphon such that it can never be emptied under any reachable marking in (N_V, M_{0V}).

Example 5.13. In Example 5.12, V_{S_1} is the monitor for $S_1 = \{p_{10}, p_{18}, p_{22}, p_{26}\}$ with $M_{0V}(V_{S_1}) = 2$ and $\mathscr{P}_{S_1} = \{p_{13}, p_{19}, p_6, p_{11}, p_{12}\}$. Lemma 5.1 indicates that $\forall M \in R(N_V, M_{0V})$, $M(V_S) + M(p_{13}) + M(p_{19}) + M(p_6) + M(p_{11}) + M(p_{12}) = 2$. Hence, the maximal number of tokens staying at $[S_1] = \{p_{13}, p_{19}\}$ is not greater than 2, i.e., $\forall M \in R(N_V, M_{0V})$, $M(p_{13}) + M(p_{19}) \leq 2$.

Recall the meanings of the complementary set of a siphon in an S^{3}PR, which contains operation places only. The places in a complementary set compete for resources with the operation places in a siphon. When the complementary set holds all the tokens initially marked in a siphon, the siphon is emptied and remains so forever, leading to occurrences of dead transitions. On the contrary, if at any reachable marking the maximal number of tokens held by the places in the complementary set is not greater than the number of tokens initially marked in it, the siphon can never be emptied.

For siphon $S_1 = \{p_{10}, p_{18}, p_{22}, p_{26}\}$ with $M_{0V}(S_1) = M_0(S_1) = 3$ and $[S_1] = \{p_{13}, p_{19}\}$, we conclude that S_1 can never be emptied due to the addition of monitor V_{S_1} that keeps $M(p_{13}) + M(p_{19}) \leq 2$, $\forall M \in R(N_V, M_{0V})$. It is easy to see that $min\{M(S_1) | M \in R(N_V, M_{0V})\} = M_0(S_1) - M_{0V}(V_{S_1}) = 1$.

Proposition 5.1. *Let V_S be a monitor for siphon S in (N_V, M_{0V}). Then $V_S + \sum_{p \in \mathscr{P}_S} p$ is a P-semiflow of N_V.*

Proof. Let I denote P-vector $V_S + \sum_{p \in \mathscr{P}_S} p$. We need to show $I^T[N_V] = \mathbf{0}^T$. $I^T[N_V] = [N_V](V_S, \cdot) + \sum_{p \in \mathscr{P}_S} [N_V](p, \cdot) = [N_V](V_S, \cdot) + [N_V](\mathscr{P}_S, \cdot) = [N_V](V_S, \cdot) + \sum_{i=1}^{n} [N_V](\mathscr{P}_S^i, \cdot)$, where n is the number of subnets by which the plant S^{3}PR (N, M_0) is composed.

Since $\mathscr{P}_S^i$ is a subset of places in state machine $\overline{N}_i$, $\forall t \in T_i$, $[N](\mathscr{P}_S^i, t) = [N_V](\mathscr{P}_S^i, t)$ is -1, 0, or 1. Note that $\forall i, j \in \mathbb{N}_n (i \neq j)$, $T_i \cap T_j = \emptyset$. We have $\forall t \in T$, $[N_V](\mathscr{P}_S, t)$ is -1, 0, or 1.

By the way of adding monitors in Definition 5.20, $\forall t \in T$, $[N_V](\mathscr{P}_S, t) = 1$ implies $[N_V](V_S, t) = -1$ and $[N_V](\mathscr{P}_S, t) = -1$ implies $[N_V](V_S, t) = 1$. Therefore, $\forall t \in T$, $[N_V](V_S, t) + [N_V](\mathscr{P}_S, t) = 0$, indicating that $I^T[N_V] = \mathbf{0}^T$. We conclude that $V_S + \sum_{p \in \mathscr{P}_S} p$ is a P-semiflow of N_V. □

Next, for a siphon S, a parameter ξ_S, called the control depth variable of the siphon, is introduced in order to establish a method to ensure the controllability of a dependent siphon by properly supervising its elementary ones.

Proposition 5.2. *Let V_S be a monitor computed by Definition 5.20 for a siphon S in an S^{3}PR (N, M_0). S is controlled if $M_{0V}(V_S) = M_0(S) - \xi_S$ where $1 \leq \xi_S \leq M_0(S) - 1$.*

Proof. By Lemma 5.1, $\forall M \in R(N_V, M_{0V})$, $M(V_S) + M(\mathscr{P}_S) = M_{0V}(V_S)$, where $\mathscr{P}_S = \cup_{i=1}^{n} \mathscr{P}_S^i$.

Hence, we have

$$M(\mathscr{P}_S) = M_0(S) - \xi_S - M(V_S).$$

Since $M(V_S) \geq 0$, $M(\mathscr{P}_S) \leq M_0(S) - \xi_S$ is true. By $[S] \subseteq \mathscr{P}_S$, we have

$$M([S]) \leq M_0(S) - \xi_S. \tag{5.5}$$

According to the definition of the complementary set of a siphon, $[S] \cup S$ is the support of a P-semiflow of N. It is also a P-semiflow of N_V. Consequently, $\forall M \in R(N_V, M_{0V})$, $M([S]) + M(S) = M_0(S) = M_{0V}(S)$, leading to the truth of

$$M(S) = M_0(S) - M([S]). \tag{5.6}$$

Consider (5.5) and (5.6); $M(S) \geq M_0(S) - (M_0(S) - \xi_S)$, i.e.,

$$M(S) \geq \xi_S \geq 1.$$

This implies that at any reachable marking in $R(N_V, M_{0V})$, S can never be unmarked. □

Increasing ξ_S intends to tighten the control of siphon S, which may degrade the control performance from the behavior permissiveness point of view. Specifically, a large ξ_S implies that some good (safe) states are possibly removed from the controlled system.

Corollary 5.1. *Let S be a siphon in an S^3PR and V_S be its monitor defined in Proposition 5.2. In (N_V, M_{0V}), $M_{min}(S) = \xi_S$.*

The relation between the controllability of elementary siphons and that of dependent ones can be established by means of the results in Chap. 3. To demonstrate this, let us first consider the Petri net in Fig. 5.3. There are three strict minimal siphons: $S_1 = \{p_5, p_9, p_{12}, p_{13}\}$, $S_2 = \{p_4, p_6, p_{13}, p_{14}\}$, and $S_3 = \{p_6, p_9, p_{12}, p_{13}, p_{14}\}$. It is verified that the rank of its characteristic T-vector matrix of these siphons equals two, indicating that there are two elementary siphons. The third one is dependent with $\eta_{S_3} = \eta_{S_1} + \eta_{S_2}$ if S_1 and S_2 are selected to be elementary ones. In other words, $\Pi_E = \{S_1, S_2\}$ implies $\Pi_D = \{S_3\}$.

Suppose that V_{S_1} and V_{S_2} are added for S_1 and S_2 with control depth variables ξ_{S_1} and ξ_{S_2}, respectively. Let (N_V, M_{0V}) denote the resultant net with V_{S_1} and V_{S_2}. We then check the controllability of S_3 in (N_V, M_{0V}). According to Corollary 3.5, S_3 is controlled if

$$M_{0V}(S_3) > M_{0V}(S_1) + M_{0V}(S_2) - M_{min}(S_1) - M_{min}(S_2).$$

i.e.,

$$M_{0V}(S_3) > M_{0V}(S_1) + M_{0V}(S_2) - \xi_{S_1} - \xi_{S_2}.$$

It is easy to see that $M_{0V}(S_1) = M_0(S_1) = 2$, $M_{0V}(S_2) = M_0(S_2) = 2$, and $M_{0V}(S_3) = M_0(S_3) = 3$. The controllability condition is true if $\xi_{S_1} = \xi_{S_2} = 1$. This indicates that S_3 is controlled if S_1 and S_2 are controlled by adding monitors V_{S_1} and V_{S_2} with $\xi_{S_1} = \xi_{S_2} = 1$, respectively. This also implies that we do not need to add a monitor for S_3 since it has been implicitly controlled due to the controllability of its elementary siphons.

Note that the controllability condition in Corollary 3.5 is sufficient but not necessary. To illustrate this, consider a dependent siphon $S_5 = \{p_2, p_4, p_8, p_{10}, p_{17}, p_{21}, p_{22}, p_{26}\}$ in the net shown in Fig. 5.8 as an example. It is easy to verify that $\eta_{S_5} = \eta_{S_1} + \eta_{S_3}$, indicating that S_5 is a strongly dependent siphon with respect to S_1 and S_3. Suppose that two monitors V_{S_1} and V_{S_3} are added with $\xi_{S_1} = \xi_{S_3} = 1$ to prevent S_1 and S_3 from being emptied, respectively. Clearly, $M_{0V}(S_5) = M_{0V}(S_1) + M_{0V}(S_3) - \xi_{S_1} - \xi_{S_3}$ when $\xi_{S_1} = \xi_{S_3} = 1$. On one hand, either $\xi_{S_1} = 2$ or $\xi_{S_3} = 2$ will surely guarantee the controllability of S_5. On the other hand, S_5 is shown below to be controlled even if $\xi_{S_1} = \xi_{S_3} = 1$.

Let (N_V, M_{0V}) denote the net that has monitors V_{S_1} and V_{S_3} with $\xi_{S_1} = \xi_{S_3} = 1$. By $M^{min}(S_5) = min\{M(S_5) | M = M_{0V} + [N_V]Y, M \geq 0, Y \geq 0\}$, we have $M^{min}(S_5) = 1$. As a result, S_5 cannot be emptied in (N_V, M_{0V}) even if $\xi_{S_1} = \xi_{S_3} = 1$. That is to say, we do not need to enlarge any siphon control depth variable in this particular case.

Motivated by the facts stated above, a deadlock prevention algorithm based on the concept of elementary siphons is proposed as follows.

Algorithm 5.1 An improved deadlock prevention policy
Input: an S^{3}PR (N, M_0)
Output: a controlled system (N_V, M_{0V})

find Π, Π_E and Π_D
/* $\Pi_D = \{S_{D1}, S_{D2}, \ldots, S_{Dn}\}$ and $\Pi_E = \{S_1, S_2, \ldots, S_m\}$ */
add monitors V_{S_1}–V_{S_m} for S_1–S_m, where $\forall i \in \mathbb{N}_m$, $\xi_{S_i} = 1$.
$i := 1$
/* check the controllability of dependent siphons */
repeat
  if S_{Di} is controlled by Corollary 3.5 or $M_{0V}^{min}(S_{Di}) > 0$ **then**
    $i := i + 1$
  else
    add a monitor $V_{S_{Di}}$ for S_{Di} with $\xi_{S_{Di}} = 1$
    $i := i + 1$
  end if
until $i \geq n + 1$
output (N_V, M_{0V})

This deadlock control policy aims to make a dependent siphon controlled by setting unit the control depth variables of its elementary siphons. When a dependent siphon cannot be controlled by its elementary siphons with their unit control depth variables, a monitor is added for it.

Example 5.14. There are 18 strict minimal siphons in the net shown in the Fig. 5.8. By the elementary siphon identification algorithm, we have $\Pi_E = \{S_1, S_2, S_3, S_4, S_9, S_{12}\}$ and $\Pi_D = \{S_5, S_6, S_7, S_8, S_{10}, S_{11}, S_{13}, S_{14}, S_{15}, S_{16}, S_{17}, S_{18}\}$. The characteristic T-vector relation between elementary and dependent siphons is shown in Table 5.5.

Table 5.5 The characteristic T-vector relation between dependent and elementary siphons

S^*	η relationship	S^*	η relationship
S_5	$\eta_{S_5} = \eta_{S_1} + \eta_{S_3}$	S_6	$\eta_{S_6} = \eta_{S_2} + \eta_{S_4}$
S_7	$\eta_{S_7} = \eta_{S_2} + \eta_{S_3}$	S_8	$\eta_{S_8} = \eta_{S_3} + \eta_{S_4}$
S_{10}	$\eta_{S_{10}} = \eta_{S_2} + \eta_{S_3} + \eta_{S_4}$	S_{11}	$\eta_{S_{11}} = \eta_{S_1} + \eta_{S_3} + \eta_{S_4}$
S_{13}	$\eta_{S_{13}} = \eta_{S_4} + \eta_{S_9}$	S_{14}	$\eta_{S_{14}} = \eta_{S_2} + \eta_{S_{12}}$
S_{15}	$\eta_{S_{15}} = \eta_{S_3} + \eta_{S_{12}}$	S_{16}	$\eta_{S_{16}} = \eta_{S_2} + \eta_{S_3} + \eta_{S_{12}}$
S_{17}	$\eta_{S_{17}} = \eta_{S_1} + \eta_{S_3} + \eta_{S_{12}}$	S_{18}	$\eta_{S_{18}} = \eta_{S_9} + \eta_{S_{12}}$

The controllability of siphon S_6 depends on whether $M_0(S_6) > M_0(S_2) + M_0(S_4) - \xi_{S_2} - \xi_{S_4}$ is true. Since $M_0(S_6) = 5$, $M_0(S_2) = 3$, and $M_0(S_4) = 3$, S_6 is controlled if monitors V_{S_2} and V_{S_4} are added when $\xi_{S_2} = \xi_{S_4} = 1$.

The controllability of S_{16} depends on the truth of $M_0(S_{16}) > M_0(S_2) + M_0(S_3) + M_0(S_{12}) - \xi_{S_2} - \xi_{S_3} - \xi_{S_{12}}$. By $M_0(S_{16}) = 10$, $M_0(S_2) = 3$, $M_0(S_3) = 3$, and $M_0(S_{12}) = 6$, this inequality holds when $\xi_{S_2} = \xi_{S_3} = \xi_{S_{12}} = 1$.

By Algorithm 5.1, it is easy to see that all dependent siphons are controlled by adding six monitors for the elementary siphons only with each siphon control depth variable being unit. That is to say, the addition of six monitors leads to a liveness-enforcing supervisor for the Petri net model of the FMS. Furthermore, the supervisor results in the same permissive behavior with the one due to the deadlock prevention policy in [16].

The number of monitors resulting from Algorithm 5.1 may be greater than that of elementary siphons. When the inequality concerning the controllability of a dependent siphon is not true, or it is not controlled by solving the corresponding LPP, a monitor needs to be added for it. As a result, the number of monitors resulting from Algorithm 5.1 is in theory exponential with respect to the net size.

A natural and fascinating problem is to find a liveness-enforcing supervisor with monitors that are added for elementary siphons only. The control of a dependent siphon is ensured by properly setting the control depth variables of its elementary siphons. If this is achieved, the size of the resulting controlled system is limited to $2n$, where n is the size of the plant Petri net model.

Theorem 5.7. *Let (N, M_0) be an S^3PR and (N_V, M_{0V}) be the net resulting from adding monitors for m elementary siphons only by Definition 5.20. (N_V, M_{0V}) is a controlled system with m monitors if the following LPP has a feasible solution:*

$$min \sum_{i=1}^{m} \xi_{S_i}$$

s.t.

$$M_0(S_{Dj}) > \sum_{i=1}^{m} a_i(M_0(S_i) - \xi_{S_i}), j = 1,2,\ldots,n,$$

$$1 \leq \xi_{S_i} \leq M_0(S_i) - 1, i = 1,2,\ldots,m,$$

where $\Pi_D = \{S_{Dj} | j = 1,2,\ldots,n\}$ *and* $\Pi_E = \{S_i | i = 1,2,\ldots,m\}$.

Proof. If the LPP has a feasible solution, it means that all dependent siphons are controlled by properly setting the control depth variables of the elementary siphons. Furthermore, the elementary siphons can be controlled by the monitors. As a result, (N_V, M_{0V}) is a controlled system with liveness. □

Example 5.15. For the Petri net shown in Fig. 5.8, monitors V_{S_1}–V_{S_4}, V_{S_9}, and $V_{S_{12}}$ are added. By solving the following LPP:

$$z = min\{\sum_{i=1}^{4} \xi_{S_i} + \xi_{S_9} + \xi_{S_{12}}\}$$

s.t.
$M_0(S_5) > M_0(S_1) + M_0(S_3) - \xi_{S_1} - \xi_{S_3}$,
$M_0(S_6) > M_0(S_2) + M_0(S_4) - \xi_{S_2} - \xi_{S_4}$,
$M_0(S_7) > M_0(S_2) + M_0(S_3) - \xi_{S_2} - \xi_{S_3}$,
$M_0(S_8) > M_0(S_3) + M_0(S_4) - \xi_{S_3} - \xi_{S_4}$,
$M_0(S_{10}) > M_0(S_2) + M_0(S_3) + M_0(S_4) - \xi_{S_2} - \xi_{S_3} - \xi_{S_4}$,
$M_0(S_{11}) > M_0(S_1) + M_0(S_3) + M_0(S_4) - \xi_{S_1} - \xi_{S_3} - \xi_{S_4}$,
$M_0(S_{13}) > M_0(S_4) + M_0(S_9) - \xi_{S_4} - \xi_{S_9}$,
$M_0(S_{14}) > M_0(S_2) + M_0(S_{12}) - \xi_{S_2} - \xi_{S_{12}}$,
$M_0(S_{15}) > M_0(S_3) + M_0(S_{12}) - \xi_{S_3} - \xi_{S_{12}}$,
$M_0(S_{16}) > M_0(S_2) + M_0(S_3) + M_0(S_{12}) - \xi_{S_2} - \xi_{S_3} - \xi_{S_{12}}$,
$M_0(S_{17}) > M_0(S_1) + M_0(S_3) + M_0(S_{12}) - \xi_{S_1} - \xi_{S_3} - \xi_{S_{12}}$,
$M_0(S_{18}) > M_0(S_9) + M_0(S_{12}) - \xi_{S_9} - \xi_{S_{12}}$,
$1 \leq \xi_{S_1} \leq M_0(S_1) - 1$,
$1 \leq \xi_{S_2} \leq M_0(S_2) - 1$,
$1 \leq \xi_{S_3} \leq M_0(S_3) - 1$,
$1 \leq \xi_{S_4} \leq M_0(S_4) - 1$,
$1 \leq \xi_{S_9} \leq M_0(S_9) - 1$,
$1 \leq \xi_{S_{12}} \leq M_0(S_{12}) - 1$.

an optimal solution $z^* = 7$ with $\xi_{S_1} = 2$, $\xi_{S_2} = 1$, $\xi_{S_3} = 1$, $\xi_{S_4} = 1$, $\xi_{S_9} = 1$, and $\xi_{S_{12}} = 1$ is found. This leads to $M_{0V}(V_{S_1}) = 1$, $M_{0V}(V_{S_2}) = 2$, $M_{0V}(V_{S_3}) = 2$, $M_{0V}(V_{S_4}) = 2$, $M_{0V}(V_{S_9}) = 5$, and $M_{0V}(V_{S_{12}}) = 5$. However, the obtained supervisor may be more restrictive than the one in Example 5.12 due to a larger control depth variable of S_1. Specially, the number of reachable states of the controlled system in Example 5.12 is 6,287 and that in this example leads to 3,506 reachable states only.

5.6 An MIP-Based Deadlock Prevention Policy

The deadlock prevention policies in [16] and in Sect. 5.5 depend on the complete siphon enumeration of a plant Petri net model. Its computation is expensive since the number of siphons grows fast with respect to the net size. For example, INA [49], a popular Petri net analysis tool, is used to find all minimal siphons in a net with 72 places and 64 transitions, the computation in a personal computer aborts due to memory overflow after several days [33]. This section introduces a deadlock prevention policy that combines the concept of elementary siphons and MIP-based deadlock detection method. More importantly, the complete siphon enumeration is avoided, leading to better computational efficiency. The following results play an important role in the establishment of the efficient deadlock prevention policy in [33].

Proposition 5.3. *Let (N_V, M_{0V}) be the net resulting from adding monitors for siphons to an S^3PR (N, M_0) by the approach in [16]. Then (N_V, M_{0V}) is an ES^3PR.*

Corollary 5.2. *An ES^3PR (N_V, M_{0V}) is live iff no siphon in it can become empty.*

Corollary 5.3. *An ES^3PR (N_V, M_{0V}) is live iff $G^{MIP}(M_{0V}) = |P_A \cup P^0 \cup P_R \cup P_V|$, where P_V is the set of monitors.*

The idea underlying the deadlock prevention policy presented in this section proceeds in an iterative way that can be stated as follows. First, the MIP-based deadlock detection method is applied to a plant S^3PR net model (N, M_0). If no unmarked siphon can be derived, it indicates that (N, M_0) itself is live since no siphon can become empty. If not, an unmarked maximal siphon S^* is hence obtained. By using the minimal siphon extraction approach in [34], a minimal siphon S_1 from S^* is then derived. Let $\Pi = \{S_1\}$.

The MIP-based method is applied to (N, M_0) with a constraint $M([S_1]) \leq M_0(S_1) - 1$, $\forall M \in R(N, M_0)$. If no unmarked maximal siphon can be found, S_1 is the only siphon that can be emptied in (N, M_0). It is then controlled by adding monitor V_{S_1} according to Definition 5.20 and the resultant net (N_V, M_{0V}) is live. Otherwise, we derive a strict minimal siphon S_2 and Π is updated by putting S_2 into it. Accordingly, a constraint $M([S_2]) \leq M_0(S_2) - 1$ is generated for the next iteration.

Suppose that at some step, a minimal siphon S_i is found. We have $\Pi = \{S_1, S_2, \ldots, S_i\}$ and constraint $M([S_i]) \leq M_0(S_i) - 1$. The MIP-based deadlock detection method is applied again to (N, M_0) with constraints $M([S_1]) \leq M_0(S_1) - 1$, $M([S_2]) \leq M_0(S_2) - 1, \ldots,$ and $M([S_i]) \leq M_0(S_i) - 1$. This process proceeds until no unmarked siphon can be found and a set of minimal siphons Π is finally obtained. With the set of siphons Π, Algorithm 5.1 can be used to compute the monitors of the supervisor for (N, M_0). Note that in general, Π computed in this way is not necessarily the set of all strict minimal siphons in (N, M_0). Even if Π happens to be the set of all strict minimal siphons in (N, M_0), it can be computed efficiently through the MIP-based deadlock detection and minimal siphon extraction method, particularly, when a large net model is dealt with. We summarize the deadlock detection and siphon extraction algorithm as follows.

Algorithm 5.2 Strict minimal siphon extraction
Input: an S^3PR (N, M_0) with $N = (P_A \cup P^0 \cup P_R, T, F)$
Output: Π
  $\Pi := \emptyset$
  repeat
    if $G^{MIP}(M_0) < |P_A| + |P^0| + |P_R|$ **then**
      find a maximal unmarked siphon S^*
      derive a minimal siphon S from S^*
      generate constraint $M(S) \leq M_0(S) - 1$ for further iteration
      $\Pi := \Pi \cup \{S\}$
    end if
  until $G^{MIP}(M_0) = |P_A| + |P^0| + |P_R|$
  output Π

Theorem 5.8. *Algorithm 5.2 can always terminate. When it terminates, the control of siphons in Π by Definition 5.20 leads to a liveness-enforcing supervisor and the controlled system (N_V, M_{0V}) is live.*

Proof. If there is no emptiable siphon in (N, M_0), the algorithm outputs $\Pi = \emptyset$ in the first iteration. Suppose that there are k strict minimal siphons in (N, M_0) and at some step, a siphon S_i is derived. The constraint $M([S_i]) \leq M_0(S_i) - 1$ ensures that in the next iteration S_i cannot be derived again, that is, a new siphon may be found in the next iteration. The algorithm terminates after at most k iterations.

Suppose that $\Pi = \{S_1, S_2, \ldots, S_m\}$ $(m \leq k)$ when the algorithm terminates. The termination also implies that there is no siphon that can be emptied in (N, M_0) under the constraints $M([S_1]) \leq M_0(S_1) - 1$, $M([S_2]) \leq M_0(S_2) - 1$, ..., and $M([S_m]) \leq M_0(S_m) - 1$. $\forall S \in \Pi$, a monitor V_S is added by Definition 5.20, which implements the constraint $M([S]) \leq M_0(S) - 1$. By Theorem 5.6, the controlled system (N_V, M_{0V}) with m monitors is live. □

Example 5.16. Consider the net shown in Fig. 5.3. The MIP-based deadlock detection method is applied to it. A maximal emptiable siphon $S^* = \{p_5, p_6, p_9, p_{12}, p_{13}, p_{14}\}$ can be distinguished, from which a minimal siphon $S_1 = \{p_5, p_9, p_{12}, p_{13}\}$ with $M_0(S_1) = 2$ is derived. We have $\Pi = \{S_1\}$.

A constraint $M([S_1]) \leq M_0(S_1) - 1$ is considered in the next iteration. We can get a maximal emptiable siphon $S^* = \{p_4, p_6, p_9, p_{12}, p_{13}, p_{14}\}$, from which a minimal siphon $S_2 = \{p_4, p_6, p_{13}, p_{14}\}$ is derived.

The MIP-based deadlock detection method is applied to (N, M_0) under the constraints $M([S_1]) \leq M_0(S_1) - 1$ and $M([S_2]) \leq M_0(S_2) - 1$. It is shown $G^{MIP}(M_0) = 15$, indicating that if S_1 and S_2 are controlled, there is no emptiable siphon in (N, M_0). Hence, we have $\Pi = \{S_1, S_2\}$.

By Algorithm 5.1, two monitors V_{S_1} and V_{S_2} are added to (N, M_0) to make S_1 and S_2 controlled, respectively. The resultant net as shown in Fig. 5.4 is the controlled system that is live.

5.7 Deadlock Prevention in S^{4}R

This section introduces a deadlock prevention policy for a class of generalized Petri nets, S^{4}R. It stands for a *system of sequential systems with shared resources*, as defined in Definition 5.10. First an important result concerning the number of the elementary siphons in an S^{4}R is given.

Theorem 5.9. *Let* $N = \bigcirc_{i=1}^{n} N_i = (P^0 \cup P_A \cup P_R, T, F, W)$ *be an* S^4R *and* N_{ES} *the number of its elementary siphons. Then,* $N_{ES} \leq |P_A|$.

Proof. It is known $N_{ES} \leq rank([N])$. As a result, if one can prove $rank([N]) = |P_A|$, the conclusion is certainly true.

Let $N_i = (P_{A_i} \cup \{p_i^0\} \cup P_{R_i}, T_i, F_i, W_i)$ and N_i' denote the resultant net after all resource places in P_{R_i} are removed from N_i. Since $\forall i \in \mathbb{N}_n$, N_i' is a strongly connected state machine, we have $rank([N_i']) = |P_{A_i}|$. In incidence matrix $[N_i]$, the row that models a resource r in net N_i is linear combinations of the rows of the idle place and the operation places that do not use resource r. We trivially have $rank([N_i]) = rank([N_i']) = |P_{A_i}|$. Noticing that $\forall i \neq j$, $P_{A_i} \cap P_{A_j} = \emptyset$, we have $rank([N]) = \sum_{i=1}^{n} rank([N_i]) = \sum_{i=1}^{n} |P_{A_i}| = |P_A|$ and conclude $N_{ES} \leq |P_A|$. □

To establish a deadlock prevention policy for S^{4}R, the following notations are also useful. Note that Π is used to denote a set of strict minimal siphons in an S^{4}R. For convenience, $[S]$ is used to denote the set of places in the complementary set $Th(S)$ of siphon S in an S^{4}R, as stated at the end of Sect. 5.2. As a multiset, $Th(S)$ can be represented by $\sum_{p \in [S]} h_S(p)p$. For example, S is a siphon in some net with $Th(S) = 3p_1 + 2p_4 + p_6$. We have $[S] = \{p_1, p_4, p_6\}$, $h_S(p_1) = 3$, $h_S(p_4) = 2$, $h_S(p_6) = 1$, and $\forall p \notin \{p_1, p_4, p_6\}$, $h_S(p) = 0$.

- $\Delta^+ : T \rightarrow 2^{\Pi}$ is a mapping defined as follows: $\forall t \in T_i$, $\Delta^+(t) = \{S \in \Pi | t <_{\overline{N}_i} [S]^i\}$. If $S \in \Delta^+(t)$ then the set $[S]^i$ is *reachable* from t, i.e., there exists a path in $\overline{N}_i$ leading from t to an operation place $p \in P_{A_i}$ that is not included in S but uses a resource of S.
- $\Delta^- : T \rightarrow 2^{\Pi}$ is a mapping defined as follows: $\forall t \in T_i$, $\Delta^-(t) = \{S \in \Pi | [S]^i <_{\overline{N}_i} t\}$.
- $\forall i \in \mathbb{N}_n$, $\forall S \in \Pi$, $\mathscr{P}_S^i = [S]^i \cup \{p \in P_{A_i} | p <_{\overline{N}_i} [S]^i\}$, and $\mathscr{P}_S = \cup_{i=1}^{n} \mathscr{P}_S^i$.

Example 5.17. There are three strict minimal siphons in the net shown in Fig. 5.5. They are $S_1 = \{p_3, p_6, p_9, p_{13}, p_{14}\}$, $S_2 = \{p_2, p_5, p_{10}, p_{12}, p_{13}\}$, and $S_3 = \{p_3, p_6, p_{10}, p_{12}{-}p_{14}\}$ with $[S_1] = \{p_2, p_5, p_8\}$, $[S_2] = \{p_1, p_9\}$, and $[S_3] = \{p_1, p_2, p_5, p_8, p_9\}$. The S^{4}R is composed of two subnets. Let $P_{A_1} = \{p_1{-}p_6\}$ and $P_{A_2} = \{p_8{-}p_{10}\}$.

$\mathscr{P}_{S_1}^1 = \{p_2, p_5\} \cup \{p_1\} = \{p_1, p_2, p_5\}$ and $\mathscr{P}_{S_1}^2 = [S_1]^2 = \{p_8\}$. As a result, we have $\mathscr{P}_{S_1} = \{p_1, p_2, p_5, p_8\}$. Similarly, $\mathscr{P}_{S_2} = \{p_1, p_8, p_9\}$ and $\mathscr{P}_{S_3} = [S_3] = \{p_1, p_2, p_5, p_8, p_9\}$.

Let S be a siphon in an S^{4}R that is composed of n state machines. A non-negative P-vector k_S for S is constructed. Without loss of generality, it is assumed that $\forall i \in$

$\mathbb{N}_l$, $[S]^i \neq \emptyset$, $\forall j \in \mathbb{N}_n \backslash \mathbb{N}_l$, $[S]^j = \emptyset$, where $\mathbb{N}_l \subseteq \mathbb{N}_n$. Define $B_S^i = \{p | p \in [S]^i, \nexists p' \in [S]^i, p <_{\overline{N}_i} p'\}$. For siphon $S_3 = \{p_3, p_6, p_{10}, p_{12}-p_{14}\}$ in Fig. 5.5, we have $B_{S_3}^1 = \{p_2, p_5\}$ and $B_{S_3}^2 = \{p_9\}$. For siphon $S_5 = \{p_2, p_4, p_8, p_{10}, p_{17}, p_{21}, p_{22}, p_{26}\}$ in the Petri net shown in Fig. 5.8, $[S_5] = \{p_{12}, p_{13}, p_{18}, p_{19}\}$. As a result, $B_{S_5} = B_{S_5}^1 \cup B_{S_5}^2 \cup B_{S_5}^3 = \{p_{13}, p_{18}\}$, where $B_{S_5}^1 = \emptyset$, $B_{S_5}^2 = \{p_{18}\}$, and $B_{S_5}^3 = \{p_{13}\}$.

Definition 5.21. A P-vector k_S for a siphon S in an S[4]R is constructed as follows:

$\forall p \notin \mathscr{P}_S$, $k_S(p) := 0$
$\forall p \in [S]$, $k_S(p) := h_S(p)$
$i := 1$
repeat
$\quad \forall p \in B_S^i$, $\alpha_p := max\{h_S(p), h_S(p') | p' <_{\overline{N}_i} p, p' \in [S]^i\}$
$\quad \forall p_x \in \{p | p \in B_S^i\} \cup \{p' | p' <_{\overline{N}_i} p, p' \in [S]^i\}$, $k_S(p_x) := \alpha_p$
$\quad \forall p_y \in \{p'' | p'' <_{\overline{N}_i} p, p \in B_S^i, p'' \in \mathscr{P}_S \setminus [S]^i\}$, $k_S(p_y) := \alpha_p$
$\quad \forall p_z \in \cap_{p_w \in B_S^i} \{p | p \in \mathscr{P}_S \setminus [S]^i, p <_{\overline{N}_i} p_w\}$, $k_S(p_z) := max\{k_S(p) | p \in B_S^i\}$
$\quad i := i+1$
until $i \geq l+1$

To further illustrate the computation of k_S for siphon S in an S[4]R, we take the net shown in Fig. 5.9 (without place V_S) as an example, where $P_{A_j} = \{p_1, p_2, \ldots, p_9, \ldots\}$ and $P_{A_{j+1}} = \{p_{11}, p_{12}, p_{13}, p_{14}, \ldots\}$. Suppose that S is a siphon with $Th(S) = 3p_5 + 2p_8 + p_9 + 4p_{13}$. Clearly, we have $[S] = \{p_5, p_8, p_9, p_{13}\}$, $h_S(p_5) = 3$, $h_S(p_8) = 2$, $h_S(p_9) = 1$, $h_S(p_{13}) = 4$, and $\forall p \notin [S]$, $h_S(p) = 0$.

It is easy to verify that $\mathscr{P}_S = \mathscr{P}_S^j \cup \mathscr{P}_S^{j+1} = \{p_1-p_5,\ p_7-p_9\} \cup \{p_{11}-p_{13}\}$. Note that $[S]^j = \{p_5, p_8, p_9\}$ and $[S]^{j+1} = \{p_{13}\}$. We have $B_S^j = \{p_5, p_9\}$ and $B_S^{j+1} = \{p_{13}\}$.

For $p_5 \in B_S^j$, we have
$\alpha_{p_5} := 3$,
$k_S(p_5) := 3$,
$k_S(p_1) := 3$, $k_S(p_2) := 3$, $k_S(p_3) := 3$, and $k_S(p_4) := 3$.
For $p_9 \in B_S^j$, we have
$\alpha_{p_8} := 2$,
$k_S(p_8) := 2$, $k_S(p_9) := 2$.
$k_S(p_1) := 2$, $k_S(p_2) := 2$, and $k_S(p_7) := 2$.
For $p_{13} \in B_S^{j+1}$, we have $\alpha_{p_{13}} := 4$,
$k_S(p_{13}) := 4$,
$k_S(p_{11}) := 4$ and $k_S(p_{12}) := 4$.

By $\{p | p \in \mathscr{P}_S^j \setminus [S]^j,\ p <_{\overline{N}_j} p_5\} \cap \{p | p \in \mathscr{P}_S^j \setminus [S]^j,\ p <_{\overline{N}_j} p_9\} = \{p_1,\ p_2\}$ and $max\{k_S(p_5),\ k_S(p_9)\} = h_S(p_5) = 3$, we have
$k_S(p_1) := 3$ and $k_S(p_2) := 3$.

In summary, we have $k_S(p_1) = k_S(p_2) = k_S(p_3) = k_S(p_4) = k_S(p_5) = 3$, $k_S(p_7) = k_S(p_8) = k_S(p_9) = 2$, and $k_S(p_{11}) = k_S(p_{12}) = k_S(p_{13}) = 4$. $\forall p \notin \mathscr{P}_S$, $k_S(p) = 0$.

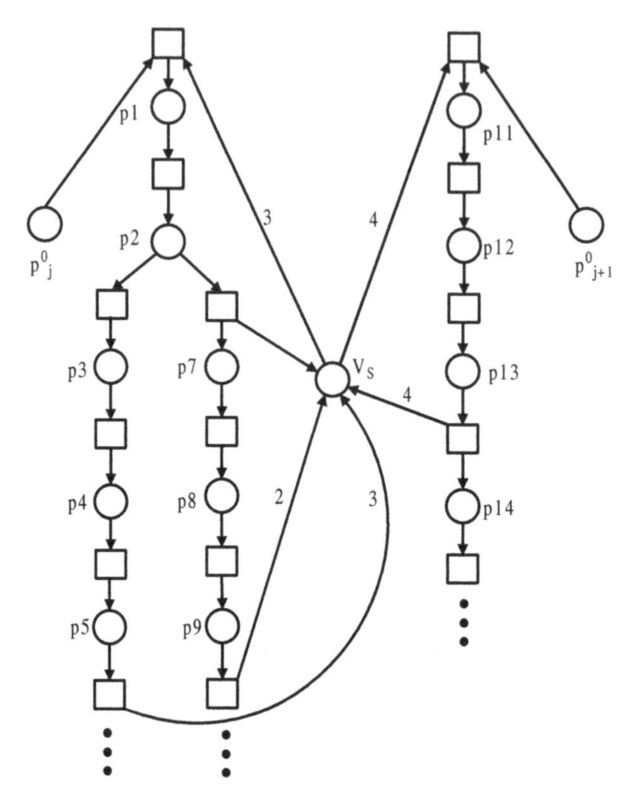

Fig. 5.9 S is a siphon with $Th(S) = 3p_5 + 2p_8 + p_9 + 4p_{13}$

Example 5.18. There are three strict minimal siphons $S_1 = \{p_3, p_6, p_9, p_{13}, p_{14}\}$, $S_2 = \{p_2, p_5, p_{10}, p_{12}, p_{13}\}$, and $S_3 = \{p_3, p_6, p_{10}, p_{12}, p_{13}, p_{14}\}$ in the net shown in Fig. 5.5 with $\mathscr{P}_{S_1} = \{p_1, p_2, p_5, p_8\}$, $\mathscr{P}_{S_2} = \{p_1, p_8, p_9\}$ and $\mathscr{P}_{S_3} = \{p_1, p_2, p_5, p_8, p_9\}$.

For S_1, $\forall p \notin \mathscr{P}_{S_1}$, $k_{S_1}(p) = 0$. $k_{S_1}(p_1) = k_{S_1}(p_2) = k_{S_1}(p_5) = k_{S_1}(p_8) = 1$.

For S_2, $Th(S_2) = p_9 + 2p_1$. $\forall p \notin \mathscr{P}_{S_2}$, $k_{S_2}(p) = 0$. We have $k_{S_2}(p_8) = k_{S_2}(p_9) = 1$ and $k_{S_2}(p_1) = 2$.

For S_3, $Th(S_3) = 2p_1 + p_2 + p_5 + p_8 + p_9$. $\forall p \notin \mathscr{P}_{S_3}$, $k_{S_3}(p) = 0$. $k_{S_3}(p_8) = k_{S_3}(p_9) = 1$, and $k_{S_3}(p_1) = k_{S_3}(p_2) = k_{S_3}(p_5) = 2$.

Next a parameterized siphon control approach is presented by adding a monitor such that a siphon is max-controlled [1, 3–5, 35, 58].

Proposition 5.4. *Let S be a strict minimal siphon in an S^4R net (N, M_0), where $N = (P^0 \cup P_A \cup P_R, T, F, W)$. A monitor V_S is added to (N, M_0) by the enforcement that $g_S = k_S + V_S$ is a P-invariant of the resultant net system (N_V, M_{0V}), where $N_V = (P^0 \cup P_A \cup P_R \cup \{V_S\}, T, F \cup F_V, W \cup W_V)$; $\forall p \in P^0 \cup P_A \cup P_R$, $M_{0V}(p) = M_0(p)$. Let*

$f_S = \sum_{r \in S^R} I_r - g_S$ *and* $M_{OV}(V_S) = M_0(S) - \xi_S$ ($\xi_S \in \mathbb{N}^+$). *Then S is max-controlled if* $\xi_S > \sum_{p \in S} f_S(p)(max_{p^\bullet} - 1)$ *and* $M_{OV}(V_S) \geq max_{V_S^\bullet}$.

Proof. Considering $g_S = k_S + V_S$ and $f_S = \sum_{r \in S^R} I_r - g_S$, we have $f_S = \sum_{r \in S^R} I_r - k_S - V_S$. Since $\sum_{r \in S^R} I_r$ and g_S are P-invariants of N_V, f_S is hence a P-invariant of N_V. Note that $\mathscr{P}_S \subseteq P_A$. Hence, we have $\forall p \in \mathscr{P}_S$, $M_{OV}(p) = M_0(p) = 0$.

$\sum_{p \in P^0 \cup P_A \cup P_R \cup \{V_S\}} f_S(p) M_{OV}(p)$
$= \sum_{p \in S} f_S(p) M_{OV}(p) - \sum_{p \in \mathscr{P}_S} k_S(p) M_{OV}(p) - M_{OV}(V_S)$
$\geq M_{OV}(S) - \sum_{p \in \mathscr{P}_S} k_S(p) M_{OV}(p) - M_{OV}(V_S)$
$= M_{OV}(S) - (M_{OV}(S) - \xi_S)$
$= \xi_S$.

Considering $\xi_S > \sum_{p \in S} f_S(p)(max_{p^\bullet} - 1)$, we conclude that
$\sum_{p \in P^0 \cup P_A \cup P_R \cup \{V_S\}} f_S(p) M_{OV}(p) > \sum_{p \in S} f_S(p)(max_{p^\bullet} - 1)$.

From the definition of f_S, it is true that $||f_S||^- \cap S = \emptyset$. Then we have $||f_S||^+ = S$. S is hence max-controlled due to Proposition 2.1. □

Similarly, ξ_S is called the control depth variable of siphon S. Clearly, we have $\sum_{p \in S} f_S(p)(max_{p^\bullet} - 1) < \xi_S < M_0(S)$. Consider the siphon S with $Th(S) = 3p_5 + 2p_8 + p_9 + 4p_{13}$ in Fig. 5.9, we have $k_S = 3p_1 + 3p_2 + 3p_3 + 3p_4 + 3p_5 + 2p_7 + 2p_8 + 2p_9 + 4p_{11} + 4p_{12} + 4p_{13}$. Let V_S be the monitor such that $k_S + V_S$ is a P-invariant of the resultant net. According to Proposition 5.4, V_S is shown in Fig. 5.9.

It is worthy of noting that in Proposition 5.4 $M_{OV}(V_S) \geq max_{V_S^\bullet}$ is not necessary for the controllability of siphon S. However, its truth ensures that the transitions in $S^\bullet$ can fire at least once. This is demonstrated by the S^{4}R net shown in Fig. 5.10(a). The siphon $S = \{p_3, p_5, p_6\}$ can lead the net to a deadlock state if it is not supervised.

By Proposition 5.4, we have $g_S = p_2 + V_S$, $\sum_{r \in S^R} I_r = p_2 + 2p_3 + 2p_5 + p_6$, $f_S = 2p_3 + 2p_5 + p_6 - V_S$. Note that $\sum_{p \in S} f_S(p)(max_{p^\bullet} - 1) = 1$ since $max_{p_6^\bullet} = 2$. As a result, $\xi_S > 1$, leading to $\xi_S \geq 2$. This means $M_{OV}(V_S) = 0$ when $\xi_S = 2$, as shown in Fig. 5.10(b). However, the addition of monitor V_S introduces an empty siphon and trap $\{p_2, V_S\}$, leading to the disablement of transition t_1 and the loss of liveness of the controlled system as shown in Fig. 5.10(b). That is to say, Proposition 5.4 fails to compute a correct monitor to control the siphon in this net.

This example shows that the siphon control approach via adding monitors as stated in [58] does not always lead to live transitions of an S^{4}R, a subclass of G-systems that are investigated in [58]. However, Fig. 5.10(c) depicts a live controlled system for the net in Fig. 5.10(a), which can be found by intuition and observation. In this sense, the siphon control approach stated in [58] cannot be applied to arbitrary S^{4}R nets and G-systems.

Example 5.19. Let us consider siphon $S_1 = \{p_3, p_6, p_9, p_{13}, p_{14}\}$ in the net shown in Fig. 5.5. By Definition 5.21, we have $k_{S_1} = p_1 + p_2 + p_5 + p_8$. As a result, $g_{S_1} = k_{S_1} + V_{S_1} = p_1 + p_2 + p_5 + p_8 + V_{S_1}$. Noticing that

$$\sum_{r \in S_1^R} I_r = I_{p_{13}} + I_{p_{14}} = p_2 + p_5 + p_9 + p_{13} + p_3 + p_6 + p_8 + p_{14},$$

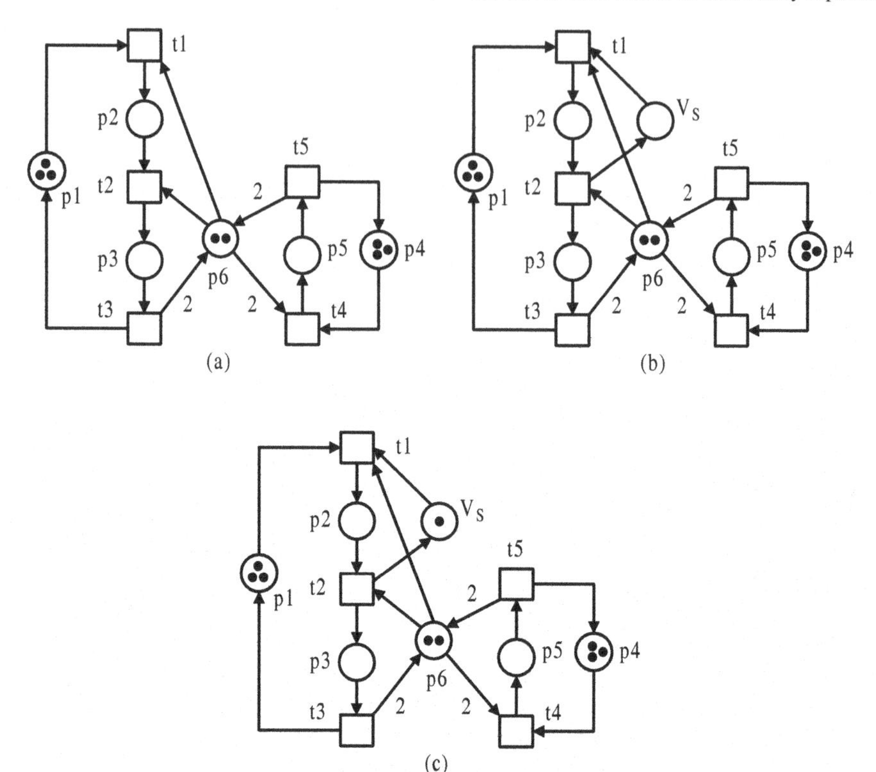

Fig. 5.10 Max-controlled siphon and its controllability

we have

$$f_{S_1} = p_3 + p_6 + p_9 + p_{13} + p_{14} - p_1 - V_{S_1}.$$

According to Proposition 5.4, f_{S_1} is a P-invariant of N_V. Let $\xi_{S_1} = 2$. Clearly, we have

$$\xi_{S_1} > \sum_{p \in S_1} f_{S_1}(p)(max_{p^\bullet} - 1) = 0.$$

S_1 is hence max-controlled by adding monitor V_{S_1} with $M_{0V}(V_{S_1}) = 3$. The resulting net is shown in Fig. 5.11. Similarly, S_2 and S_3 are max-controlled by monitors V_{S_2} and V_{S_3}, respectively, with $\xi_{S_2} = 2$, $\xi_{S_3} = 2$, $M_{0V}(V_{S_2}) = 2$, and $M_{0V}(V_{S_3}) = 5$. This means that Proposition 5.4 can compute monitors for the example.

In an S^4R, the controllability of a dependent siphon can also be ensured by properly supervising its elementary siphons. To achieve this, Corollary 3.16 is useful.

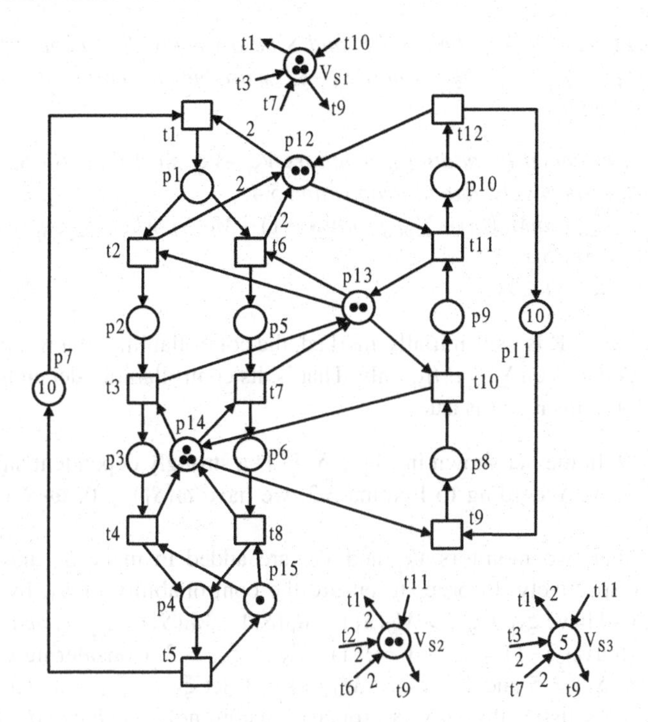

Fig. 5.11 Three siphons are max-controlled by monitors

Theorem 5.10. *Let (N,M_0) be an S^4R and S be a strongly dependent siphon with $\eta_S = \sum_{i=1}^{n} a_i \eta_{S_i}$, where $S_1 - S_n$ are the elementary siphons of S. S is max-controlled if:*

1. *(N,M_0) is extended by adding n monitors $V_{S_1} - V_{S_n}$ such that $S_1 - S_n$ are max-controlled, respectively, due to Proposition 5.4;*
2. *$\sum_{i=1}^{n} a_i \xi_{S_i} > \sum_{i=1}^{n} a_i M_0(S_i) + \sum_{p\in S}(max_{p^\bullet} - 1) - M_0(S)$, where $\forall i \in \mathbb{N}_n$, $\sum_{p\in S_i} f_{S_i}(p)(max_{p^\bullet} - 1) < \xi_{S_i} < M_0(S_i)$;*
3. *$M^{min}(S_i) \geq \xi_{S_i}$, $\forall i \in \mathbb{N}_n$.*

Proof. By Corollary 3.16, S is max-controlled if

$$M_0(S) > \sum_{i=1}^{n} a_i (M_0(S_i) - M^{min}(S_i)) + \omega(S),$$

where $\omega(S) = \sum_{p\in S}(max_{p^\bullet} - 1)$.

Considering that $\forall i \in \mathbb{N}_n$, $M^{min}(S_i) \geq \xi_{S_i}$ if monitor V_{S_i} is added to make S_i max-controlled, S is max-controlled if $\sum_{i=1}^{n} a_i \xi_{S_i} > \sum_{i=1}^{n} a_i M_0(S_i) + \sum_{p\in S}(max_{p^\bullet} - 1) - M_0(S)$. □

Theorem 5.11. *Let (N, M_0) be an S^4R and S be a weakly dependent siphon with $\eta_S = \sum_{i=1}^{n} a_i \eta_{S_i} - \sum_{j=n+1}^{n+m} a_j \eta_{S_j}$, where $S_1 - S_{n+m}$ are the elementary siphons of S. S is max-controlled if:*

1. *(N, M_0) is extended by adding n monitors $V_{S_1} - V_{S_n}$ such that $S_1 - S_n$ are max-controlled, respectively, due to Proposition 5.4;*
2. *$\sum_{i=1}^{n} a_i \xi_{S_i} > \sum_{i=1}^{n} a_i M_0(S_i) + \sum_{p \in S}(max_{p^\bullet} - 1) - M_0(S)$, where $\sum_{p \in S_i} f_{S_i}(p)\ (max_{p^\bullet} - 1) < \xi_{S_i} < M_0(S_i)$;*
3. *$M^{min}(S_i) \geq \xi_{S_i}$, $\forall i \in \mathbb{N}_n$.*

Proof. Since an S[4]R is well-initially-marked, the controllability of a weakly dependent siphon depends on $\sum_{i=1}^{n} a_i \eta_{S_i}$ only. That is, its controllability depends on S_1-S_n only. Therefore, this result is true. □

Example 5.20. In the net shown in Fig. 5.5, S_3 is a strongly dependent siphon with $\eta_{S_3} = \eta_{S_1} + \eta_{S_2}$. According to Lemma 3.2, we have $\omega(S_1) = 0$, $\omega(S_2) = 1$, and $\omega(S_3) = 1$.

Suppose that two monitors V_{S_1} and V_{S_2} are added to make S_1 and S_2 max-controlled, respectively. In order to ensure the controllability of S_3, by Theorem 5.10, we should have $\xi_{S_1} + \xi_{S_2} > M_0(S_1) + M_0(S_2) - M_0(S_3) + \sum_{p \in S_3}(max_{p^\bullet} - 1) = 5 + 4 - 7 + (max_{p_{12}^\bullet} - 1) = 2 + 1 = 3$, i.e., $\xi_{S_1} + \xi_{S_2} > 3$. Considering constraints $0 < \xi_{S_1} < M_0(S_1) = 5$ and $1 < \xi_{S_2} < M_0(S_2) = 4$, let $\xi_{S_1} = \xi_{S_2} = 2$. This leads to the fact that S_3 satisfies the max cs-property. Accordingly, we have $M_{0V}(V_{S_1}) = 3$ and $M_{0V}(V_{S_2}) = 2$, as shown in Fig. 5.11.

Next a deadlock prevention policy is presented for an S[4]R. Its output depends on the applicability of Proposition 5.4.

Algorithm 5.3 Deadlock prevention for S[4]R

Input: An S[4]R (N, M_0)
Output: a controlled system (N_V, M_{0V}) with liveness or "undecided"
$\Pi_E := \{S_i | S_i$ is an elementary siphon of $N\}$
$\Pi_D := \{S_{Dj} | S_{Dj}$ is a dependent siphon of $N\}$
$i := 1$
$flag := 0$
repeat
  add V_{S_i} to (N, M_0) to make S_i max-controlled by Proposition 5.4 with $M_{0V}(V_{S_i}) = M_0(S_i) - \xi_{S_i}$, where $\xi_{S_i} = \sum_{p \in S_i} f_{S_i}(p)(max_{p^\bullet} - 1) + 1$
  if $M_{0V}(V_{S_i}) < max_{V_{S_i}^\bullet}$ **then**
    $flag := 1$
  end if
until $i \geq |\Pi_E| + 1$
$j := 1$
repeat
  if S_{Dj} is max-controlled with respect to its elementary siphons **then**
    $j := j + 1$
  else

add monitor V_{Dj} for S_{Dj}
if $M_{0V}(V_{Dj}) < max_{V^{\bullet}_{Dj}}$ **then**
$flag := 1$
end if
$j := j+1$
end if
until $j \geq |\Pi_D|+1$
if flag=1 **then**
Output "Undecided"
else
Output (N_V, M_{0V})
end if

This algorithm first adds monitors for the elementary siphons. The addition of monitor V_{Dj} for a dependent siphon S_{Dj} depends on its controllability under the control depth variables of its elementary siphons. Obviously, it cannot always lead to a controlled system with liveness since the siphon control method that is based on the max-cs property does not work for arbitrary S^{4}R nets. That is to say, this algorithm can generate a controlled system with liveness for some particular S^{4}R nets.

Theorem 5.12. *Algorithm 5.3 always terminates and its termination gives a liveness-enforcing supervisor and a controlled system* (N_V, M_{0V}) *if for any monitor* V, $M_{0V}(V) \geq max_{V^{\bullet}}$.

Its proof is left as an exercise for the reader.

The existence of a liveness-enforcing supervisor with monitors added for the elementary siphons can be decided by solving the following LPP:

$$min\{\sum_{i=1}^{|\Pi_E|} \xi_{S_i}\} \tag{5.7}$$

s.t.

$$\sum_{i=1}^{n} a_i \xi_{S_i} > \sum_{i=1}^{n} a_i M_0(S_i) + \omega(S) - M_0(S), \forall S \in \Pi_D,$$

$$\sum_{p \in S_i} f_{S_i}(p)(max_{p^{\bullet}} - 1) < \xi_{S_i} \leq M_0(S_i) - 1, \forall i \in \mathbb{N}_{|\Pi_E|},$$

$$M_{0V}(V_{S_i}) \geq max_{V^{\bullet}_{S_i}}, M_{0V}(V_{S_i}) = M_0(S_i) - \xi_{S_i}, \forall i \in \mathbb{N}_{|\Pi_E|}.$$

If LPP (5.7) has a feasible solution and $\forall i \in \mathbb{N}_{|\Pi_E|}$, $M^{min}(S_i) \geq \xi_{S_i}$, it implies that there exists a liveness-enforcing supervisor with monitors that are added for the elementary siphons only.

Example 5.21. Consider an FMS with its layout shown in Fig. 5.12 and production routes in Fig. 5.13. It consists of four robots R1–R4 and three machines M1–M3. Each of R1–R3 can hold one product and R4 can hold three products every time. Machines M1 can process two products and each of M2 and M3 can process three products every time. There are three loading buffers I1–I3 and three unloading buffers O1–O3 to load and unload the FMS. There are three raw product types, namely J1, J2, and J3, to be processed. For these raw product types the production cycles are shown in Fig. 5.13 in which r/r' means a conjunctive requirement of resources r and r' in some operation. According to the production cycles, a raw product J1 is taken from I1 by R1 and R2 and put in M1. After being processed by M1, it is then moved to M3 by R4. Finally, after being processed by M3, it is processed by M2 and R3 and then moved to O1. A raw product J2 is taken from I2 and processed by M2 and R4, and then processed by M2 only. After being processed by M2, it is then moved from M2 to O2 by R4. A raw product J3 is taken from I3, processed by M3 and R3, and then by M3 and R4. After that, it is then processed by M1. Finally, after being processed by M1 it is moved from M1 to O3 by R1 and R2 sequentially.

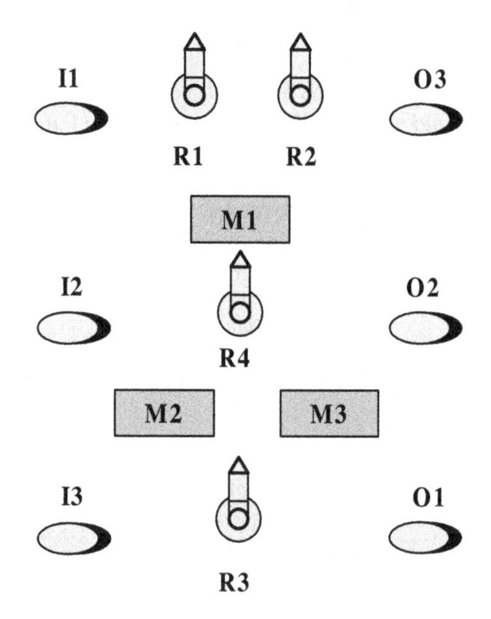

Fig. 5.12 An FMS layout

Figure 5.14 shows the net model of the FMS that may use a multiset of resources at a processing step. The system net is an S^{4}R where $P^0 = \{p_8, p_{12}, p_{20}\}$, $P_{A_1} = \{p_1 - p_7\}$, $P_{A_2} = \{p_9 - p_{11}\}$, $P_{A_3} = \{p_{13} - p_{17}\}$, $P_{R_1} = \{p_{18}, p_{19}, p_{21} - p_{25}\}$, $P_{R_2} = \{p_{24}, p_{25}\}$, and $P_{R_3} = \{p_{18}, p_{19}, p_{21} - p_{24}\}$. Places p_{21}, p_{25}, p_{22}, p_{18}, p_{19}, p_{23} and p_{24} denote M1, M2, M3, R1, R2, R3, and R4, respectively. Initially, it is assumed

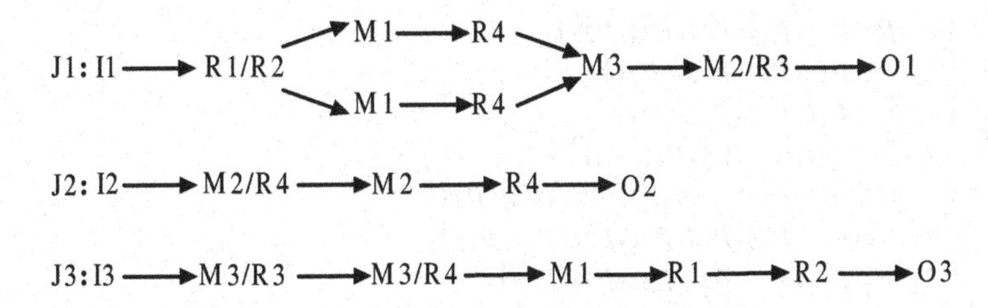

Fig. 5.13 The production routings of an FMS

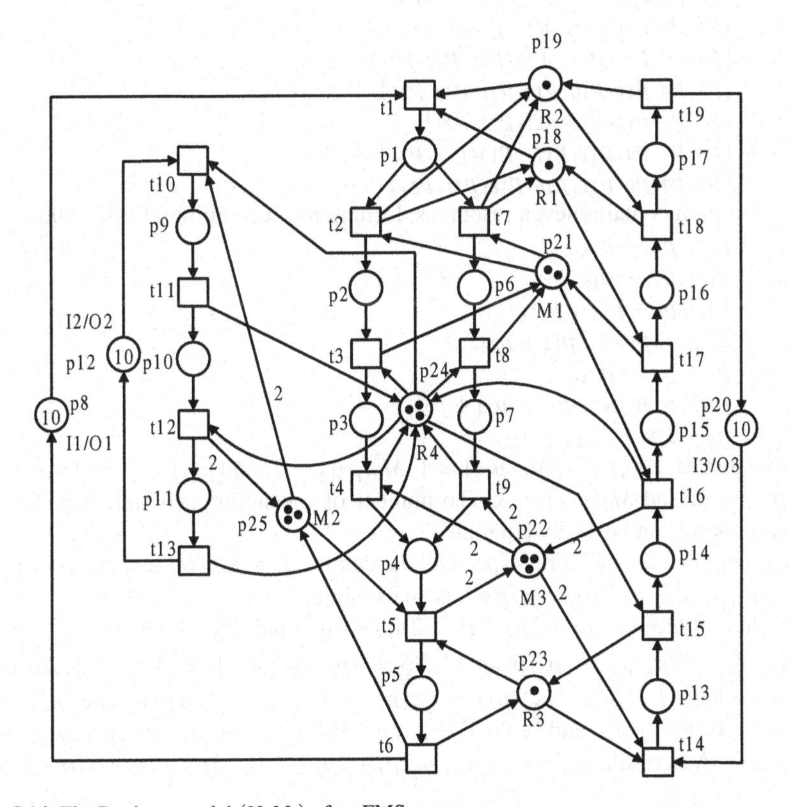

Fig. 5.14 The Petri net model (N, M_0) of an FMS

that there are no parts in the system. $M_0(p_8) = M_0(p_{12}) = M_0(p_{20}) = 10$ represents the maximal number of concurrent jobs that can be processed for part types J1, J2, and J3, respectively.

The net system contains deadlocks. There are 17 strict minimal siphons as shown below, among which $S_1 - S_6$ are elementary and $S_7 - S_{17}$ are strongly dependent:

$S_1 = \{p_2, p_6, p_{17} - p_{19}, p_{21}\}$,

$S_2 = \{p_2, p_6, p_{16}, p_{18}, p_{21}\}$,

$S_3=\{p_5,p_9,p_{11},p_{14},p_{22},p_{24},p_{25}\}$,
$S_4=\{p_5,p_9,p_{11},p_{14},p_{22}-p_{24}\}$,
$S_5=\{p_4,p_9,p_{11},p_{14},p_{22},p_{24}\}$,
$S_6=\{p_3,p_7,p_9,p_{11},p_{15},p_{21},p_{24}\}$,
$S_7=\{p_5,p_9,p_{11},p_{17}-p_{19},p_{21},p_{22},p_{24},p_{25}\}$,
$S_8=\{p_5,p_9,p_{11},p_{16},p_{18},p_{21},p_{22},p_{24},p_{25}\}$,
$S_9=\{p_5,p_9,p_{11},p_{15},p_{21},p_{22},p_{24},p_{25}\}$,
$S_{10}=\{p_5,p_9,p_{11},p_{17}-p_{19},p_{21}-p_{24}\}$,
$S_{11}=\{p_5,p_9,p_{11},p_{16},p_{18},p_{21}-p_{24}\}$,
$S_{12}=\{p_5,p_9,p_{11},p_{15},p_{21}-p_{24}\}$,
$S_{13}=\{p_4,p_9,p_{11},p_{17}-p_{19},p_{21},p_{22},p_{24}\}$,
$S_{14}=\{p_4,p_9,p_{11},p_{16},p_{18},p_{21},p_{22},p_{24}\}$,
$S_{15}=\{p_4,p_9,p_{11},p_{15},p_{21},p_{22},p_{24}\}$,
$S_{16}=\{p_3,p_7,p_9,p_{11},p_{17}-p_{19},p_{21},p_{24}\}$,
$S_{17}=\{p_3,p_7,p_9,p_{11},p_{16},p_{18},p_{21},p_{24}\}$.

This system contains seven resources, leading to seven minimal P-semiflows:
$I_{p_{18}}=p_1+p_{16}+p_{18}$,
$I_{p_{19}}=p_1+p_{17}+p_{19}$,
$I_{p_{21}}=p_2+p_6+p_{15}+p_{21}$,
$I_{p_{22}}=2p_4+2p_{13}+2p_{14}+p_{22}$,
$I_{p_{23}}=p_5+p_{13}+p_{23}$,
$I_{p_{24}}=p_3+p_7+p_9+p_{11}+p_{14}+p_{24}$,
$I_{p_{25}}=p_5+2p_9+2p_{10}+p_{25}$.

Note that $M_0(p_{18})=1$, $M_0(p_{19})=1$, $M_0(p_{21})=2$, $M_0(p_{22})=3$, $M_0(p_{23})=1$, $M_0(p_{24})=3$, and $M_0(p_{25})=3$. The number of tokens in each siphon under the initial marking can be easily computed.

Six monitors $V_{S_1}-V_{S_6}$ are respectively added for S_1-S_6. We take the control of $S_3=\{p_5,p_9,p_{11},p_{14},p_{22},p_{24},p_{25}\}$ as an example.

We have $Th(S_3)=p_3+2p_4+p_7+2p_{10}+2p_{13}$ and $\mathscr{P}_{S_3}=\mathscr{P}^1_{S_3}\cup\mathscr{P}^2_{S_3}\cup\mathscr{P}^3_{S_3}$, where $\mathscr{P}^1_{S_3}=\{p_1,p_2,p_3,p_4,p_6,p_7\}$, $\mathscr{P}^2_{S_3}=\{p_9,p_{10}\}$, and $\mathscr{P}^3_{S_3}=\{p_{13}\}$. Accordingly, we have $k_{S_3}(p_1)=k_{S_3}(p_2)=k_{S_3}(p_3)=k_{S_3}(p_6)=k_{S_3}(p_7)=k_{S_3}(p_4)=2$, $k_{S_3}(p_9)=k_{S_3}(p_{10})=2$, and $k_{S_3}(p_{13})=2$. $\forall p\in P_R\cup P^0\cup\{p_5,p_{11},p_{14},p_{15},p_{16},p_{17}\}$, $k_{S_3}(p)=0$. As a result, $g_{S_3}=k_{S_3}+V_{S_3}=2p_1+2p_2+2p_3+2p_4+2p_6+2p_7+2p_9+2p_{10}+2p_{13}+V_{S_3}$.

By

$\sum_{r\in S_3^R}I_r=I_{p_{22}}+I_{p_{24}}+I_{p_{25}}=(2p_4+2p_{13}+2p_{14}+p_{22})+(p_9+p_{11}+p_3+p_7+p_{14}+p_{24})+(2p_9+2p_{10}+p_5+p_{25})$,

we have

$f_{S_3}=\sum_{r\in S_3^R}I_r-g_{S_3}=p_5+p_9+p_{11}+3p_{14}+p_{22}+p_{24}+p_{25}-V_{S_3}-2p_1-2p_2-p_3-2p_6-p_7$.

Clearly, $||f_{S_3}||^-\cap S_3=\emptyset$ and $||f_{S_3}||^+=S_3$ are true. Considering

$M_0(S_3)=M_0(p_{22})+M_0(p_{24})+M_0(p_{25})=9$

and

$\omega(S_3)=\sum_{p\in S}f_{S_3}(max_{p^\bullet}-1)=(max_{p_{22}^\bullet}-1)+(max_{p_{25}^\bullet}-1)=2$, we have $2 < \xi_{S_3} < 9$. In an analogous way, monitors V_{S_1}, V_{S_2}, V_{S_4}, V_{S_5}, and V_{S_6} can be accordingly added to the plant net model. The controlled system is denoted by (N_V, M_{0V}) as shown in Fig. 5.15. The details concerning the control of elementary siphons are given as follows:

(1) S_1:
$\sum_{r \in S_1^R} I_r = 2p_1 + p_2 + p_6 + p_{15} + p_{16} + p_{17} + p_{18} + p_{19} + p_{21}$;
$Th(S_1) = 2p_1 + p_{15} + p_{16}$;
$k_{S_1} = 2p_1 + p_{13} + p_{14} + p_{15} + p_{16}$;
$g_{S_1} = 2p_1 + p_{13} + p_{14} + p_{15} + p_{16} + V_{S_1}$;
$f_{S_1} = p_2 + p_6 + p_{17} + p_{18} + p_{19} + p_{21} - V_{S_1} - p_{13} - p_{14}$;
$M_{0V}(V_{S_1}) = 4 - \xi_{S_1}$;
$0 < \xi_{S_1} \leq 3$.

(2) S_2:
$\sum_{r \in S_2^R} I_r = p_1 + p_2 + p_6 + p_{15} + p_{16} + p_{18} + p_{21}$;
$Th(S_2) = p_1 + p_{15}$;
$k_{S_2} = p_1 + p_{13} + p_{14} + p_{15}$;
$g_{S_2} = p_1 + p_{13} + p_{14} + p_{15} + V_{S_2}$;
$f_{S_2} = p_2 + p_6 + p_{16} + p_{18} + p_{21} - V_{S_2} - p_{13} - p_{14}$;
$M_{0V}(V_{S_2}) = 3 - \xi_{S_2}$;
$0 < \xi_{S_2} \leq 2$.

(3) S_3:
$\sum_{r \in S_3^R} I_r = p_3 + 2p_4 + p_5 + p_7 + 3p_9 + 2p_{10} + p_{11} + 2p_{13} + 3p_{14} + p_{22} + p_{24} + p_{25}$;
$Th(S_3) = p_3 + 2p_4 + p_7 + 2p_{10} + 2p_{13}$;
$k_{S_3} = 2p_1 + 2p_2 + 2p_3 + 2p_4 + 2p_6 + 2p_7 + 2p_9 + 2p_{10} + 2p_{13}$;
$g_{S_3} = 2p_1 + 2p_2 + 2p_3 + 2p_4 + 2p_6 + 2p_7 + 2p_9 + 2p_{10} + 2p_{13} + V_{S_3}$;
$f_{S_3} = p_5 + p_9 + p_{11} + 3p_{14} + p_{22} + p_{24} + p_{25} - V_{S_3} - 2p_1 - 2p_2 - p_3 - 2p_6 - p_7$;
$M_{0V}(V_{S_3}) = 9 - \xi_{S_3}$;
$2 < \xi_{S_3} \leq 8$.

(4) S_4:
$\sum_{r \in S_4^R} I_r = p_3 + 2p_4 + p_5 + p_7 + p_9 + p_{11} + 3p_{13} + 3p_{14} + p_{22} + p_{23} + p_{24}$;
$Th(S_4) = p_3 + 2p_4 + p_7 + 3p_{13}$;
$k_{S_4} = 2p_1 + 2p_2 + 2p_3 + 2p_4 + 2p_6 + 2p_7 + 3p_{13}$;
$g_{S_4} = 2p_1 + 2p_2 + 2p_3 + 2p_4 + 2p_6 + 2p_7 + 3p_{13} + V_{S_4}$;
$f_{S_4} = p_5 + p_9 + p_{11} + 3p_{14} + p_{22} + p_{23} + p_{24} - V_{S_4} - 2p_1 - 2p_2 - p_3 - 2p_6 - p_7$;
$M_{0V}(V_{S_4}) = 7 - \xi_{S_4}$;
$1 < \xi_{S_4} \leq 6$.

(5) S_5:
$\sum_{r \in S_5^R} I_r = p_3 + 2p_4 + p_7 + p_9 + p_{11} + 2p_{13} + 3p_{14} + p_{22} + p_{24}$;
$Th(S_5) = p_3 + p_7 + 2p_{13}$;
$k_{S_5} = p_1 + p_2 + p_3 + p_6 + p_7 + 2p_{13}$;
$g_{S_5} = p_1 + p_2 + p_3 + p_6 + p_7 + 2p_{13} + V_{S_5}$;
$f_{S_5} = 2p_4 + p_9 + p_{11} + 3p_{14} + p_{22} + p_{24} - V_{S_5} - p_1 - p_2 - p_6$;
$M_{0V}(V_{S_5}) = 6 - \xi_{S_5}$;
$1 < \xi_{S_5} \leq 5$.

(6) S_6:
$\sum_{r\in S_6^R} I_r = p_2+p_3+p_6+p_7+p_9+p_{11}+p_{14}+p_{15}+p_{21}+p_{24}$;
$Th(S_6) = p_2+p_6+p_{14}$;
$k_{S_6} = p_1+p_2+p_6+p_{13}+p_{14}$;
$g_{S_6} = p_1+p_2+p_6+p_{13}+p_{14}+V_{S_6}$;
$f_{S_6} = p_3+p_7+p_9+p_{11}+p_{15}+p_{21}+p_{24}-V_{S_6}-p_1-p_{13}$;
$M_{OV}(V_{S_6}) = 5-\xi_{S_6}$;
$0<\xi_{S_6}\leq 4$.

Table 5.6 shows the controllability of all dependent siphons by Theorem 5.10. Algorithm 5.3 adds six monitors for the elementary siphons with $\xi_{S_1}=1$, $\xi_{S_2}=1$, $\xi_{S_3}=3$, $\xi_{S_4}=2$, $\xi_{S_5}=2$, and $\xi_{S_6}=4$, which can ensure the controllability of all dependent siphons. Accordingly, we have $M_{OV}(V_{S_1})=3$, $M_{OV}(V_{S_2})=2$, $M_{OV}(V_{S_3})$ $=6$, $M_{OV}(V_{S_4})=5$, $M_{OV}(V_{S_5})=4$ and $M_{OV}(V_{S_6})=1$, as shown in Fig. 5.15.

Table 5.6 Controllability of dependent siphons

Siphon	T-vector relationship	Controllability
S_7	$\eta_{S_7}=\eta_{S_1}+\eta_{S_3}+\eta_{S_6}$	$\xi_{S_1}+\xi_{S_3}+\xi_{S_6}>7$
S_8	$\eta_{S_8}=\eta_{S_2}+\eta_{S_3}+\eta_{S_6}$	$\xi_{S_2}+\xi_{S_3}+\xi_{S_6}>7$
S_9	$\eta_{S_9}=\eta_{S_3}+\eta_{S_6}$	$\xi_{S_3}+\xi_{S_6}>5$
S_{10}	$\eta_{S_{10}}=\eta_{S_1}+\eta_{S_4}+\eta_{S_6}$	$\xi_{S_1}+\xi_{S_4}+\xi_{S_6}>6$
S_{11}	$\eta_{S_{11}}=\eta_{S_2}+\eta_{S_4}+\eta_{S_6}$	$\xi_{S_2}+\xi_{S_4}+\xi_{S_6}>6$
S_{12}	$\eta_{S_{12}}=\eta_{S_4}+\eta_{S_6}$	$\xi_{S_4}+\xi_{S_6}>4$
S_{13}	$\eta_{S_{13}}=\eta_{S_1}+\eta_{S_5}+\eta_{S_6}$	$\xi_{S_1}+\xi_{S_5}+\xi_{S_6}>6$
S_{14}	$\eta_{S_{14}}=\eta_{S_2}+\eta_{S_5}+\eta_{S_6}$	$\xi_{S_2}+\xi_{S_5}+\xi_{S_6}>6$
S_{15}	$\eta_{S_{15}}=\eta_{S_5}+\eta_{S_6}$	$\xi_{S_5}+\xi_{S_6}>4$
S_{16}	$\eta_{S_{16}}=\eta_{S_1}+\eta_{S_6}$	$\xi_{S_1}+\xi_{S_6}>2$
S_{17}	$\eta_{S_{17}}=\eta_{S_2}+\eta_{S_6}$	$\xi_{S_2}+\xi_{S_6}>2$

Alternatively, a set of control depth variables of the elementary siphons can be found by solving the following LPP:

$$z = min\{\sum_{i=1}^{6} \xi_{S_i}\}$$

s.t.
$\xi_{S_1}+\xi_{S_3}+\xi_{S_6}>7$,
$\xi_{S_2}+\xi_{S_3}+\xi_{S_6}>7$,
$\xi_{S_3}+\xi_{S_6}>5$,
$\xi_{S_1}+\xi_{S_4}+\xi_{S_6}>6$,
$\xi_{S_2}+\xi_{S_4}+\xi_{S_6}>6$,
$\xi_{S_4}+\xi_{S_6}>4$,
$\xi_{S_1}+\xi_{S_5}+\xi_{S_6}>6$,
$\xi_{S_2}+\xi_{S_5}+\xi_{S_6}>6$,
$\xi_{S_5}+\xi_{S_6}>4$,
$\xi_{S_1}+\xi_{S_6}>2$,

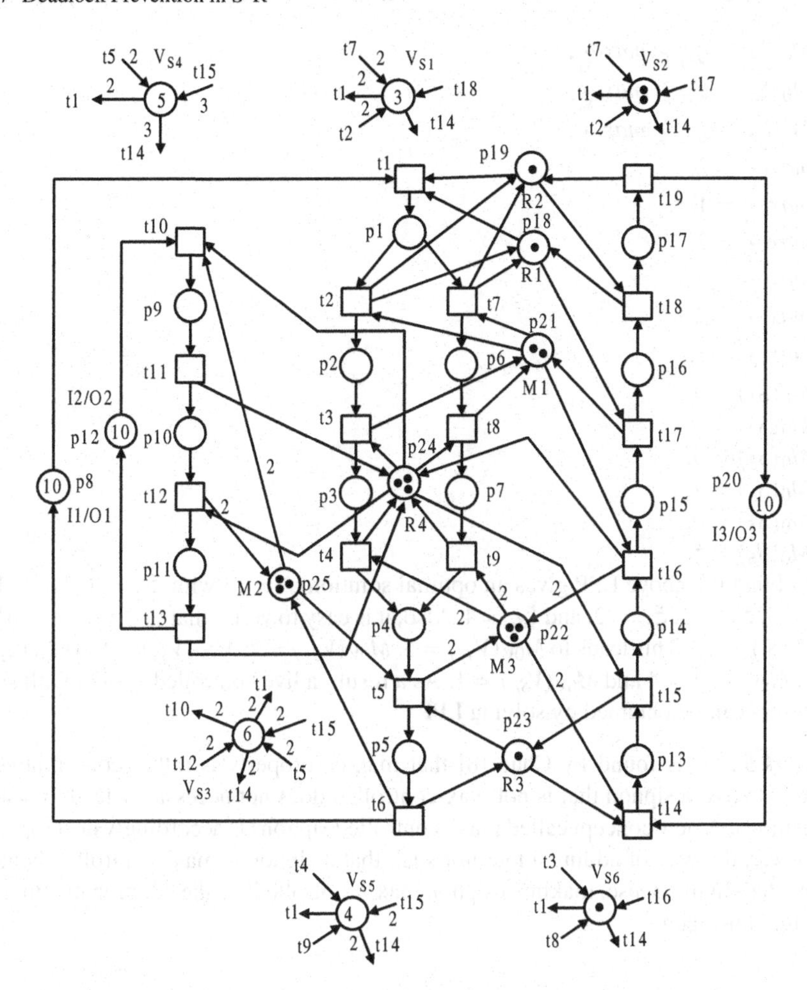

Fig. 5.15 The controlled system (N_V, M_{0V})

$\xi_{S_2} + \xi_{S_6} > 2,$
$0 < \xi_{S_1} \leq M_0(S_1) - 1,$
$0 < \xi_{S_2} \leq M_0(S_2) - 1,$
$2 < \xi_{S_3} \leq M_0(S_3) - 1,$
$1 < \xi_{S_4} \leq M_0(S_4) - 1,$
$1 < \xi_{S_5} \leq M_0(S_5) - 1,$
$0 < \xi_{S_6} \leq M_0(S_6) - 1,$
$M_0(S_1) - \xi_{S_1} \geq max_{V^{\bullet}_{S_1}},$
$M_0(S_2) - \xi_{S_2} \geq max_{V^{\bullet}_{S_2}},$
$M_0(S_3) - \xi_{S_3} \geq max_{V^{\bullet}_{S_3}},$

$M_0(S_4) - \xi_{S_4} \geq max_{V^\bullet_{S_4}}$,
$M_0(S_5) - \xi_{S_5} \geq max_{V^\bullet_{S_5}}$,
$M_0(S_6) - \xi_{S_6} \geq max_{V^\bullet_{S_6}}$,
$max_{V^\bullet_{S_1}} = 2$,
$max_{V^\bullet_{S_2}} = 1$,
$max_{V^\bullet_{S_3}} = 2$,
$max_{V^\bullet_{S_4}} = 3$,
$max_{V^\bullet_{S_5}} = 2$,
$max_{V^\bullet_{S_6}} = 1$,
$M_0(S_1) = 4$,
$M_0(S_2) = 3$,
$M_0(S_3) = 9$,
$M_0(S_4) = 7$,
$M_0(S_5) = 6$,
$M_0(S_6) = 5$.

Solving the above LPP gives an optimal solution $z^* = 13$ with $\xi_{S_1} = 1$, $\xi_{S_2} = 1$, $\xi_{S_3} = 3$, $\xi_{S_4} = 2$, $\xi_{S_5} = 2$, and $\xi_{S_6} = 4$. Also, it is easy to verify that $\forall i \in \{1, 2, \ldots, 6\}$, $M^{min}(S_i) \geq \xi_{S_i}$. This leads to $M_{0V}(V_{S_1}) = 3$, $M_{0V}(V_{S_2}) = 2$, $M_{0V}(V_{S_3}) = 6$, $M_{0V}(V_{S_4}) = 5$, $M_{0V}(V_{S_5}) = 4$ and $M_{0V}(V_{S_6}) = 1$. As a result, a live controlled system with six monitors can be obtained by solving LPP.

Remark 5.2. It is found by Chao [8] that max-cs property is rather conservative. That is to say, a siphon that is not max-controlled does not necessarily lead to dead transitions. A new concept called max'-controlled siphons is accordingly developed. However, the way of adding a monitor such that a siphon is max'-controlled is not presented. In this sense, making a siphon max'-controlled via the addition of a monitor remains open. □

5.8 Bibliographical Remarks

Deadlock is a major issue that must be addressed in contemporary resource allocation systems. Deadlocks and related blocking phenomena not only degrade the productivity of a system but also lead to catastrophic results in some highly automated systems, e.g., semiconductor manufacturing systems [17]. Their efficient handling becomes a necessary condition for a system to gain high throughput and safety. This is the reason why this problem is extensively investigated in the literature, particularly within the area of flexible manufacturing systems [14, 21, 56, 57].

Deadlock is first addressed by computer scientists [10, 22, 23, 26, 48]. Some important results about deadlocks in Petri nets are obtained by Commoner in 1972 [13]. Handling deadlock prevention in manufacturing dates back to 1990 due to the seminal work by Roszkowska and Viswanadham et al. [2, 43–46, 55]. The last two decades have seen much effort for deadlock control in flexible manufacturing [15].

The work of Fanti and Zhou [17] surveys a variety of deadlock control approaches that are based on graph theory, automata, and Petri nets in the literature. A survey and comparison of Petri net-based deadlock prevention policies for FMS can be found in [36]. Also readers are referred to the books [28] and [42].

Problems and Discussions

5.1. For an S^{3}PR, the existence of a liveness-enforcing supervisor with monitors that are added for elementary siphons only depends on the presence of a feasible solution to the following LPP:

$$min \sum_{i=1}^{m} \xi_{S_i} \tag{5.8}$$

s.t.

$$M_0(S_{Dj}) > \sum_{i=1}^{m} (M_0(S_i) - \xi_{S_i}), j = 1, 2, \ldots, n,$$

$$1 \leq \xi_{S_i} \leq M_0(S_i) - 1, i = 1, 2, \ldots, m,$$

where $\Pi_D = \{S_{Dj} | j = 1, 2, \ldots, n\}$ and $\Pi_E = \{S_i | i = 1, 2, \ldots, m\}$ are the sets of dependent and elementary siphons, respectively.

Can it be proved that LPP (5.8) definitely has a feasible solution for an S^{3}PR with acceptable initial markings?

5.2. Design a liveness-enforcing supervisor for the net shown in Fig. 5.8 by the deadlock prevention policy presented in Sect. 5.6.

5.3. The net (N, M_0) shown in Fig. 5.16 is the model of an FMS consisting of four machine tools and two robots. Two types of parts can be produced in this system. It is an S^{3}PR where p_1 and p_8 are idle process places, $p_{14}-p_{19}$ are resource places, and the others are operation places.

1. Find all the strict minimal siphons.
2. By the elementary siphon identification algorithm, compute the set of elementary siphons. Check whether there exists a liveness-enforcing supervisor by explicitly controlling its elementary siphons only.
3. According to the definition of the elementary siphons in a Petri net, they are not unique. Suppose that a net has n siphons and $rank([\eta]) = m$. We can at most find $n!/(m!(n-m)!)$ different sets of elementary siphons. By using this example, compare the permissive behavior of the supervisors resulting from controlling different sets of elementary siphons.

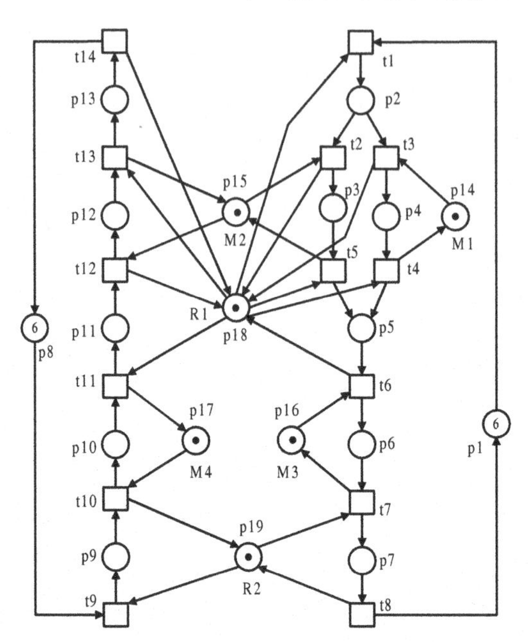

Fig. 5.16 An S^{3}PR net (N, M_0)

5.4. The development of a deadlock prevention policy without the complete siphon enumeration of a plant Petri net model can be first traced to the work by Huang et al. [24] in which an algorithm that extracts a strict minimal siphon from an unmarked maximal siphon is proposed. However, the algorithm is problematic. The corrections are given in [34].

The deadlock prevention policy in [24] consists of two phases: siphon control and control-induced siphon control. However, in each control phrase, the concept of elementary siphons is not considered. Design a novel deadlock control policy based on the policy in [24] by considering the existence of elementary siphons in each phrase. Take the nets shown in Figs. 5.8 and 5.16 as examples.

5.5. One of the most important contributions in the development of deadlock control strategies is due to Park and Reveliotis. In [40], a deadlock prevention policy is proposed for S^{4}R, which is of polynomial complexity. By this deadlock prevention policy, design a liveness-enforcing supervisor for the net shown in Fig. 5.5.

5.6. Iterative siphon control for deadlock prevention is a good but old idea. A basic iterative siphon control approach [27, 39] for plant net model (N, M_0) can be generalized as follows:

$i := 0$
$\Pi^{(0)} = \emptyset$
$N^{(0)} := N$
$M^{(0)} := M_0$

$i := i+1$
Find the set $\Pi^{(i)}$ of minimal siphons of $N^{(i-1)}$ not belonging to $\cup_{k=0}^{i-1}\Pi^{(k)}$
repeat
Implement on $(N^{(i-1)}, M^{(i-1)})$ the set of GMECs [20] $\{\lambda_S^T M \geq 1, \forall S \in \Pi^{(i)}\}$ and denote the resultant net by $(N^{(i)}, M^{(i)})$
$i := i+1$
Find the set $\Pi^{(i)}$ of minimal siphons of $N^{(i-1)}$ not belonging to $\cup_{k=0}^{i-1}\Pi^{(k)}$
until $\Pi^{(i)} = \emptyset$
output $(N^{(i)}, M^{(i)})$

Using the iterative siphon control scheme mentioned above, design a liveness-enforcing supervisor for the net shown in Fig. 5.8. It is worth noting that this algorithm is effective if at each iteration step, the net that is handled is ordinary. In the process of iterations, a generalized net may be produced. We refer the reader to the work in [27,29], and [41] to complete the iteration for this example. Decide whether an optimal liveness-enforcing supervisor can be found.

5.7. Implicit places [11, 12, 18, 19, 47] are a class of elements in a Petri net whose removal does not change the behavior of the Petri net. For example, the deadlock prevention policy proposed in [16] computes 18 monitors for the Petri net model shown in Fig. 5.8. By considering the elementary siphons, however, the addition of six monitors for the elementary siphons can lead to a live controlled system. Both nets with the different number of monitors have the same behavior. Identify the implicit places from the net that has 18 monitors and compare with the one that has six monitors. The readers are suggested to pay attention to the computational complexity of identifying implicit places and controllability decision of dependent siphons [37].

5.8. The redundancy problem of the monitors that are added for siphon control is also noticed by Iordache, Moody, and Antsaklis. Use the redundant monitor identification method proposed in [27] to decide whether the monitors in the live controlled system resulting from the deadlock prevention policy in [16] can be removed.

5.9. A redundant monitor identification method is proposed in [53] by using the complete state enumeration. Discuss its pros and cons.

5.10. Compare the computational burden by using INA [49] and the MIP-based deadlock detection method [9] to find all the strict minimal siphons in differently sized S^3PR nets. Supposedly, the superiority of the MIP-based deadlock detection method is more and more evident as a net size grows.

5.11. By Definitions 5.19 and 5.20, the deadlock prevention policy developed in [16] guarantees that, at any reachable marking, the token count in the adjoint set of a siphon does not exceed the initial number of tokens marked in the siphon. To be formal, the policy ensures that $\forall S \in \Pi$, $\forall M \in R(N, M_0)$, $M(\mathscr{P}_S) \leq M_0(S) - 1$. That is to say, deadlocks in an S^3PR can be prevented by enforcing the following set of GMECs by monitors: $(L, B) = \{(l_i, b_i) | l_i = \sum_{p \in \mathscr{P}_{S_i}} p, b_i = M_0(S_i) - 1\}$.

By using the results presented in Chap. 4, a set of monitors that are added for elementary constraints only may enforce all GMECs in the set.

The Petri net (N,M_0) in Fig. 5.3 is an S[3]PR if $P^0 = \{p_1, p_{10}\}$, $P_R = \{p_{11}, p_{12}, p_{13}, p_{14}, p_{15}\}$, and others are operation places. There are three strict minimal siphons $S_1 = \{p_5, p_9, p_{12}, p_{13}\}$, $S_2 = \{p_4, p_6, p_{13}, p_{14}\}$, and $S_3 = \{p_6, p_9, p_{12}, p_{13}, p_{14}\}$. The adjoint sets of the three siphons are $\mathscr{P}_{S_1} = \{p_3, p_4, p_7, p_8\}$, $\mathscr{P}_{S_2} = \{p_3, p_5, p_7, p_8\}$, and $\mathscr{P}_{S_3} = \{p_3, p_4, p_5, p_7, p_8\}$.

1. List the set of GMECs for this Petri net to implement the deadlock prevention policy in [16].
2. Find a monitor solution in which only elementary constraints are explicitly controlled.
3. Compare the two supervisors found by this GMEC-based method and the policy in [16].
4. Check whether there exist implicit places in the supervisor that is computed by the GMEC-based method.
5. By the MIP-based deadlock detection method, check whether there exist redundant monitors whose removal keeps the liveness of the resultant net systems.

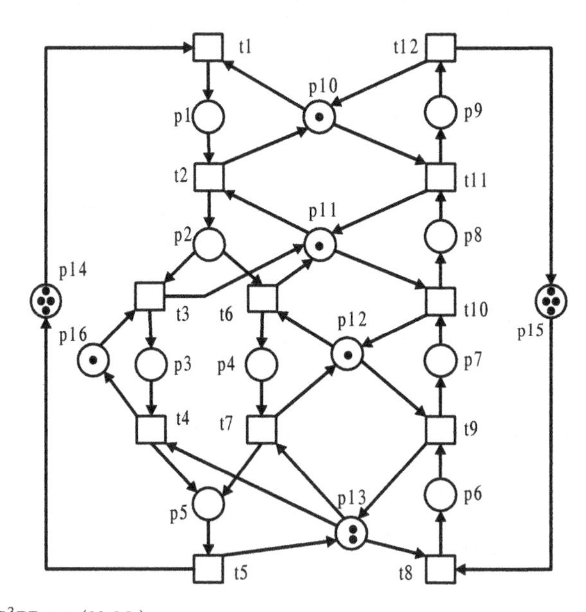

Fig. 5.17 An S[3]PR net (N,M_0)

5.12. Redo (1)–(5) of Problem 5.11 for the Petri net model of the FMS in Example 5.12.

5.13. Let $S_1 = \{p_5, p_9, p_{10}, p_{11}, p_{12}, p_{13}\}$, $S_2 = \{p_3, p_4, p_9, p_{10}, p_{11}, p_{12}\}$, $S_3 = \{p_2, p_3, p_9, p_{10}, p_{11}\}$, and $S_4 = \{p_5, p_7, p_{12}, p_{13}\}$. Construct an S^{3}PR N such that $\Pi = \{S_1, S_2, S_3, S_4\}$ is the set of strict minimal siphons of N. Verify whether the net shown in Fig. 5.18 is such an S^{3}PR.

5.14. Let $S_1 = \{p_5, p_9, p_{10}, p_{11}, p_{12}, p_{13}, p_{16}\}$, $S_2 = \{p_5, p_8, p_{11}, p_{12}, p_{13}, p_{16}\}$, $S_3 = \{p_2, p_9, p_{10}, p_{11}\}$, and $S_4 = \{p_5, p_7, p_{12}, p_{13}\}$. Construct an S^{3}PR N such that $\Pi = \{S_1, S_2, S_3, S_4\}$ is the set of strict minimal siphons of N. Verify whether the net shown in Fig. 5.17 is such an S^{3}PR.

5.15. Let $S_1 = \{p_1, p_2, r_1, r_2\}$ and $S_2 = \{p_3, p_4, r_2, r_3, r_4\}$. Find an S^{3}PR $N = (P^0 \cup P_A \cup P_R, T, F)$ such that $\Pi_E = \{S_1, S_2\}$, where $\{p_1 - p_4\} \subseteq P_A$ and $\{r_1 - r_4\} \subseteq P_R$.

5.16. Let $\Pi_E = \{S_i | i \in \mathbb{N}_n\}$. Develop an algorithm that can construct an S^{3}PR $N = (P^0 \cup P_A \cup P_R, T, F)$ such that Π_E is a set of elementary siphons of N.

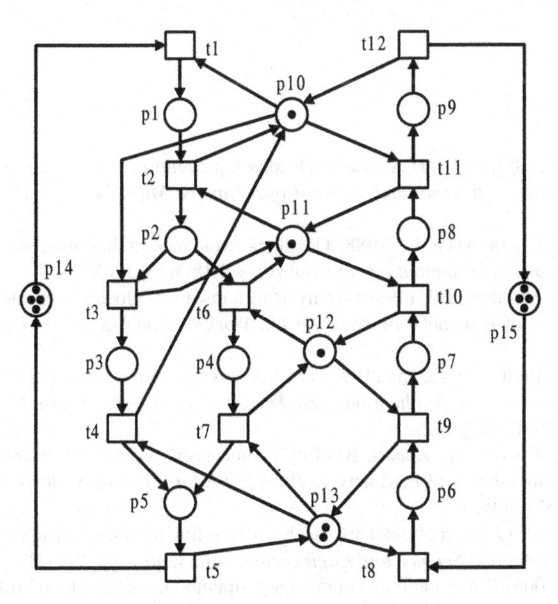

Fig. 5.18 An S^{3}PR net (N, M_0)

5.17. Properly set parameters $w_1 - w_{12}$, x, y, and z for the net in Fig. 5.19 such that its liveness is independent from the non-zero initial markings d_1 and d_2 of places p_4 and p_5, respectively.

5.18. Prove Theorem 5.12. Hint: We refer the reader to the proof of Theorem VI.I in [16].

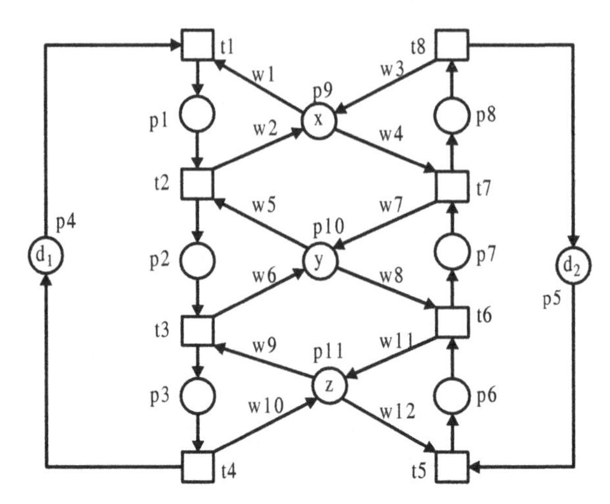

Fig. 5.19 An S^4R net (N, M_0)

References

1. Abdallah, I.B., ElMaraghy, H.A. (1998) Deadlock prevention and avoidance in FMS: A Petri net based approach. *International Journal of Advanced Manufacturing Technology*, vol.14, no.10, pp.704–715.
2. Banaszak, Z., Roszkowska, E. (1988) Deadlock avoidance in pipeline concurrent processes. *Podstawy Sterowania (Foundations of Control)*, vol.18, no.1, pp.3–17.
3. Barkaoui, K., Abdallah, I.B. (1996) Analysis of a resource allocation problem in FMS using structure theory of Petri nets. In *Proc. 1st Int. Workshop on Manufacturing and Petri Nets*, pp.62–76.
4. Barkaoui, K., Pradat-Peyre, J.F. (1996) On liveness and controlled siphons in Petri nets. In *Proc. 17th Int. Conf. on Applications and Theory of Petri Nets Lecture Notes in Computer Science*, vol.1091, pp.57–72.
5. Barkaoui, K., Chaoui, A., Zouari, B. (1997) Supervisory control of discrete event systems based on structure theory of Petri nets. In *Proc. IEEE Int. Conf. on Systems, Man, and Cybernetics*, pp.3750–3755.
6. Chao, D.Y. (2006) An incremental approach to extracting minimal bad siphons, *International Journal of Information Science and Engineering*, vol.23, no.1, pp.203–214.
7. Chao, D.Y. (2006) Computation of elementary siphons for deadlock control. *The Computer Journal*, vol.49, no.4, pp.470–479.
8. Chao, D.Y. (2007) Max′-controlled siphons for liveness of S^3PGR^2. *IET Control Theory and Applications*, vol.1, no.4, pp.933–936.
9. Chu, F., Xie, X.L. (1997) Deadlock analysis of Petri nets using siphons and mathematical programming. *IEEE Transactions on Robotics and Automation*, vol.13, no.6, pp.793–804.
10. Coffman, E.G., Elphick, M.J., Shoshani, A. (1971) Systems deadlocks. *ACM Computing Surveys*, vol.3, no.2, pp.67–78.
11. Colom, J.M., Silva, M. (1989) Improving the linearly based characterization of P/T nets. In *Proc. 10th Int. Conf. on Applications and Theory of Petri Nets*, G. Rozenberg (Ed.), *Lecture Notes in Computer Science*, vol.483, pp.113–145.
12. Colom, J.M., Campos, J., Silva, M. (1990) On liveness analysis through linear algebraic techniques. In *Proc. of Annual General Meeting of ESPRIT Basic Research Action 3148 Design Methods Based on Nets DEMON.*

13. Commoner, F. (1972) Deadlocks in Petri nets. Report CA-7206-2311, Massachusetts Computer Associates, Wakefield, Massachusetts.
14. D'souza, K.A., Khator, S.K. (1994) A survey of Petri nets in automated manufacturing systems control. *Computers in Industry Engineering*, vol.24, no.1, pp.5–16.
15. D'souza, K.A., Khator, S.K. (1997) System reconfiguration to avoid deadlocks in automated manufacturing systems. *Computers in Industry Engineering*, vol.32, no.2, pp.455–465.
16. Ezpeleta, J., Colom, J.M., Martinez, J. (1995) A Petri net based deadlock prevention policy for flexible manufacturing systems. *IEEE Transactions on Robotics and Automation*, vol.11, no.2, pp.173–184.
17. Fanti, M.P., Zhou, M.C. (2004) Deadlock control methods in automated manufacturing systems. *IEEE Transactions on Systems, Man, and Cybernetics, Part A*, vol.34, no.1, pp.5–22.
18. García-Vallés, F., Colom, J.M. (1999) Implicit places in net systems. In *Proc. 8th Int. Workshop on Petri Nets and Performance Models*, pp.104–113.
19. García-Vallés, F., Colom, J.M. (2002) Checking redundancy in supervisory control. A complexity result. In *Proc. 15th IFAC World Congress on Automatic Control*.
20. Giua, A., DiCesare, F., Silva, M. (1992) Generalized mutual exclusion constraints on nets with uncontrollable transitions. In *Proc. IEEE Int. Conf. on Systems, Man, and Cybernetics*, pp.974–979.
21. Giua, A., Seatzu, C. (2007) A systems theory view of Petri nets. In *Advances in Control Theory and Applications, Lecture Notes in Control and Information Science*, vol.353, C. Bonivento et al. (Eds.), pp.99–127.
22. Gold, E.M. (1978) Deadlock predication: Easy and difficult cases. *SIAM Journal of Computing*, vol.7, no.3, pp.320–336.
23. Haberman, A. (1969) Prevention of system deadlocks. *Communications of the ACM*, vol.12, no.7, pp.373–377.
24. Huang, Y.S., Jeng, M.D., Xie, X.L., Chung, S. L. (2001) Deadlock prevention policy based on Petri nets and siphons. *International Journal of Production Research*, vol.39, no.2, pp.283–305.
25. Huang, Y.S., Jeng, M.D., Xie, X.L., Chung, S.L. (2001) A deadlock prevention policy for flexible manufacturing systems using siphons. In *Proc. IEEE Int. Conf. on Robotics and Automation*, pp.541–546.
26. Isloor, S.S., Marsland, T.A. (1980) The deadlock problem: An overview. *Computer*, vol.13, no.9, pp.58–77.
27. Iordache, M.V., Moody, J.O., Antsaklis, P.J. (2002) Synthesis of deadlock prevention supervisors using Petri nets. *IEEE Transactions on Robotics and Automation*, vol.18, no.1, pp.59–68.
28. Iordache, M.V., Antsaklis, P.J. (2006) *Supervisory Control of Concurrent Systems: A Petri Net Structural Approach*. Berlin: Springer.
29. Lautenbach, K., Ridder, H. (1996) The linear algebra of deadlock avoidance–a Petri net approach. No.25-1996, Technical Report, Institute of Software Technology, University of Koblenz-Landau, Koblenz, Germany.
30. Li, Z.W., Zhou, M.C. (2004) Elementary siphons of Petri nets and their application to deadlock prevention in flexible manufacturing systems. *IEEE Transactions on Systems, Man, and Cybernetics, Part A*, vol.34, no.1, pp.38–51.
31. Li, Z.W., Uzam, M., Zhou, M.C. (2004) Comments on "Deadlock prevention policy based on Petri nets and siphons". *International Journal of Production Research*, vol.42, no.24, pp.5253–5254.
32. Li, Z.W., Zhou, M.C. (2006) Two-stage method for synthesizing liveness-enforcing supervisors for flexible manufacturing systems using Petri nets. *IEEE Transactions on Industrial Informatics*, vol.2, no.4, pp.313–325.
33. Li, Z.W., Hu, H.S., Wang, A.R. (2007) Design of liveness-enforcing supervisors for flexible manufacturing systems using Petri nets. *IEEE Transactions on Systems, Man, and Cybernetics, Part C*, vol.37, no.4, pp.517–526.
34. Li, Z.W., Liu, D. (2007) A correct minimal siphons extraction algorithm from a maximal unmarked siphon of a Petri net. *International Journal of Production Research*, vol.45, no.9, pp.2163–2167.

35. Li, Z.W., Zhao, M. (2008) On controllability of dependent siphons for deadlock prevention in generalized Petri nets. *IEEE Transactions on Systems, Man, and Cybernetics, Part A*, vol.38, no.2, pp.369–384.
36. Li, Z.W., Zhou, M.C. (2008) A survey and comparison of Petri net-based deadlock prevention policies for flexible manufacturing systems. *IEEE Transactions on Systems, Man, and Cybernetics, Part C*, vol.38, no.2, pp.172–188.
37. Li, Z.W. (2009) On systematic methods to remove redundant monitors from liveness-enforcing net supervisors. To appear in *Computer and Industrial Engineering*.
38. Lindo, Premier Optimization Modeling Tools, http://www.lindo.com/.
39. Moody, J.O., Antsaklis, P.J. (1998) *Supervisory Control of Discrete Event Systems Using Petri Nets*. Boston, MA: Kluwer.
40. Park, J., Reveliotis, S.A. (2001) Deadlock avoidance in sequential resource allocation systems with multiple resource acquisitions and flexible routings. *IEEE Transactions on Automatic Control*, vol.46, no.10, pp.1572–1583.
41. Piroddi, L., Cordone, R., Fumagalli, I. (2008) Selective siphon control for deadlock prevention in Petri nets. *IEEE Transactions on Systems, Man, and Cybernetics, Part A*. vol. 38, no. 6, pp.1337–1348.
42. Reveliotis, S.A. (2005) *Real-time Management of Resource Allocation Systems: A Discrete Event Systems Approach*. New York: Springer.
43. Roszkowska, E. (1990) Deadlock avoidance in concurrent compound pipeline processes, *Archives of Theoretical and Engineering Informatics*, vol.2, no.3–4, pp. 227–242.
44. Roszkowska, E. (1991) Application of Petri nets to the modelling and efficiency evaluation of FMS, Ph.D. thesis (in Polish), Report 4/91, Institute of Engineering Cybernetics, Wroclaw University of Technology, Poland.
45. Roszkowska, E., Wojcik, R. (1993) Problems of process flow feasibility in FAS. In *CIM in Process and Manufacturing Industries*, Oxford: Pergamon Press, pp.115–120.
46. Roszkowska, E., Jentink, J. (1993) Minimal restrictive deadlock avoidance in FMSs. In *Proc. European Control Conf.*, J. W. Nieuwenhuis, C. Pragman, and H. L. Trentelman, Eds., vol.2, pp. 530–534.
47. Silva, M., Teruel, E., Colom, J.M. (1998) Linear algebraic and linear programming techniques for the analysis of place/transition net systems. In *Lectures on Petri Nets I: Basic Models, Lectures Notes in Computer Science*, vol.1491, W. Reisig and G. Rozenberg (Eds.), pp.309–373.
48. Singhal, M. (1989) Deadlock detection in distributed systems. *IEEE Computer*, vol.22, no.11, pp.37–48.
49. Starke, P.H. (2003) *INA: Integrated Net Analyzer*. http://www2.informatik.hu-berlin.de/~starke/ina.html.
50. Tricas, F., García-Vallés, F., Colom, J.M., Ezpeleta, J. (1998) A structural approach to the problem of deadlock prevention in processes with shared resources. In *Proc. 4th Workshop on Discrete Event Systems*, pp.273–278.
51. Tricas, F., García-Vallés, F., Colom, J.M., Ezpeleta, J. (2000) An iterative method for deadlock prevention in FMSs. In *Proc. 5th Workshop on Discrete Event Systems*, R. Boel and G.Stremersch (Eds.), pp.139–148.
52. Uzam, M., Zhou, M.C. (2006) An improved iterative synthesis method for liveness enforcing supervisors of flexible manufacturing systems. *International Journal of Production Research*, vol.44, no.10, pp.1987–2030.
53. Uzam, M., Li, Z.W., Zhou, M.C. (2007) Identification and elimination of redundant control places in Petri net based liveness enforcing supervisors of FMS. *International Journal of Advanced Manufacturing Technology*, vol.35, no.1–2, pp.150–168.
54. Uzam, M., Zhou, M.C. (2007) An iterative synthesis approach to Petri net based deadlock prevention policy for flexible manufacturing systems. *IEEE Transactions on Systems, Man, and Cybernetics, Part A*, vol.37, no.3, pp.362–371.
55. Viswanadham, N., Narahari, Y., Johnson, T. (1990) Deadlock prevention and deadlock avoidance in flexible manufacturing systems using Petri net models. *IEEE Transactions on Robotics and Automation*, vol.6, no.6, pp.713–723.

56. Wu, N.Q. (1999) Necessary and sufficient conditions for deadlock-free operation in flexible manufacturing systems using a colored Petri net model. *IEEE Transactions on Systems, Man, and Cybernetics, Part C*, vol.29, no.2, pp.192–204.
57. Wu, N.Q., Zhou, M.C. (2001) Avoiding deadlock and reducing starvation and blocking in automated manufacturing systems. *IEEE Transactions on Robotics and Automation*, vol.17, no.5, pp.658–669.
58. Zouari, B., Barkaoui, K. (2003) Parameterized supervisor synthesis for a modular class of discrete event systems. In *Proc. IEEE Int. Conf. on Systems, Man, and Cybernetics*, pp.1874–1879.

Chapter 6
Optimal Liveness-Enforcing Supervisors

Abstract This chapter considers the design of optimal, i.e., maximally permissive, liveness-enforcing (Petri net) supervisors for automated manufacturing systems. It first reviews a deadlock prevention policy that is based on theory of regions, which is optimal. Then, based on the elementary siphon theory, sufficient conditions are presented under which there exists an optimal liveness-enforcing supervisor for a class of Petri nets, S^3PR (system of simple sequential processes with resources). A synthesis method of such a supervisor is given if it exists. This chapter also shows that an optimal liveness-enforcing supervisor can be computed in polynomial time if all the siphons in an S^3PR are elementary. Moreover, there exists an optimal liveness-enforcing supervisor if the capacity of every resource is greater than one.

6.1 Background

The quality of a deadlock prevention policy can be shown by evaluating its resultant controlled net system from a number of aspects. Usually, in addition to its structural complexity, behavior permissiveness is one of the most important criteria in evaluating the performance of a liveness-enforcing (Petri net) supervisor. In supervisory control theory [22], the existence and synthesis approach of an optimal, i.e., maximally permissive, supervisor for a discrete-event system is well addressed in the framework of formal languages and finite-state automata. When the control specification languages are controllable, the resultant automaton serves as a maximally permissive supervisor by properly trimming the automaton that represents the behavior of a plant. Its computational complexity depends on the complete state enumeration. Results on the existence and synthesis of a liveness-enforcing marking-based supervisor are reported by Sreenivas in [24–26]. It is shown that the existence of a liveness-enforcing marking-based supervisor is undecidable for an arbitrary Petri net. In case of bounded Petri nets, it is decidable. However, the decision procedure requires the KM-tree of a Petri net, which is basically the reachability graph of the net.

The theory of regions [2, 9] can be used to synthesize pure Petri nets from automaton-based models and is an important method for supervisory control of discrete-event systems. By using the theory of regions, Uzam [28] develops an optimal liveness-enforcing supervisor synthesis method for FMS. Later, in terms of plain and popular linear algebra notions, Ghaffari et al. [11] explore the sufficient and necessary conditions on the existence of a liveness-enforcing supervisor that is optimal, and develop a methodology to synthesize such a supervisor. The most attractive advantage of these approaches is that such an optimal supervisor can be always obtained if it exists. This chapter explores the existence of an optimal liveness-enforcing supervisor, $\mathcal{M}^*$ for short, for S^3PR nets in terms of elementary siphons through pure structural analysis.

6.2 Optimal Supervisor Design by the Theory of Regions

Deadlock prevention or liveness enforcement is a special class of forbidden state problems that are typical in supervisory control of discrete-event systems. In a Petri net formalism, $\mathcal{M}_F$ is usually used to denote the set of markings for which control specifications do not hold in a Petri net (N, M_0). Set $\mathcal{M}_F$ is also called the set of forbidden markings. The markings in it are hence unsafe [14]. To find a Petri net supervisor for the given control specifications, the objective is to determine a set of monitors that, once added to a given plant net model, prevent the whole system from reaching these states. For example, the GMECs in Chap. 4 are also a typical class of forbidden state problems.

Definition 6.1. The set $\mathcal{M}_L$ of legal or admissible markings is the maximal set of reachable markings such that (1) $\mathcal{M}_L \cap \mathcal{M}_F = \emptyset$, and (2) it is possible to reach initial marking M_0 from any legal marking without leaving $\mathcal{M}_L$.

By Definition 6.1 we have $\mathcal{M}_L = R(N, M_0) \setminus \mathcal{M}_F$. Let R_c be the reachability graph containing all legal markings for the given control specification in a plant Petri net model (N, M_0). It is clear that every element in $\mathcal{M}_L$ can be found in R_c and every node in R_c is an element in $\mathcal{M}_L$. At any marking in $\mathcal{M}_L$, the system cannot be led outside $\mathcal{M}_L$. A marking in $\mathcal{M}_L$ is called dangerous if an unsafe marking (in $\mathcal{M}_F$) can be possibly reached due to improper firing of an enabled transition. To solve the control problem, one has to identify the set of state/event separation instances (or marking/transition separation instances in net terminology) from an admissible marking to a non-admissible one. The additional monitors are used to prevent these transitions from occurring in order to keep the state space of the controlled system in the set of legal markings. Formally, the set of marking/transition separation instances that the supervisor has to disable is $\Omega = \{(M,t) | M[t\rangle M' \wedge M \in \mathcal{M}_L \wedge M' \notin \mathcal{M}_L\}$, where M is a dangerous marking and M' is bad. Let $\mathcal{M}_D$ be the set of dangerous markings. Clearly, we have $\mathcal{M}_D = \{M | M \in \mathcal{M}_L \wedge \exists t \in T, \ni M[t\rangle M' \wedge M' \in \mathcal{M}_F\}$.

An optimal supervisor is the one that ensures the reachability of all markings in $\mathcal{M}_L$ and forbids all marking/transition separation instances in Ω. An algorithm is proposed in [11], which is of polynomial complexity with respect to the number of states in the reachability graph of a plant net model. It can give legal behavior R_c, $\mathcal{M}_L$, Ω, $\mathcal{M}_D$, and the set of transitions leading outside R_c.

Next we illustrate the above concepts. Let us consider an FMS with two machine tools M1 and M2, each of which can process only one part at a time, and one robot R that can hold one part at a time. Parts enter the FMS through input/output buffers I1/O1 and I2/O2. The system can repeatedly produce two part-types J1 and J2. The production sequences are J1: M1→R→M2 and J2: M2→R→M1.

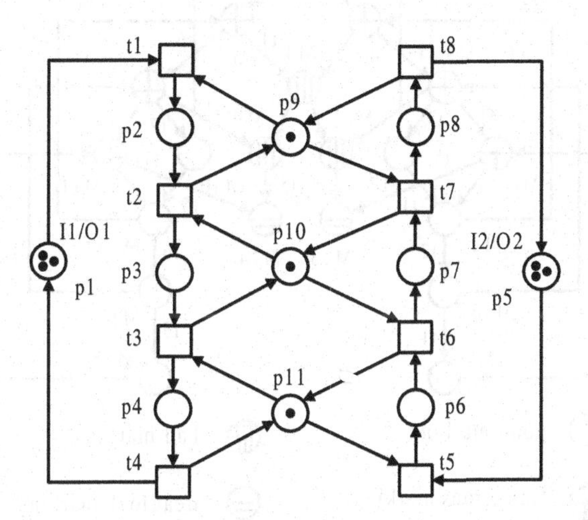

Fig. 6.1 The Petri net model (N, M_0) of an FMS

Figure 6.1 shows the Petri net model, denoted by (N, M_0), of the system, where tokens in p_1 and p_5 indicate the maximal number of parts of J1 and J2, which can be concurrently processed in the system, p_2, p_3, p_4, p_6, p_7, and p_8 represent the operations on J1 and J2, and $p_9 - p_{11}$ denote the availability of M1, R, and M2, respectively. Its reachability graph is shown in Fig. 6.2 given $M_0 = 3p_1 + 3p_5 + p_9 + p_{10} + p_{11}$.

In the reachability graph of a Petri net, markings are generally categorized into four classes in the sense of deadlock control: deadlock, bad, dangerous, and good markings. A deadlock one means a dead system state by which no successive marking is followed. A bad one is the one that has subsequent markings, from which the initial marking cannot be reached. A dangerous one can reach a good, bad or deadlock one depending on supervisory control. It is a node of the maximal strongly connected component containing the initial marking and its son nodes are either bad or deadlock ones. Good markings are the ones except deadlock, bad, and dangerous ones. Good and dangerous markings are included in the maximal strongly connected component containing the initial marking and bad and deadlock ones are not included in the component.

In Fig. 6.2, M_{13} and M_{14} are deadlock and M_4, M_8, and M_9 are bad, while M_1, M_2, M_3, M_5, M_6, and M_{11} are dangerous. All the other markings are hence good. The set of good and dangerous markings in $R(N_0, M_0)$, denoted by $\mathcal{M}_L$, should constitute the maximum legal behavior if a supervisory controller is optimally designed.

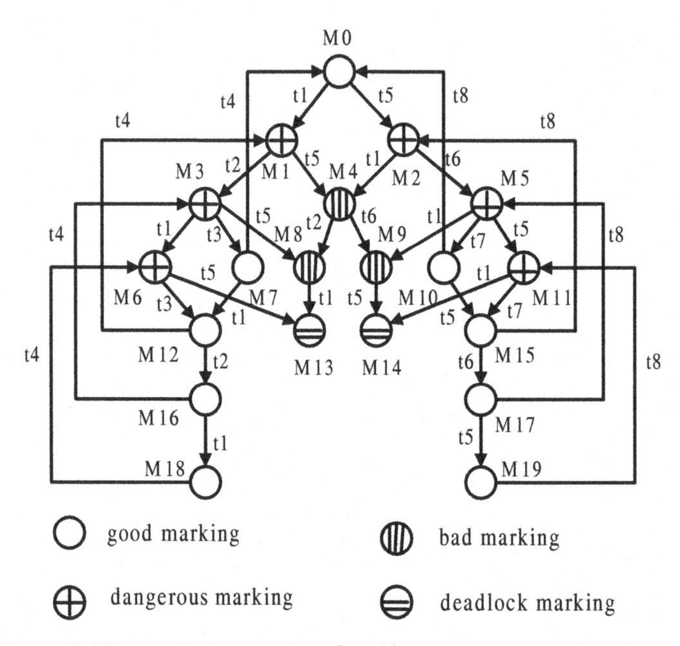

Fig. 6.2 The reachability graph of the Petri net (N, M_0)

As stated previously, a marking/transition separation instance has the form of (M, t), where M is a dangerous marking, and t is a transition whose firing results in marking $M[t\rangle$, which is a bad or deadlock marking. In Fig. 6.2, we have $\Omega = \{(M_1, t_5), (M_2, t_1), (M_3, t_5), (M_5, t_1), (M_6, t_5), (M_{11}, t_1)\}$.

The theory of regions is proposed for the synthesis of pure nets from given finite transition systems [2], which can be adopted to synthesize the liveness-enforcing supervisor for a plant net model [11].

Given a plant model (N, M_0) of a system with n transitions to control and its R_c, the theory of regions designs a monitor p_m and then adds it to the original Petri net for each marking/transition separation instance (M, t) such that t is disabled at M. Let (N^α, M_0^α) be the resultant net with monitors. We need to compute $M_0^\alpha(p_m)$ and $[N^\alpha](p_m, T)$.

Note that it shall be ensured that the addition of p_m does not exclude the markings in R_c, which implies that p_m has to satisfy the following reachability condition (6.1) and circuit equation (6.2):

$$M(p_m) = M_0^\alpha(p_m) + [N^\alpha](p_m, \cdot)\overrightarrow{\Gamma}_M \geq 0, \forall M \in \mathcal{M}_L \tag{6.1}$$

where Γ_M is any non-oriented path in R_c from M_0 to M, and $\overrightarrow{\Gamma}_M$ is the counting vector of the path Γ_M, which denotes the algebraic sum of all occurrences of transitions in Γ_M.

Consider any non-oriented circuit γ in R_c. Applying the state equation to the nodes in γ and summing them up give the following circuit equation:

$$\sum_{t \in T} [N^{\alpha}](p_m, t) \cdot \overrightarrow{\gamma}[t] = 0, \forall \gamma \in \mathscr{C} \tag{6.2}$$

where $\overrightarrow{\gamma}[t]$ denotes the algebraic sum of all occurrences of t in γ and $\mathscr{C}$ is the set of non-oriented circuits in R_c. $\overrightarrow{\gamma}$ is called the counting vector of γ. Note that different circuits in R_c may have the same circuit equation when their counting vectors are identical.

Furthermore, the addition of p_m necessarily forbids the firing of t for marking/transition separation instance (M,t). That is to say, $M(p_m) + [N^{\alpha}](p_m,t) \leq -1$. Thanks to $M(p_m) = M_0^{\alpha}(p_m) + [N^{\alpha}](p_m,\cdot)\overrightarrow{\Gamma}_M$, for every separation instance (M,t), we have the marking/transition separation equation

$$M_0^{\alpha}(p_m) + [N^{\alpha}](p_m,\cdot)\overrightarrow{\Gamma}_M + [N^{\alpha}](p_m,t) \leq -1. \tag{6.3}$$

We now use Figs. 6.1 and 6.2 to illustrate the above procedure.

Example 6.1. In Fig. 6.2, we have $\Omega = \{(M_1,t_5), (M_2,t_1), (M_3,t_5), (M_5,t_1), (M_6,t_5), (M_{11},t_1)\}$. First consider (M_1,t_5). We need to find a monitor p_{m1} to achieve this marking/transition separation. Note that $M_0^{\alpha}(p_{m1}) \geq 0$ is trivial and for simplicity, let $x = M_0^{\alpha}(p_{m1})$ and $x_i = [N^{\alpha}](p_{m1},t_i)$, where $i \in \{1,2,\ldots,8\}$. $\mathscr{M}_L$ has a total of 15 markings, i.e., $\mathscr{M}_L = \{M_0-M_3,\ M_5-M_7,\ M_{10}-M_{12},\ M_{15}-M_{19}\}$. Also, different circuits in Fig. 6.2 may have the same circuit equations. For example, $M_0t_1M_1t_2M_3t_3M_7t_4M_0$ and $M_{12}t_2M_{16}t_1M_{18}t_4M_6t_3M_{12}$ are two circuits. It is easy to verify that they have the same circuit equations.

For (M_1,t_5), we have the following reachability condition equations (6.4)–(6.18), two circuit equations (6.19) and (6.20), and marking/transition separation equation (6.21).

$$x \geq 0, \tag{6.4}$$

$$x + x_1 \geq 0, \tag{6.5}$$

$$x + x_1 + x_2 \geq 0, \tag{6.6}$$

$$x + x_1 + x_2 + x_1 \geq 0, \tag{6.7}$$

$$x + x_1 + x_2 + x_3 \geq 0, \tag{6.8}$$

$$x + x_1 + x_2 + x_3 + x_1 \geq 0, \tag{6.9}$$

$$x + x_1 + x_2 + x_3 + x_1 + x_2 \geq 0, \tag{6.10}$$

$$x + x_1 + x_2 + x_3 + x_1 + x_2 + x_1 \geq 0, \tag{6.11}$$

$$x + x_5 \geq 0, \tag{6.12}$$

$$x + x_5 + x_6 \geq 0, \tag{6.13}$$

$$x + x_5 + x_6 + x_5 \geq 0, \tag{6.14}$$

$$x + x_5 + x_6 + x_7 \geq 0, \tag{6.15}$$

$$x + x_5 + x_6 + x_7 + x_5 \geq 0, \tag{6.16}$$

$$x + x_5 + x_6 + x_7 + x_5 + x_6 \geq 0, \tag{6.17}$$

$$x + x_5 + x_6 + x_7 + x_5 + x_6 + x_5 \geq 0, \tag{6.18}$$

$$x_1 + x_2 + x_3 + x_4 = 0, \tag{6.19}$$

$$x_5 + x_6 + x_7 + x_8 = 0, \tag{6.20}$$

$$x + x_1 + x_5 \leq -1. \tag{6.21}$$

Solving inequalities (6.4)–(6.21), we get $x = 1, x_1 = -1, x_2 = 1, x_5 = -1, x_6 = 1$, and $x_3 = x_4 = x_7 = x_8 = 0$. Hence monitor p_{m1} is accordingly added, as shown in Fig. 6.3, to the original Petri net.

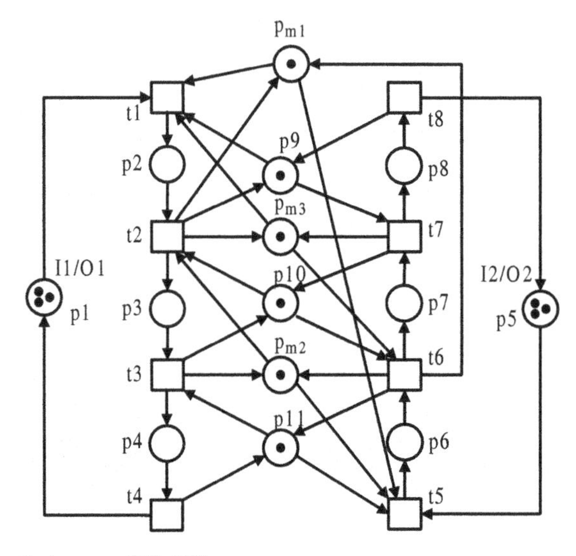

Fig. 6.3 A controlled system (N^α, M_0^α)

Now we deal with marking/transition separation instance (M_2, t_1). It is obvious to see that the reachability and circuit equations for all marking/transition separation instances are the same, i.e., (6.4)–(6.20), in this example. When we deal with a new marking/transition separation instance, what we shall do is just to replace (6.21) with the corresponding separation equation. The separation equation for (M_2, t_1) happens to be (6.21) as well. Hence we can say that p_{m1} has implemented (M_2, t_1) as well. Similarly, we can find monitor p_{m2} that implements (M_3, t_5) and (M_6, t_5) and monitor p_{m3} that implements (M_5, t_1) and (M_{11}, t_1). Consequently, the resultant controlled system with three monitors is live, as shown in Fig. 6.3.

Note that although the number of monitors to add is theoretically at most equal to, practically much smaller than, the number of marking/transition separation instances in the reachability graph of a Petri net, the number of the sets of inequalities that we have to solve is actually equal to that of marking/transition separation instances. This is so since we do not know whether a monitor can implement two or more marking/transition separation instances until all the sets of inequalities are already solved.

For example, there are 20 reachable markings in Fig. 6.2, where two of them are deadlock markings and six are dangerous markings. Hence there are six marking/transition separation instances. We know that three monitors actually implement all these six marking/transition separation instances only after we have solved all the six sets of inequalities. This is one of the major disadvantages of the method.

The next result from [11] indicates the existence of an optimal liveness-enforcing supervisor. A controlled system is said to be optimal if it results from the synchronous synthesis of a plant net model and an optimal supervisor. For example, the controlled net system shown in Fig. 6.3 is optimal since the supervisor with three monitors is optimal.

Theorem 6.1. *There exists an optimal liveness-enforcing supervisor for a plant Petri net model* (N,M_0) *iff there exist a set of monitors that implement all marking/transition separation instances of* (N,M_0).

Corollary 6.1. $R(N^\alpha,M_0^\alpha) = \{M|M \text{ is a marking in } R_c\} = \mathcal{M}_L$ *if* (N^α,M_0^α) *is an optimal controlled system for* (N,M_0).

Remark 6.1. The synthesis methods of optimal liveness-enforcing supervisors in [11] and [28] need the complete state enumeration of a plant net model, whose computation is expensive or impossible when we deal with either a large-sized net model, or a small-sized one with a large initial marking. The work of Ghaffari et al. [11] shows that, in general, the existence of an optimal liveness-enforcing supervisor $\mathcal{M}^*$ can be determined by first generating the reachability graph, and then solving an LPP for each marking/transition separation instance. Unfortunately, the number of LPPs that need to be solved is in theory exponential with respect to the size of the plant net and the initial marking. Worst of all, the number of constraints in such an LPP is in theory exponential with respect to the net size and the initial marking. Although an LPP can be solved in polynomial time, the complexity of the supervisor synthesis approach based on the theory of regions makes it actually impractical.

The next section presents the existence conditions and synthesis method of an optimal liveness-enforcing supervisor for a class of Petri nets, S^3PR, by pure structural analysis.

6.3 Existence of an Optimal Liveness-Enforcing Supervisor

In what follows, $N=(P^0 \cup P_A \cup P_R, T, F)$ is used to denote an S^{3}PR where there is no confusion. We suppose that N is composed of n subnets N_i's, which is denoted by $N=\bigcirc_{i=1}^{n} N_i$. Accordingly, we have $P_A=\cup_{i=1}^{n} P_{A_i}$.

Proposition 6.1. *Let S be a strict minimal siphon in N. If $p \in S^A$, then $r_p \in S$, where r_p denotes the resource place that is required by the operation represented by p.*

Proof. We prove it by contradiction. Let $p_k \in S \cap P_{A_i}$ such that $r_{p_k} \notin S$. Then $\exists t_k \in T_i$, $t_k \in {}^\bullet p_k$. Since $p_k \in S$ and S is a siphon, we have $t_k \in {}^\bullet S$. Therefore, t_k must be in $S^\bullet$. Note that $r_{p_k} \notin S$ and each operation place needs only one resource in N. We have $\exists p_{k-1} \in P_{A_i} \cap S$, $t_k \in p_{k-1}^\bullet$. Thus, we have two cases (a) $r_{p_{k-1}} \notin S$ and (b) $r_{p_{k-1}} \in S$.

(a) $r_{p_{k-1}} \notin S$.
In this case, there exists $t_{k-1} \in T_i$ such that $t_{k-1} \in {}^\bullet p_{k-1}$. Since $p_{k-1} \in S$ and S is a siphon, we have $t_{k-1} \in {}^\bullet S$. Therefore, t_{k-1} must be in $S^\bullet$. Note that $r_{p_{k-1}} \notin S$ and each operation place needs only one resource in N. We have $\exists p_{k-2} \in P_{A_i} \cap S$, $t_{k-1} \in p_{k-2}^\bullet$, leading to two subcases depending on whether or not $r_{p_{k-2}} \in S$ is true. Without loss of generality, in this case, we assume that $r_{p_{k-2}}, r_{p_{k-3}}, \cdots$ do not belong to siphon S. The case that S contains resource places $r_{p_{k-2}}, r_{p_{k-3}}, \cdots$ is considered in (b).
The same reasoning is applied to $r_{p_{k-2}}, r_{p_{k-3}}, \cdots$, and r_{p_1}. Finally, we have $p_1 \in S \cap P_{A_i}$, $r_{p_1} \notin S$, $t_1 \in {}^\bullet p_1$, and $t_1 \in (p_i^0)^\bullet$. Since t_1 must be in $S^\bullet$ and r_{p_1} is not in S, we conclude that $p_i^0 \in S$ is true. This contradicts the fact that $\forall S$ (a strict minimal siphon) $\in \Pi$, $\forall p \in P^0$, $p \notin S$, where Π is the set of strict minimal siphons in N.

(b) $r_{p_{k-1}} \in S$.
In this case, $\exists t_{k+1} \in T_i$, $\exists p_{k+1} \in P_{A_i}$ such that $t_{k+1} \in p_k^\bullet \cap {}^\bullet p_{k+1}$. Without loss of generality we assume that t_{k+1} is not a sink transition of the S^{3}PR. We hence have two subcases. (b.1) $t_{k+1} \notin {}^\bullet S$ and (b.2) $t_{k+1} \in {}^\bullet S$.

(b.1) $t_{k+1} \notin {}^\bullet S$: $t_{k+1} \notin {}^\bullet S$ implies that ${}^\bullet(S \setminus \{p_k\}) \subseteq {}^\bullet S \subseteq S^\bullet \setminus \{t_{k+1}\} = (S \setminus \{p_k\})^\bullet$, leading to the fact that S is not minimal.

(b.2) $t_{k+1} \in {}^\bullet S$: It is easy to see that $\exists r_{p_{k+1}} \in P_R$ such that $t_{k+1} \in r_{p_{k+1}}{}^\bullet$. Depending on $r_{p_{k+1}}$, we hence have two subcases (b.2.1) $r_{p_{k+1}} \notin S$ and (b.2.2) $r_{p_{k+1}} \in S$.

(b.2.1) $r_{p_{k+1}} \notin S$ and $t_{k+1} \in {}^\bullet S$ imply $p_{k+1} \in S$. The same reasoning about p_k can be applied to p_{k+1}, similar to Case (a). Here we assume that resource places $r_{p_{k+2}}, r_{p_{k+3}}, \ldots$ do not belong to S. Finally, we have $S = P_{A_i} \cup \{p_i^0\} \cup \{r_{p_{k-1}}\}$. This contradicts the minimality of S.

(b.2.2) $r_{p_{k+1}} \in S$ and $t_{k+1} \in {}^\bullet S$ lead to $p_{k+1} \in S$. As a result, in this case, we have the following facts: $p_{k-1} \in S$, $r_{p_{k-1}} \in S$, $p_{k+1} \in S$, $r_{p_{k+1}} \in S$, and $p_k \in S$. We are led to infer ${}^\bullet(S \setminus \{p_k\}) = {}^\bullet S \subseteq S^\bullet = (S \setminus \{p_k\})^\bullet$. This implies that $S \setminus \{p_k\}$ is a siphon, which contradicts the minimality of S.

It is easy to see that r_p must be a shared resource place. If r_p is unshared, then $\{p, r_p\}$ is a siphon included in S, which contradicts its minimality. □

Property 6.1. [3,7] Let S be a minimal siphon in a Petri net. The subnet G_S derived from $S \cup {}^{\bullet}S$ is strongly connected.

Corollary 6.2. *Let S be a strict minimal siphon in N. $\exists t \in {}^{\bullet}S$, $|t^{\bullet} \cap S| = 2$ with $t^{\bullet} \cap S = \{r, p\}$, where $r \in P_R$, $p \notin H(r)$, and $p \in P_A$.*

Proof. A strict minimal siphon in an S[3]PR contains at least two resource places and a number of operation places. Without loss of generality, we assume $S = \{r, r_p, \ldots, p, \ldots\}$. By Proposition 6.1, $\forall p \in S \cap P_A$, $\exists r_p \in P_R$, $r_p \in S$ and $p \in H(r_p)$. From Property 6.1, the subnet derived from $S \cup {}^{\bullet}S$ is strongly connected. That is to say, for any two nodes x_1 and x_2 in G_S, there exists a directed path from x_1 to x_2. Let t be an input transition of p. Clearly, $t \in r_p{}^{\bullet}$ is true, as shown in Fig. 6.4. There exists a resource place r such that $t \in {}^{\bullet}r$ since otherwise r and r_p are not strongly connected. Hence, we have $t^{\bullet} \cap S = \{r, p\}$. Since $p \in H(r_p)$, we conclude $p \notin H(r)$. □

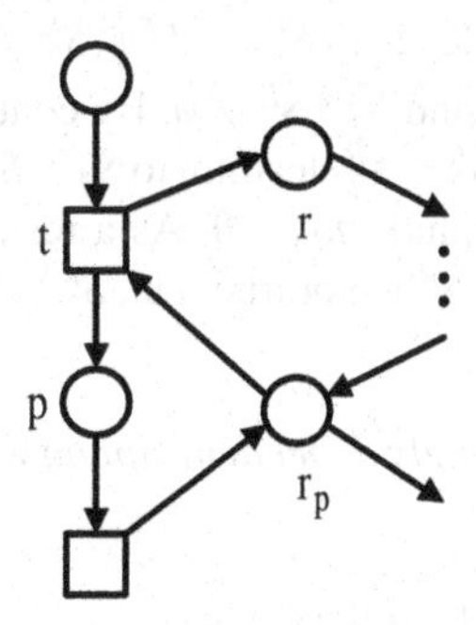

Fig. 6.4 A subnet of G_S derived from $S \cup {}^{\bullet}S$

Corollary 6.2 indicates that $\{r, p\} \subseteq S$. From Proposition 6.1, there exists a resource $r' \in S$ such that $p \in H(r')$ is true. As a result, any strict minimal siphon in an S[3]PR contains at least two resource places. The following result is from [15].

Theorem 6.2. *Let S be a strongly dependent siphon in an S[3]PR N with $\eta_S = \sum_{i=1}^{n} a_i \eta_{S_i}$. Then $\forall i \in \mathbb{N}_n$, $a_i = 1$.*

A multiset is useful in the synthesis of an optimal liveness-enforcing supervisor for an S[3]PR. In this chapter, the complementary set of a siphon in an S[3]PR is treated as a multiset. We have the following results.

Corollary 6.3. *Let S_0 be a strongly dependent siphon with $\eta_{S_0} = \eta_{S_1} + \eta_{S_2}$ in N. Then (1) $S_0^R = S_1^R \cup S_2^R$ and (2) $[S_0] = [S_1] + [S_2]$.*

Proof. (1) can be easily proved by contradiction. We next prove (2). $\forall i \in \{0, 1, 2\}$, $\lambda_{S_i \cup [S_i]}$ is a P-semiflow. As a result, we have $(\lambda_{S_0 \cup [S_0]} - \lambda_{S_1 \cup [S_1]} - \lambda_{S_2 \cup [S_2]})^T [N] = \mathbf{0}^T$. That is to say, $(\lambda_{S_0} - \lambda_{S_1} - \lambda_{S_2})^T [N] + (\lambda_{[S_0]} - \lambda_{[S_1]} - \lambda_{[S_2]})^T [N] = \mathbf{0}^T$. Since

$\eta_{S_0} = \eta_{S_1} + \eta_{S_2}$, we have $(\lambda_{[S_0]} - \lambda_{[S_1]} - \lambda_{[S_2]})^T[N] = \mathbf{0}^T$. Let $\rho = \lambda_{[S_0]} - \lambda_{[S_1]} - \lambda_{[S_2]}$. Since $||\rho|| \subseteq P_A$ is a subset of operation places and the support of any P-invariant of N contains either a resource place or an idle place, ρ is not a P-invariant of N. We hence have $\lambda_{[S_0]} - \lambda_{[S_1]} - \lambda_{[S_2]} = \mathbf{0}$, which implies the truth of this corollary. □

Proposition 6.2. *Let S_0, S_1, $S_2 \in \Pi$ be strict minimal siphons in N. If $\eta_{S_0} = \eta_{S_1} + \eta_{S_2}$, then $S^R_{1,2} \neq \emptyset$, $S^R_1 \backslash S^R_2 \neq \emptyset$, and $S^R_2 \backslash S^R_1 \neq \emptyset$, where $S^R_{1,2} = S^R_1 \cap S^R_2$.*

Proof. Let S be a strict minimal siphon, $T^{in}_S = \{t \in S^\bullet \,|\, |t^\bullet \cap S| > |^\bullet t \cap S|\}$, $T^{out}_S = \{t \in S^\bullet \,|\, |t^\bullet \cap S| < |^\bullet t \cap S|\}$, and $T^{equ}_S = \{t \in S^\bullet \,|\, |t^\bullet \cap S| = |^\bullet t \cap S|\}$. Clearly, we have $S^\bullet = T^{in}_S \cup T^{out}_S \cup T^{equ}_S$.

By contradiction, we suppose that $S^R_1 \cap S^R_2 = \emptyset$. By Proposition 6.1, $p \in S$ implies $r_p \in S$, where S is a strict minimal siphon. Hence, we have $S^A_1 \cap S^A_2 = \emptyset$, i.e., $S_1 \cap S_2 = \emptyset$. By Corollary 6.2, we have $T^{in}_{S_1} \cap T^{in}_{S_2} = \emptyset$, $T^{out}_{S_1} \cap T^{out}_{S_2} = \emptyset$, and $T^{equ}_{S_1} \cap T^{equ}_{S_2} = \emptyset$.

Therefore, $\eta_{S_0} = \eta_{S_1} + \eta_{S_2}$ means the truth of $S_0 = S_1 \cup S_2$, which contradicts the minimality of siphon S_0. This leads to $S^R_{1,2} = S^R_1 \cap S^R_2 \neq \emptyset$.

Next we prove $S^R_1 \setminus S^R_2 \neq \emptyset$ and $S^R_2 \setminus S^R_1 \neq \emptyset$. By contradiction, suppose that $S^R_1 \subseteq S^R_2$. By $S^R_0 = S^R_1 \cup S^R_2$, we have $S^R_0 = S^R_2$, leading to $S_0 = S_2$ and furthermore $\eta_{S_0} = \eta_{S_2}$. Since S_1 is a strict minimal siphon, $\eta_{S_1} \neq \mathbf{0}$. As a result, $\eta_{S_0} = \eta_{S_2}$ contradicts the known result. Therefore, $S^R_1 \subseteq S^R_2$ is not true, i.e., $S^R_1 \setminus S^R_2 \neq \emptyset$. Similarly, $S^R_2 \setminus S^R_1 \neq \emptyset$ can be shown. □

Corollary 6.4. *Let S_0 be a strongly dependent siphon with $\eta_{S_0} = \sum^n_{i=1} \eta_{S_i}$ in N. Then $[S_0] = [S_1] + [S_2] + \cdots + [S_n]$.*

Proof. Similar to the proof of Corollary 6.3. □

Example 6.2. The Petri net shown in Fig. 6.5(a) has three strict minimal siphons: $S_1 = \{p_5, p_9, p_{12}, p_{13}\}$, $S_2 = \{p_4, p_6, p_{13}, p_{14}\}$, and $S_3 = \{p_6, p_9, p_{12}, p_{13}, p_{14}\}$. We have $\lambda_{S_1} = p_5 + p_9 + p_{12} + p_{13}$, $\lambda_{S_2} = p_4 + p_6 + p_{13} + p_{14}$, $\lambda_{S_3} = p_6 + p_9 + p_{12} + p_{13} + p_{14}$, $\eta_{S_1} = -t_2 + t_3 - t_9 + t_{10}$, $\eta_{S_2} = -t_3 + t_4 - t_8 + t_9$, and $\eta_{S_3} = -t_2 + t_4 - t_8 + t_{10}$. It is easy to verify that $\eta_{S_3} = \eta_{S_1} + \eta_{S_2}$, and $N_{ES} = 2$, which means that there are two elementary siphons. Note that $|S^R_1| = |S^R_2| = 2$ and $|S^R_3| = 3$. Accordingly, S_1 and S_2 are elementary siphons and S_3 is a strongly dependent one. In addition, $[S_1] = \{p_3, p_4\}$, $[S_2] = \{p_5, p_8\}$, and $[S_3] = \{p_3, p_4, p_5, p_8\}$. If they are treated to be multisets, we have $[S_3] = [S_1] + [S_2]$. □

Corollary 6.5. *Let S_0 be a weakly dependent siphon in a Petri net with $\eta_{S_0} = \sum^n_{i=1} \eta_{S_i} - \sum^m_{j=n+1} \eta_{S_j}$. Then $[S_0] + ([S_{n+1}] + [S_{n+2}] + \cdots + [S_m]) = [S_1] + [S_2] + \cdots + [S_n]$.*

As a typical class of control specifications in supervisory control of discrete-event systems, generalized mutual exclusion constraints (GMECs) play an important role in the synthesis of an optimal liveness-enforcing supervisor for an S[3]PR. Their definitions [12] are recalled.

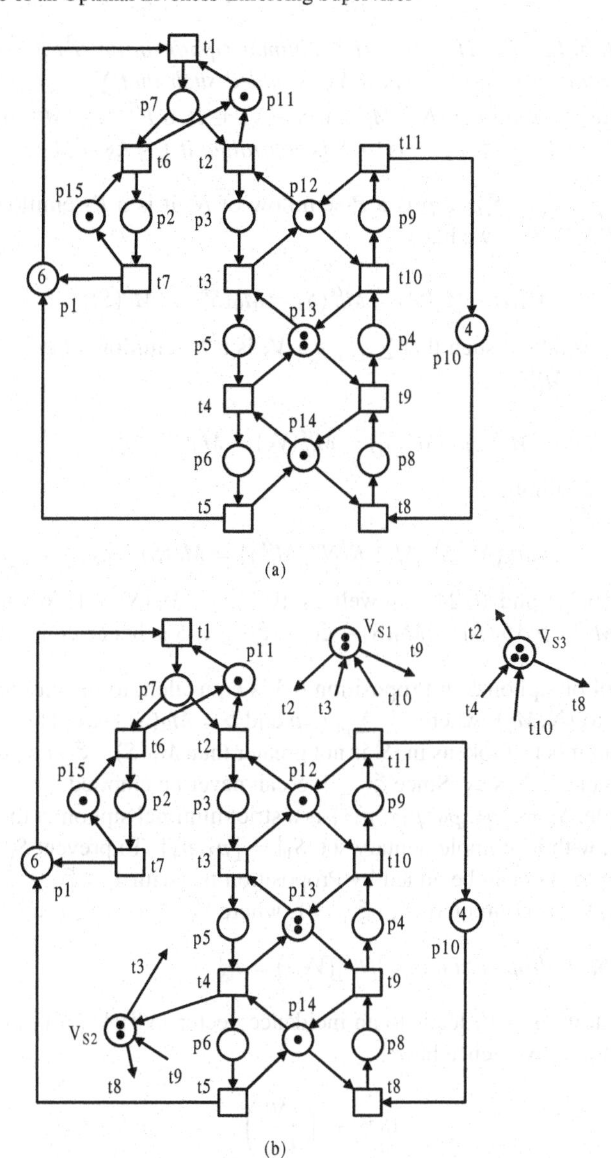

Fig. 6.5 (a) An S^3PR (N, M_0) and (b) its controlled system (N^α, M_0^α)

Definition 6.2. A single GMEC (l, b) in a net system, with place set P, defines a set of legal markings $\mathcal{M}(l,b) = \{M \in \mathbb{N}^{|P|} | l^T M \leq b\}$, where $l : P \to \mathbb{N}$ is a P-vector and $b \in \mathbb{N}^+$ is a constant.

Proposition 6.3. *Let $S \in \Pi$ be a strict minimal siphon in an S^3PR (N, M_0) with its complementary set $[S]$. A monitor V_S is added such that $\sum_{p\in[S]} p + V_S$ be a P-semiflow of the resultant net (N^α, M_0^α), where $\forall p \in P_A \cup P^0 \cup P_R$, $M_0^\alpha(p) = M_0(p)$, and $M_0^\alpha(V_S) = M_0(S) - \xi_S$ ($\xi_S \in \mathbb{N}^+$). S is controlled if $1 \le \xi_S \le M_0(S) - 1$.*

Proof. Since $\sum_{p\in S} p + \sum_{p\in[S]} p$ is a P-semiflow of N, it is a P-semiflow of N^α as well. $\forall M \in R(N^\alpha, M_0^\alpha)$, we have

$$M(S) + M([S]) = M_0^\alpha(S) = M_0(S^R) = M_0(S). \tag{6.22}$$

Monitor V_S is added such that $\sum_{p\in[S]} p + V_S$ is a P-semiflow of N^α. This means that $\forall M \in R(N^\alpha, M_0^\alpha)$,

$$M(V_S) + M([S]) = M_0^\alpha(V_S) = M_0(S) - \xi_S. \tag{6.23}$$

Equation 6.23 implies

$$max\{M([S]) | M \in R(N^\alpha, M_0^\alpha)\} = M_0(S) - \xi_S. \tag{6.24}$$

Consider (6.22) and (6.24), as well as $1 \le \xi_S \le M_0(S) - 1$; We have, $\forall M \in R(N^\alpha, M_0^\alpha)$, $M(S) \ge M_0(S) - (M_0(S) - \xi_S) = \xi_S \ge 1$. S is hence controlled. □

The control of siphon S in Proposition 6.3 is equivalent to the enforcement of a GMEC (l, b) to (N, M_0), where $l = \sum_{p\in[S]} p$ and $b = M_0(S) - \xi_S$. This implies that the maximal number of tokens in $[S]$ is not greater than $M_0(S) - \xi_S$, i.e., the minimal number of tokens in S is ξ_S. Since $\xi_S \ge 1$, S can never be emptied.

For example, $S_1 = \{p_5, p_9, p_{12}, p_{13}\}$ is a strict minimal siphon in the net shown in Fig. 6.5(a), with its complementary set $[S_1] = \{p_3, p_4\}$. To prevent S_1 from being emptied, monitor V_{S_1} can be added by Proposition 6.3 with $1 \le \xi_{S_1} \le 2$. As shown in Fig. 6.5(b), V_{S_1} is computed with $\xi_{S_1} = 1$, where $V_{S_1} + p_3 + p_4$ is a P-semiflow.

Corollary 6.6. *In Proposition 6.3, $[N^\alpha](V_S, \cdot) = \eta_S^T$.*

Proof. The addition of V_S leads to an incidence vector $[N^\alpha](V_S, \cdot)$ in $[N^\alpha]$. Let L_{V_S} denote $[N^\alpha](V_S, \cdot)$. We hence have

$$[N^\alpha] = \begin{pmatrix} [N] \\ L_{V_S} \end{pmatrix}.$$

Let $I_1 = \sum_{p\in S} p + \sum_{p\in[S]} p$, $I_2 = V_S + \sum_{p\in[S]} p$, and $I_3 = I_1 - I_2$. Clearly, $I_3 = \sum_{p\in S} p - V_S = \lambda_S - V_S$ is a P-invariant of N^α. From $I_3^T[N^\alpha] = \mathbf{0}^T$, one has $\lambda_S^T[N] - L_{V_S} = \mathbf{0}^T$, which implies the truth of this corollary. □

In Fig. 6.5(b), $[N^\alpha](V_{S_1}, \cdot) = -t_2 + t_3 - t_9 + t_{10} = \eta_{S_1}$.

Corollary 6.7. *(N^α, M_0^α) is an ES^3PR.*

Proof. Suppose that there are n monitors V_{S_1}, V_{S_2}, $\cdots$, and V_{S_n} in (N^α, M_0^α). The resulting net from deleting all monitors and their related arcs from (N^α, M_0^α) is (N, M_0). $\forall i \in \mathbb{N}_n$, V_{S_i} is added such that $V_{S_i} + [S_i]$ is a P-semiflow $I_{V_{S_i}}$ of N^α. Note that each element in $I_{V_{S_i}}$ is either one or zero and $[S_i]$ is a subset of operation places. In this sense, V_{S_i} behaves as a resource place in (N^α, M_0^α). By the definition of ES^{3}PR, this corollary is true. □

Corollary 6.8. *In Proposition 6.3, $\xi_S = 1$ implies that monitor V_S is a GMEC implementation of $(\lambda_{[S]}, M_0(S) - 1)$.*

Proof. Note that $\sum_{p \in S} p + \sum_{p \in [S]} p$ is a *P*-semiflow of (N, M_0). In order to keep S always marked at any reachable marking, the number of tokens contained in its complementary set $[S]$ at any reachable marking should be less than $M_0(S)$, i.e., $\forall M \in R(N, M_0)$, $M([S]) < M_0(S)$. As a GMEC [12], $(\lambda_{[S]}, M_0(S) - 1)$ can be implemented by an additional monitor V_S with $M_0^\alpha(V_S) = (M_0(S) - 1) - \lambda_{[S]}^T M_0$. Considering that $\forall p \in [S]$, $p \in P_A$, we have $M_0([S]) = \lambda_{[S]}^T M_0 = 0$, leading to $M_0^\alpha(V_S) = M_0(S) - 1$. □

Corollary 6.9. *The addition of V_S for S with $\xi_S = 1$ minimally restricts the behavior of (N^α, M_0^α).*

Proof. It immediately follows from Proposition 2 in [12]. □

Corollary 6.9 indicates that the addition of V_S prevents only the transition firings that the yield forbidden markings that do not satisfy the GMEC $(\lambda_{[S]}, M_0(S) - 1)$.

Theorem 6.3. *Let S be a strongly dependent siphon in an S^{3}PR (N, M_0) with $\eta_S = \eta_{S_1} + \eta_{S_2}$. Monitors V_{S_1} and V_{S_2} are added by Proposition 6.3, which leads to an augmented net system (N^α, M_0^α). Then $S_c = (\{V_{S_1}, V_{S_2}\} \cup [S_1] \cup [S_2]) \setminus P_{1,2}$ is a strict minimal siphon in N^α, where $P_{1,2} = \{p \in [S_1] \cup [S_2] | \forall t \in p^\bullet, |^\bullet t \cap (\{V_{S_1}\} \cup [S_1] \cup \{V_{S_2}\} \cup [S_2])| = 2\}$.*

Proof. It is known that $\forall i \in \{1, 2\}$, $D_i = \{V_{S_i}\} \cup [S_i]$ is a minimal siphon and trap that is initially marked in (N^α, M_0^α). Therefore, $D_1 \cup D_2 = \{V_{S_1}, V_{S_2}\} \cup [S_1] \cup [S_2]$ is also a siphon but clearly not minimal. We claim that $(D_1 \cup D_2) \setminus P_{1,2}$ is a strict minimal siphon. By $P_{1,2} \subseteq [S_1] \cup [S_2]$, S_c contains V_{S_1} and V_{S_2}. Furthermore, the strictness of S_c is ensured by removing $P_{1,2}$ from $D_1 \cup D_2$. We need to show that S_c is a minimal siphon.

Note that $\eta_S = -\eta_{[S]}$, $\eta_{S_1} = -\eta_{[S_1]}$, and $\eta_{S_2} = -\eta_{[S_2]}$. $\eta_S = \eta_{S_1} + \eta_{S_2}$ implies $\eta_{[S]} = \eta_{[S_1]} + \eta_{[S_2]}$.

Let $T_\alpha^1 = \{t | \eta_{[S]}(t) > 0\}$, $T_\alpha^2 = \{t | \eta_{[S]}(t) = 0\}$, and $T_\alpha^3 = \{t | \eta_{[S]}(t) < 0\}$. Let $T_\alpha^A = \{t \in T_\alpha^2 | \eta_{[S_1]}(t) \neq 0, \eta_{[S_2]}(t) \neq 0\}$ and $T_\alpha^B = \{t \in T_\alpha^2 | \eta_{[S_1]}(t) = 0, \eta_{[S_2]}(t) = 0\}$. According to the definition of an S^{3}PR and Proposition 6.3, we have $T_\alpha^A = \{t | t \in T, |^\bullet t \cap (D_1 \cup D_2)| = 2\}$. As a result, the removal of $P_{1,2}$ from $D_1 \cup D_2$ does not change the postset of $D_1 \cup D_2$, i.e., $(D_1 \cup D_2)^\bullet = ((D_1 \cup D_2) \setminus P_{1,2})^\bullet = S_c^\bullet$. Considering that ${}^\bullet S_c \subseteq {}^\bullet(D_1 \cup D_2) = (D_1 \cup D_2)^\bullet = S_c^\bullet$, we conclude that S_c is a siphon. Next its minimality is shown by the fact that the removal of any place p makes $S_c \setminus \{p\}$ a non-siphon. Two subcases are considered: (a) p is an operation place and (b) p is a monitor.

(a) Without loss of generality, suppose that $p \in [S_1]$. $\forall t \in p^\bullet$, $|^\bullet t \cap S_c| = 1$. Obviously, $t \in S_c^\bullet$ but the removal of p falsifies $t \in {}^\bullet S_c$. Therefore, the removal of any operation place p leads to the fact that $S_c \setminus \{p\}$ is not a siphon any more.

(b) Without loss of generality, suppose that $p = V_{S_1}$. It is proved with the aid of Fig. 6.6 showing a general case in which $p_1 \in S_c$ and $V_{S_1} \in S_c$. It is easy to see that $p_2 \notin S_c$ since otherwise we have $|^\bullet t_1 \cap S_c| = 2$. Clearly, we have $t_1 \in {}^\bullet S_c$ since $t_1 \in {}^\bullet p_1$ and $p_1 \in S_c$. However, the removal of V_{S_1} falsifies $t_1 \in S_c^\bullet$.

From the two subcases, S_c is minimal. In summary, S_c is a strict minimal siphon that contains V_{S_1} and V_{S_2}. □

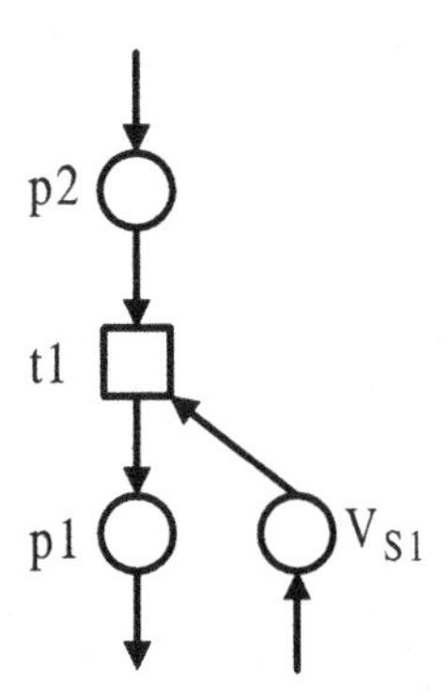

Fig. 6.6 The case of removing a monitor from S_c

A siphon that contains monitors is called a control-induced siphon. For example, there are three strict minimal siphons S_1, S_2, and S_3 with $\eta_{S_3} = \eta_{S_1} + \eta_{S_2}$ in Fig. 6.5(a). The addition of V_{S_1} and V_{S_2} leads to a strict minimal siphon $\{p_4, p_5, V_{S_1}, V_{S_2}\}$ in the net depicted in Fig. 6.5(b).

Theorem 6.4. *Let S be a strongly dependent siphon in an S^3PR (N, M_0) with $\eta_S = \eta_{S_1} + \eta_{S_2}$. Monitors V_S, V_{S_1}, and V_{S_2} are added by Proposition 6.3. The augmented net system is denoted by (N^α, M_0^α). S_c cannot be emptied if $M_0(S_{1,2}^R) > \xi_{S_1} + \xi_{S_2} - \xi_S$, where S_c is the strict minimal siphon containing V_{S_1} and V_{S_2} as stated in Theorem 6.3.*

Proof. Note that $V_S + \sum_{p \in [S]} p$, $V_{S_1} + \sum_{p \in [S_1]} p$, and $V_{S_2} + \sum_{p \in [S_2]} p$ are P-invariants of N^α. By $[S] = [S_1] + [S_2]$ and $S^R = S_1^R \cup S_2^R$, S_c cannot be emptied if $M_0^\alpha(V_S) < M_0^\alpha(V_{S_1}) + M_0^\alpha(V_{S_2})$.

$\forall i \in \{1,2\}$, $M_0^\alpha(V_{S_i}) = M_0(S_i) - \xi_{S_i} = M_0(S_i^R \setminus S_{1,2}^R) + M_0(S_{1,2}^R) - \xi_{S_i}$. We have

$$M_0^\alpha(V_{S_1}) = M_0(S_1^R \backslash S_{1,2}^R) + M_0(S_{1,2}^R) - \xi_{S_1}, \tag{6.25}$$

$$M_0^\alpha(V_{S_2}) = M_0(S_2^R \backslash S_{1,2}^R) + M_0(S_{1,2}^R) - \xi_{S_2}. \tag{6.26}$$

$(6.25)+(6.26) \Rightarrow M_0^\alpha(V_{S_1}) + M_0^\alpha(V_{S_2}) = M_0(S_1^R \backslash S_{1,2}^R) + M_0(S_2^R \backslash S_{1,2}^R) + 2M_0(S_{1,2}^R) - \xi_{S_1} - \xi_{S_2} \Rightarrow M_0^\alpha(V_{S_1}) + M_0^\alpha(V_{S_2}) = [M_0(S_1^R \backslash S_{1,2}^R) + M_0(S_2^R \backslash S_{1,2}^R) + M_0(S_{1,2}^R)] + M_0(S_{1,2}^R) - \xi_{S_1} - \xi_{S_2}$.

Since $M_0^\alpha(V_S) + \xi_S = M_0(S) = M_0(S_1^R \backslash S_{1,2}^R) + M_0(S_2^R \backslash S_{1,2}^R) + M_0(S_{1,2}^R)$, one can have

$M_0^\alpha(V_{S_1}) + M_0^\alpha(V_{S_2}) = M_0^\alpha(V_S) + M_0(S_{1,2}^R) + \xi_S - \xi_{S_1} - \xi_{S_2}$.

$M_0(S_{1,2}^R) > \xi_{S_1} + \xi_{S_2} - \xi_S$ implies the truth of $M_0^\alpha(V_S) < M_0^\alpha(V_{S_1}) + M_0^\alpha(V_{S_2})$. □

Example 6.3. There are three strict minimal siphons S_1, S_2, and S_3 in the net shown in Fig. 6.5(a), where $S_{1,2}^R = \{p_{13}\}$ and $M_0(S_{1,2}^R) = M_0(p_{13}) = 2$. Accordingly, three monitors V_{S_1}, V_{S_2}, and V_{S_3} are added as shown in Fig. 6.5(b), where $\xi_{S_1} = \xi_{S_2} = \xi_{S_3} = 1$. $M_0(S_{1,2}^R) = M_0(p_{13}) = 2 > \xi_{S_1} + \xi_{S_2} - \xi_{S_3}$ means that siphon $S_c = \{p_4, p_5, V_{S_1}, V_{S_2}\}$ cannot be emptied in (N^α, M_0^α).

Corollary 6.10. In (N^α, M_0^α), any siphon that contains V_{S_1} and V_{S_2} is controlled if $M_0(S_{1,2}^R) > \xi_{S_1} + \xi_{S_2} - \xi_S$.

Proof. $M_0(S_{1,2}^R) > \xi_{S_1} + \xi_{S_2} - \xi_S$ implies the truth of $M_0^\alpha(V_S) < M_0^\alpha(V_{S_1}) + M_0^\alpha(V_{S_2})$.

Note that $V_S + \sum_{p \in [S]} p$, $V_{S_1} + \sum_{p \in [S_1]} p$, and $V_{S_2} + \sum_{p \in [S_2]} p$ are P-invariants of N^α and $[S] = [S_1] + [S_2]$. Thus, $M_0^\alpha(V_S) < M_0^\alpha(V_{S_1}) + M_0^\alpha(V_{S_2})$ implies that any reachable marking in (N^α, M_0^α) cannot unmark both V_{S_1} and V_{S_2}. That is to say, $\forall M \in R(N^\alpha, M_0^\alpha)$, $M(\{V_{S_1}, V_{S_2}\}) > 0$. As a result, any siphon containing V_{S_1} and V_{S_2} cannot be unmarked. □

A siphon is said to be optimally controlled if the addition of its monitor does not exclude any legal behavior of the plant model. In an S^{3}PR (N, M_0), if the control of a siphon S by adding a monitor V_S is considered to be a GMEC problem ($\forall M \in R(N, M_0), M([S]) \leq M_0(S) - 1$), it is optimally controlled when $\xi_S = 1$.

It is trivial that Theorems 6.3, 6.4 and Corollary 6.10 are true even if S_1 and S_2 are not elementary siphons.

Corollary 6.11. *(1) S, S_1, and S_2 are optimally controlled if $\xi_S = \xi_{S_1} = \xi_{S_2} = 1$. (2) S_c is optimally controlled if (i) $M_0(S_{1,2}^R) > 1$ and (ii) $\xi_S = \xi_{S_1} = \xi_{S_2} = 1$.*

Proof. (1) If $\xi_S = \xi_{S_1} = \xi_{S_2} = 1$, the control of S, S_1, and S_2 is a GMEC problem. (2) According to Theorem 6.4 and Corollary 6.9, $\xi_S = \xi_{S_1} = \xi_{S_2} = 1$ leads to the optimal controllability of S_c. □

Definition 6.3. T-vector $\eta = \eta_{S_1} + \eta_{S_2}$ in an S^{3}PR (N, M_0) is said to be optimally controlled if:

1. S_1 and S_2 are optimally controlled;
2. Any control-induced siphon containing V_{S_1} and V_{S_2} is optimally controlled;

3. If there exists a siphon $S \in \Pi$ such that $\eta_S = \eta_{S_1} + \eta_{S_2}$, then S is optimally controlled.

Example 6.4. In Fig. 6.5(a), we have $M_0(S^R_{1,2}) = 2 > 1$. As a result, all the siphons in the net in Fig. 6.5(b) are optimally controlled, i.e., $\eta_{S_3} = \eta_{S_1} + \eta_{S_2}$ is optimally controlled.

Let $\eta^{[n]} = \sum_{i=1}^{n} \eta_{S_i}$, where $\forall i \in \mathbb{N}_n$, $S_i \in \Pi_E$.

Definition 6.4. The optimal controllability of $\eta^{[n]}$ is recursively defined as follows:

1. $\eta^{[2]}$ is optimally controlled if T-vector $\eta_{S_1} + \eta_{S_2}$ is optimally controlled, as stated in Definition 6.3.
2. $\eta^{[i+1]} = \eta^{[i]} + \eta_{S_{i+1}}$ is optimally controlled if
   a. $\eta^{[i]}$ is optimally controlled;
   b. S_{i+1} is optimally controlled;
   c. If $\exists S \in \Pi$ such that $\eta_S = \eta^{[i+1]}$, S is optimally controlled and any siphon containing V_S and $V_{S_{i+1}}$ is optimally controlled.

Next we discuss the optimal controllability of the T-vector of a weakly dependent siphon S with $\eta_S = \sum_{i=1}^{n} \eta_{S_i} - \sum_{j=n+1}^{m} \eta_{S_j}$. Since an S^{3}PR is well-initially-marked, the controllability of S is independent from $\Gamma^-(S) = \sum_{j=n+1}^{m} \eta_{S_j}$. Let $\eta = \sum_{i=1}^{n} \eta_{S_i} = \eta_S + \sum_{j=n+1}^{m} \eta_{S_j}$.

Definition 6.5. $\eta_S = \sum_{i=1}^{n} \eta_{S_i} - \sum_{j=n+1}^{m} \eta_{S_j}$ is optimally controlled if both $\eta = \sum_{i=1}^{n} \eta_{S_i}$ and $\eta = \eta_S + \sum_{j=n+1}^{m} \eta_{S_j}$ are optimally controlled.

The optimal controllability of $\eta_S = \sum_{i=1}^{n} \eta_{S_i} - \sum_{j=n+1}^{m} \eta_{S_j}$ implies that S, $S_1 - S_m$, and the ones containing two or more monitors in $\{V_S, V_{S_1} - V_{S_m}\}$ are optimally controlled.

Corollary 6.12. *Let (N, M_0) be an S^{3}PR. $\forall i \in \mathbb{N}_{|\Pi|}$, a monitor V_{S_i} is added for $S_i \in \Pi$ by Proposition 6.3 and the resultant net is denoted by (N^α, M_0^α). If $\Pi_E = \Pi$, any minimal siphon containing V_{S_i} is controlled in (N^α, M_0^α).*

Proof. First, all strict minimal siphons in (N, M_0) are controlled due to the addition of monitors. $\forall i \in \mathbb{N}_{|\Pi|}$, $V_{S_i} + \sum_{p \in [S_i]} p$ is a minimal P-semiflow, implying that $\{V_{S_i}\} \cup [S_i]$ is a minimal siphon and trap that is marked at M_0^α. We claim that no strict minimal siphon in N^α contains two or more monitors. This is proved by contradiction.

First, suppose that in N^α there is a strict minimal siphon $S_\alpha = \{V_{S_1}, V_{S_2}\} \cup S_\alpha^A$, where $S_\alpha^A \subseteq P_A$ is a subset of operation places in N. Furthermore, we have $S_\alpha^A \subseteq [S_1] \cup [S_2]$ since otherwise S_α is not a siphon.

$\forall i \in \{1, 2\}$, let $D_i = \{V_{S_i}\} \cup [S_i]$, $P_{1,2} = \{p \in [S_1] \cup [S_2] | \forall t \in p^\bullet, |^\bullet t \cap (D_1 \cup D_2)| = 2\}$, and $E = S_\alpha^A \cup P_{1,2}$. Clearly, $E = [S_1] \cup [S_2]$ or, equivalently, $E = [S_1] + [S_2]$.

Let $S = \cup_{p \in E} ||I_{r_p}|| \setminus E$, where $r_p \in P_R$ is a resource place with $p \in H(r_p)$ and I_{r_p} is a minimal P-semiflow associated with resource place r_p in N. If S is a minimal siphon in N, then E is its complementary set, i.e., $[S] = E$.

Similar to the proof of Theorem 6.3, it can be shown that S is a strict minimal siphon in N with $[S] = [S_1] + [S_2]$. This leads to $\eta_{[S]} = \eta_{[S_1]} + \eta_{[S_2]}$. Since $\forall S' \in \Pi$ in N, $\eta_{S'} = -\eta_{[S']}$, we have $\eta_S = \eta_{S_1} + \eta_{S_2}$, implying that S is a strongly dependent siphon with respect to strict minimal siphons S_1 and S_2. This contradicts $\Pi_E = \Pi$ in N.

In summary, we conclude that if $\Pi_E = \Pi$ in N, N^α does not have a strict minimal siphon containing two monitors. The cases involving more monitors can be similarly proved. □

Theorem 6.5. *Let (N, M_0) be an S^3PR. $\forall S \in \Pi$, a monitor V_S is added by Proposition 6.3 with $\xi_S = 1$, and the resultant net is denoted by (N^m, M_0^m). (N^m, M_0^m) is an optimal controlled system for (N, M_0) if $\Pi_E = \Pi$.*

Proof. By Corollary 6.12, $\Pi_E = \Pi$ in (N, M_0) leads to the fact that the addition of monitors used to control siphons in (N, M_0) cannot lead to new emptiable siphons in (N^m, M_0^m). Proposition 2 in [12] indicates that monitor V_S minimally restricts the behavior of (N^m, M_0^m), in the sense that it disables only transitions whose firings yield forbidden marking M such that $\lambda_{[S]}^T M > M_0(S) - 1$ holds. As a result, if $\Pi_E = \Pi$ in (N, M_0), its siphons are therefore optimally controlled, which leads to the truth of this corollary. □

Example 6.5. The net shown in Fig. 6.7 is an S^{3}PR in which p_1 and p_6 are idle places, $p_{11} - p_{15}$ are resource places, and the others are operation places. The net has two strict minimal siphons $S_1 = \{p_5, p_{12}, p_{14}\}$ and $S_2 = \{p_7, p_{13}, p_{15}\}$ that are independent, i.e., both are elementary siphons. Hence, there exists an optimal controlled system resulting from adding two monitors V_{S_1} and V_{S_2} for S_1 and S_2, respectively, as shown in Fig. 6.8.

Theorem 6.6. *An $\mathscr{M}^*$ of an S^3PR can be computed in polynomial time if $\Pi_E = \Pi$.*

Proof. As shown in [18], a resource circuit in an S^{3}PR can derive a place set that is either a strict minimal siphon or a minimal siphon that is also an initially marked trap. Let $N = (P^0 \cup P_A \cup P_R, T, F)$ denote an S^{3}PR with $\Pi_E = \Pi$. According to the properties of an S^{3}PR, each idle or resource place is associated with a minimal siphon that is also an initially marked trap. We conclude that the number of minimal siphons in N is $|P^0| + |P_R| + |\Pi|$, not greater than $|P^0| + |P_R| + min\{|P^0| + |P_A| + |P_R|, |T|\}$.

Consider the directed graph $G_N = (V, E)$ derived from N: (1) $V = P_R$, and (2) let $r, r' \in P_R$; there is an edge in E from r to r' iff $r^\bullet \cap {}^\bullet r' \neq \emptyset$. By the definition of a resource circuit [18], finding a resource circuit in net N is equivalent to finding a cycle in G_N [21], and this can be done in $O(|P_R| + |P_R^\bullet \cap {}^\bullet P_R|)$ time [8].

Let $m = |P^0| + |P_R| + min\{|P^0| + |P_A| + |P_R|, |T|\}$ and $n = |P_R| + |P_R^\bullet \cap {}^\bullet P_R|$. In the worst case Π can be found in $O(mn)$ time that is polynomial with respect to the size of N.

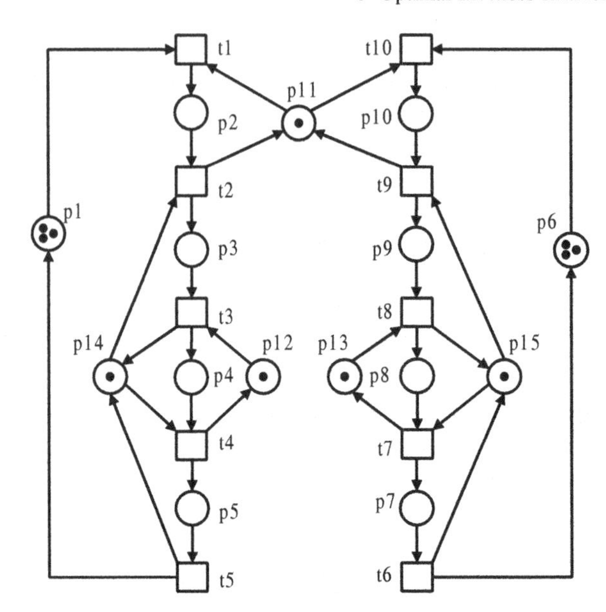

Fig. 6.7 An S^{3}PR (N, M_0) without dependent siphons

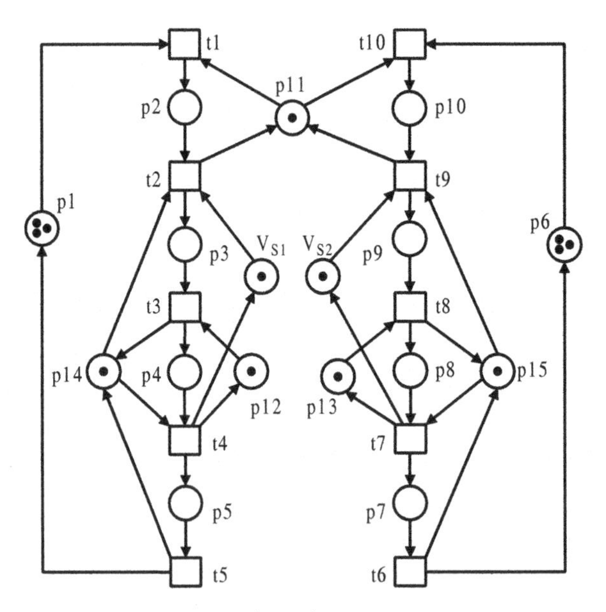

Fig. 6.8 An optimal controlled system for (N, M_0)

For each siphon S, V_S can be computed by $M_0^\alpha(V_S) = M_0(S) - 1$ and $[N^\alpha](V_S, \cdot) = \eta_S^T$. Thus, the complexity of finding an $\mathscr{M}^*$ is $O(m^2 n)$. □

Theorems 6.5 and 6.6 present the existence and computational complexity of $\mathscr{M}^*$ when there are no dependent siphons in an S^{3}PR.

Theorem 6.7. *Let (N, M_0) be an S^{3}PR. For each strict minimal siphon S, a monitor V_S is added to (N, M_0) and the resultant net system is denoted by (N^m, M_0^m). (N^m, M_0^m) is an optimal controlled system for (N, M_0) if each siphon in (N^m, M_0^m) is optimally controlled.*

Proof. This result follows due to the optimal controllability of the siphons in (N^m, M_0^m). □

Definition 6.6. *Let S be a strongly (weakly) dependent siphon with $\eta_S = \sum_{i=1}^n \eta_{S_i}$ ($\eta_S = \sum_{i=1}^n \eta_{S_i} - \eta_{j=n+1}^m \eta_{S_j}$) in N. $\forall i \in \mathbb{N}_n$ ($\forall i \in \mathbb{N}_m$), η_{S_i} is called a component of η_S.*

Definition 6.7. *An elementary siphon is said to be essential if its T-vector is not a component of the T-vector of any dependent siphon.*

Corollary 6.13. Let $N = (P_A \cup P^0 \cup P_R, T, F)$ be an S^{3}PR with initial marking M_0 and Π_E^e the set of essential siphons. There exists an $\mathscr{M}_P$ if $\forall S \in \Pi \setminus \Pi_E^e$, $\forall r \in S$, $M_0(r) \geq 2$.

Proof. Any siphon in Π_E^e can be optimally controlled. $\forall S_1, S_2 \in \Pi \setminus \Pi_E^e$, $S_1^R \cap S_2^R \neq \emptyset$ implies the truth of $M_0(S_{1,2}^R) \geq 2$. As a result, Proposition 6.3 ensures the optimal controllability of siphons in the resulting net, and thus $\mathscr{M}^*$ exists. □

Example 6.6. As stated previously, all siphons in the net shown in Fig. 6.5(b) are optimally controlled, which leads to the fact that it is an optimal controlled system for the net model in Fig. 6.5(a). Specifically, the plant net model in Fig. 6.5(a) has 188 reachable states, 168 of which are permissive states from the deadlock control point of view. That is to say, an $\mathscr{M}^*$ should lead to 168 reachable states in an optimal controlled system. It is easy to verify by using INA [27] that the net in Fig. 6.5(b) does so.

To illustrate the deadlock control of an S^{3}PR with weakly dependent siphons, we consider the net shown in Fig. 6.9. It has four emptiable minimal siphons $S_1 = \{p_4, p_{12} - p_{15}\}$, $S_2 = \{p_4, p_{11}, p_{14}, p_{15}\}$, $S_3 = \{p_5, p_{11}, p_{14} - p_{16}\}$, and $S_4 = \{p_5, p_{12} - p_{16}\}$. It is easy to find that $\eta_{S_4} + \eta_{S_2} = \eta_{S_1} + \eta_{S_3}$. We have $\eta_{S_4} = \eta_{S_1} + \eta_{S_3} - \eta_{S_2}$, implying that S_4 is weakly dependent if S_1, S_2, and S_3 are selected as elementary siphons.

The controllability of S_4 does not depend on that of S_2. First we add monitors V_{S_1}, V_{S_3}, and V_{S_4}, respectively. Note that $S_1^R \cap S_3^R = S_{1,3}^R = \{p_{14}, p_{15}\}$ and hence $M_0(S_{1,3}^R) = 2$. The addition of V_{S_1}, V_{S_3}, and V_{S_4} guarantees the optimal controllability of S_1, S_3, and V_{S_4}, respectively, with $\xi_{S_1} = \xi_{S_3} = \xi_{S_4} = 1$. Then, we add a monitor V_{S_2} for S_2 with $\xi_{S_2} = 1$.

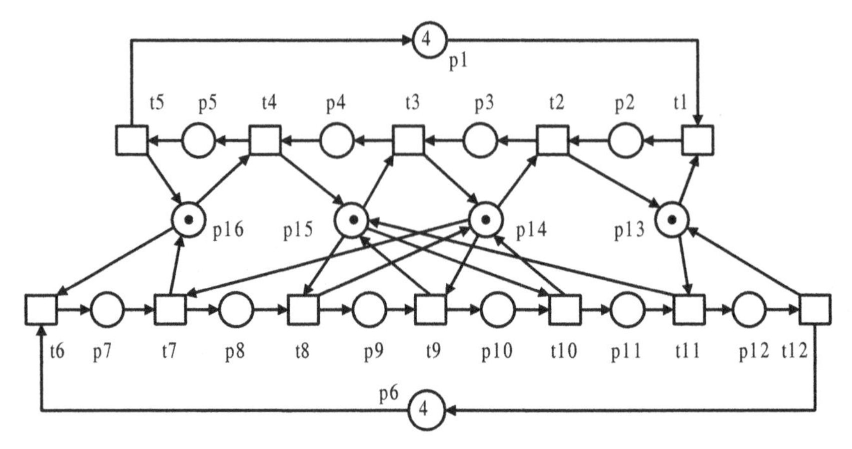

Fig. 6.9 An S^3PR (N, M_0) with a weakly dependent siphon

Table 6.1 The monitors added for the plant net in Fig. 6.9

Monitor	$M_0^m(\cdot)$	Preset	Postset
V_{S_1}	2	t_3, t_{11}	t_1, t_7
V_{S_2}	1	t_3, t_{10}	t_2, t_7
V_{S_3}	2	t_4, t_{10}	t_2, t_6
V_{S_4}	3	t_4, t_{11}	t_1, t_6

Considering $M_0(S^R_{2,4}) = M_0(S^R_{1,3}) = M_0(p_{14}) + M_0(p_{15}) = 2 > 1$, we conclude that T-vectors $\eta = \eta_{S_1} + \eta_{S_3}$ and $\eta = \eta_{S_2} + \eta_{S_4}$ are optimally controlled. These monitors are shown in Table 6.1. By Theorem 6.7, the addition of the monitors leads to an $\mathscr{M}^*$ for the plant model in Fig. 6.9. It can be verified that the plant model has 99 reachable states and its $\mathscr{M}^*$ leads to 77 ones.

6.4 Synthesis of Optimal Supervisors

This section develops an algorithm to synthesize an optimal liveness-enforcing supervisor $\mathscr{M}^*$ and the controlled system (N^m, M_0^m) for an S^3PR (N, M_0) when such a supervisor exists. With the aid of the MIP-based deadlock detection method [7], this synthesis algorithm enjoys high computational efficiency. It generates an optimal controlled system with liveness represented by (N^m, M_0^m) or reports "Undecided". The latter means that the algorithm cannot sufficiently decide whether there exists an $\mathscr{M}^*$ for a plant model (N, M_0).

Algorithm 6.1 Synthesis of an $\mathscr{M}^*$

Input: an S^3PR plant model (N, M_0)

Output: (N^m, M_0^m) or "Undecided"

```
N_1 := N
M_1 := M_0
L := |P_A| + |P^0| + |P_R|
Π_E := ∅
Π_D := ∅
Π := ∅
P_V := ∅
flag_a := 1
flag_b := 1
while (G^{MIP}(M_1) < L and flag_b = 1) do
  find a maximal unmarked siphon S_max
  Π_{S_max} := {S|S is a minimal siphon derived from S_max and ∀p ∈ S, p ∉ P_V}
  find a siphon S from Π_{S_max}\Π such that ∀S' ∈ Π_{S_max}, |S'_R| ≥ |S_R|
  Π := Π ∪ {S}
  if S is elementary with respect to Π_E then
    Π_E := Π_E ∪ {S}
    add monitor V_S for S by Proposition 6.3 with ξ_S = 1
    P_V := P_V ∪ {V_S}
    L := L+1
    the augmented net is denoted by (N^m, M_0^m)
    N_1 := N^m
    M_1 := M_0^m
  else
    Π_D := Π_D ∪ {S}
    add monitor V_S for S by Proposition 6.3 with ξ_S = 1
    if η_S is optimally controlled then
      P_V := P_V ∪ {V_S}
      L := L+1
      the augmented net is denoted by (N^m, M_0^m)
      N_1 := N^m
      M_1 := M_0^m
    else
      flag_a := 0
      flag_b := 0
    end if
  end if
end while
if (flag_a = 1) then
  N^m := N_1
  M_0^m := M_1
  output (N^m, M_0^m)
else
  output "Undecided"
end if
```

This algorithm can synthesize an $\mathcal{M}^*$ for a plant S^{3}PR model (N, M_0) if some conditions are satisfied. Note that if there is no emptiable siphon in the plant, (N, M_0) itself can be considered to be optimally controlled already. An intuitive but naive algorithm to synthesize such a supervisor is first to compute the set of strict minimal siphons Π in (N, M_0). Then, a monitor V_S is added for each siphon S in Π with $\xi_S = 1$. For each dependent siphon, if its T-vector is optimally controlled, the intuitive algorithm can output an $\mathcal{M}^*$. Note that the number of additional monitors exactly equals that of strict minimal siphons in (N, M_0), i.e., $|P_V| = |\Pi|$, which is in theory exponential with respect to the size of N.

The naive algorithm suffers from two problems: computational complexity and structure complexity. For example, as shown in [16], it takes more than 6 hours to compute all 169 strict minimal siphons in an S^{3}PR with 68 places and 54 transitions by using INA [27]. However, it takes 178 CPU seconds only for the MIP-based deadlock detection method to find 24 strict minimal siphons among the 169, where Lindo [19] is used to solve the MIP problems. The high efficiency of the MIP-based method is also fully shown in [7] by case studies.

The structural complexity of a supervisor is usually measured by the number of monitors. The intuitive algorithm adds exactly $|\Pi|$ monitors. However, the potential redundancy of some monitors motivates us to develop a method to systematically remove them.

Algorithm 6.2 Structural reduction of $\mathcal{M}^*$

Input: (N^m, M_0^m), an optimal controlled system
Output (N^{Rm}, M_0^{Rm}), a structurally reduced controlled system

suppose that $P_V = \{V_{S_i} | i \in \mathbb{N}_n\}$ is the set of monitors
$L := |P_A| + |P^0| + |P_R| + |P_V|$
$N_1 := N^m$
$M_1 := M_0^m$
$i := 1$
while $(i \leq n)$ **do**
  remove V_{S_i} and its related arcs from (N_1, M_1)
  denote the resultant net as $G = (N_2, M_2)$
  if $G^{MIP}(M_2) = L - 1$ **then**
    $L := L - 1$
    $N_1 := N_2$
    $M_1 := M_2$
    $i := i + 1$
    $P_V := P_V \setminus \{V_{S_i}\}$
  else
    $i := i + 1$
  end if
end while
$N^{Rm} := N_1$
$M_0^{Rm} := M_1$
output (N^{Rm}, M_0^{RM})

Algorithm 6.2 can sufficiently distinguish monitors whose removal does not change the liveness property of the optimal controlled system computed by Algorithm 6.1. It depends on the order in which monitors are taken into consideration. Furthermore it requires that all monitors are generated first. Note that a redundancy test algorithm for monitors is also proposed in [30], which needs the complete state enumeration.

To illustrate the supervisor synthesis and reduction algorithms, the net in Fig. 6.5(a) is taken as an example.

Example 6.7. Applying the MIP-based deadlock detection method to the net shown in Fig. 6.5(a), we can first obtain a maximal unmarked siphon $S_{max} = \{p_5, p_6, p_9, p_{12} - p_{14}\}$ from which a strict minimal siphon $S_1 = \{p_5, p_9, p_{12}, p_{13}\}$ with $M_0(S_1) = 3$ is derived. It is easy to see that $[S_1] = \{p_3, p_4\}$. By Proposition 6.3, a monitor V_{S_1} is added such that $V_{S_1} + p_3 + p_4$ is a *P*-invariant of the resultant net that is denoted by (N^α, M_0^α) with $\xi_{S_1} = 1$ and $M_0^\alpha(V_{S_1}) = 2$, as shown in Fig. 6.5(b).

The MIP-based deadlock detection method is applied to (N^α, M_0^α) and we can obtain a maximal unmarked siphon $S_{max} = \{p_4, p_6, p_9, p_{12} - p_{14}, V_{S_1}\}$. From S_{max}, we can derive a minimal siphon $S_2 = \{p_4, p_6, p_{13}, p_{14}\}$ with $[S_2] = \{p_5, p_8\}$. Since η_{S_1} and η_{S_2} are linearly independent, we add monitor V_{S_2}. The resultant net system, as shown in Fig. 6.5(b), is denoted by (N^α, M_0^α), where $M_0^\alpha(V_{S_2}) = 2$.

The same method is applied to (N^α, M_0^α). A maximal unmarked siphon $S_{max} = \{p_4, p_6, p_9, p_{12} - p_{14}, V_{S_1}, V_{S_2}\}$ is detected, from which a strict minimal siphon $S_3 = \{p_6, p_9, p_{12} - p_{14}\}$ is found. The fact $\eta_{S_3} = \eta_{S_1} + \eta_{S_2}$ indicates that S_3 is a strongly dependent siphon. A monitor V_{S_3} is added with $\xi_{S_3} = 1$. The resultant net is denoted by (N^α, M_0^α). Clearly, we have $M_0^\alpha(V_{S_3}) = 3$. Since $M_0(S_{1,2}^R) = M_0(p_{13}) = 2 > 1$, $\eta_{S_3} = \eta_{S_1} + \eta_{S_2}$ is optimally controlled. When the MIP-based deadlock detection method is applied to $G = (N^\alpha, M_0^\alpha)$ that has three monitors, we have $G^{MIP}(M_0^\alpha) = 18$, implying that no unmarked siphon can be derived. As a result, the net shown in Fig. 6.5(b) is an optimal controlled system for the plant model in Fig. 6.5(a).

For this example, Algorithm 6.2 indicates that the removal of any monitor deteriorates the liveness of the controlled system. Hence, we have the monitor set $P_V = \{V_{S_1}, V_{S_2}, V_{S_3}\}$. By $\Pi = \{S_1, S_2, S_3\}$, we conclude $|P_V| = |\Pi|$.

Remark 6.2. A hybrid policy that combines deadlock prevention and avoidance is developed in [1, 4], and [6], respectively. It consists of two stages. First, a monitor V_S is added by Proposition 6.3 for each strict minimal siphon S with $\xi_S = 1$ on condition that all strict minimal siphons in a plant net model (N, M_0) are known. Then, the exertion of an online deadlock avoidance policy depends on whether the net (N^α, M_0^α) resulting from the first stage contains deadlocks, which is verified via computing the existence of emptiable siphons in (N^α, M_0^α).

A two-stage deadlock prevention policy is developed in [13]. The first stage adds monitors to plant model (N, M_0) such that no siphons in it can be unmarked, leading to an augmented net (N^α, M_0^α). The second stage, control-induced siphon control, becomes necessary via adding monitors if (N^α, M_0^α) contains deadlocks, which is verified by solving an MIP problem.

The deadlock control strategy presented in [5] is also of two stages. The first is identical to those proposed in [1,4], and [6], leading to an augmented net (N^α, M_0^α). The second stage aims to eliminate deadlocks in (N^α, M_0^α) by properly modifying the initial markings of the monitors added in the first stage. Likewise, the second stage is put into execution provided that (N^α, M_0^α) contains deadlocks, which is verified by finding the emptiable siphons in it.

The two-stage policies mentioned above suffer from the siphon computation problem in the augmented net (N^α, M_0^α) resulting from the first stage. As is known, the number of siphons in a net is exponential with respect to its size. Since each strict minimal siphon needs a monitor to add to prevent it from being emptied or insufficiently marked, in theory, the size of (number of the monitors in) (N^α, M_0^α) is exponential with respect to the size of (N, M_0). Therefore, siphon computation in (N^α, M_0^α) is time-consuming or impossible by the above methods in the case of a large-sized plant model.

However, the results in this chapter indicate that once $\Pi_E = \Pi$ in an S^3PR plant model is true, the first stage of these two-stage policies can result in an $\mathscr{M}^*$, leading to the fact that the second stage is no longer needed.

6.5 An Example

This section presents an example that is a modified version of the FMS investigated in [10].

Example 6.8. The Petri net (N, M_0) shown in Fig. 6.10 is an S^3PR where $P^0 = \{p_1\} \cup \{p_5\} \cup \{p_{14}\} = \{p_1, p_5, p_{14}\}$, $P_A = P_{A_1} \cup P_{A_3} \cup P_{A_2} = \{p_2 - p_4\} \cup \{p_6 - p_{13}\} \cup \{p_{15} - p_{19}\}$, $P_R = \{p_{20} - p_{26}\}$, and $M_0 = 10p_1 + 15p_5 + 15p_{14} + p_{20} + 2p_{21} + p_{22} + 2p_{23} + 2p_{24} + 2p_{25} + 2p_{26}$.

The MIP-based deadlock detection method is applied to (N, M_0). A maximal unmarked siphon $S_{max} = \{p_4, p_8 - p_{12}, p_{15}, p_{17}, p_{18}, p_{20} - p_{26}\}$ can be found. Using the minimal siphon extraction algorithm [17], a minimal siphon $S_1 = \{p_{10}, p_{18}, p_{22}, p_{26}\}$ can be derived from S_{max}. By Definition 5.7, $[S_1] = \{p_{13}, p_{19}\}$. Due to Proposition 6.3, a monitor V_{S_1} is added such that $V_{S_1} + p_{13} + p_{19}$ is a P-invariant of the resultant net that is denoted by (N_1, M_1) with $\xi_{S_1} = 1$, where $M_1(V_{S_1}) = M_0(S_1) - 1 = 3 - 1 = 2$. We hence have $\Pi_E = \{S_1\}$, $\Pi_D = \emptyset$, and $\Pi = \{S_1\}$.

The MIP-based deadlock detection method is applied to (N_1, M_1). We can find a maximal unmarked siphon $S_{max} = \{p_4, p_8 - p_{10}, p_{12}, p_{15} - p_{17}, p_{20} - p_{26}, V_{S_1}\}$. A minimal siphon $S_2 = \{p_4, p_9, p_{12}, p_{17}, p_{21}, p_{24}\}$ can be derived. Since $[S_2] = \{p_2, p_3, p_8\}$, monitor V_{S_2} is accordingly added and the resultant net is denoted by (N_1, M_1). As a result, one gets $M_1(V_{S_2}) = M_0(S_2) - \xi_{S_2} = M_0(S_2) - 1 = 3$. Noticing that η_{S_1} and η_{S_2} are linearly independent, we have $\Pi_E = \{S_1, S_2\}$, $\Pi_D = \emptyset$, and $\Pi = \{S_1, S_2\}$.

Similarly, we can derive strict minimal siphons $S_3 = \{p_2, p_4, p_8, p_{13}, p_{17}, p_{21}, p_{26}\}$ and $S_4 = \{p_2, p_4, p_8, p_{12}, p_{16}, p_{21}, p_{25}\}$. Note that η_{S_1}, η_{S_2}, η_{S_3}, and η_{S_4} are lin-

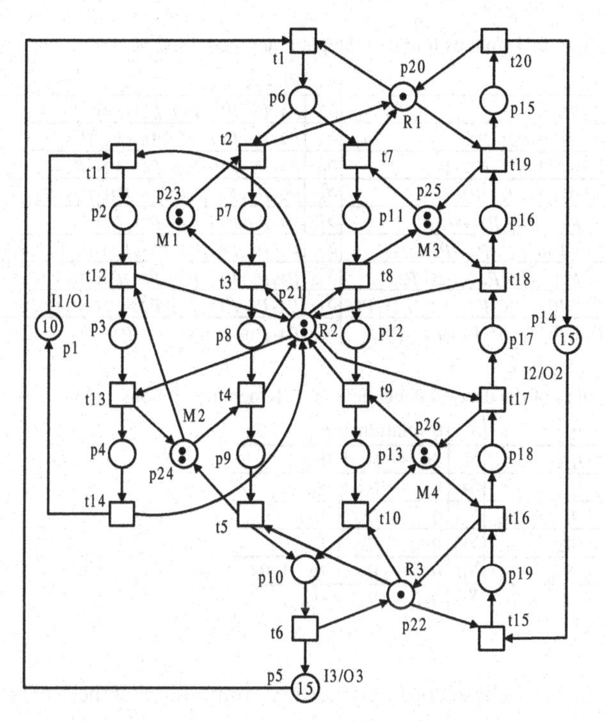

Fig. 6.10 An S³PR (N, M_0)

early independent. We have $\Pi_E = \{S_1, S_2, S_3, S_4\}$, $\Pi_D = \emptyset$, and $\Pi = \{S_1, S_2, S_3, S_4\}$. Monitors V_{S_3} and V_{S_4} are added with $\xi_{S_3} = \xi_{S_4} = 1$ and the resultant net system is denoted by (N_1, M_1).

The next strict minimal siphon that can be derived is $S_5 = \{p_2, p_4, p_8, p_{10}, p_{17}, p_{21}, p_{22}, p_{26}\}$. It is easy to verify that $\eta_{S_5} = \eta_{S_1} + \eta_{S_3}$ and S_5 is a strongly dependent siphon with respect to S_1 and S_3. Noticing $M_0(S^R_{1,3}) = M_0(p_{26}) = 2 > 1$, a monitor V_{S_5} is added with $\xi_{S_5} = 1$. Therefore, $\eta_{S_5} = \eta_{S_1} + \eta_{S_3}$ is optimally controlled.

The other strict minimal siphons derived and controlled are shown in Table 6.2 where dependent siphons are marked by stars. Linear relationships among the T-vectors of the dependent and elementary siphons are shown in Table 6.3.

To further illustrate the proposed optimal supervisor synthesis approach, we take dependent siphon S_{16} as an example, where $\eta_{S_{16}} = \eta_{S_2} + \eta_{S_3} + \eta_{S_{12}}$. Let $S_\alpha = S_7$ and $S_\beta = S_{12}$. Note that $\eta_{S_7} = \eta_{S_2} + \eta_{S_3}$ and $M_0(S^R_{2,3}) = M_0(p_{21}) = 2 > 1$. Thus, $\eta_{S_7} = \eta_{S_2} + \eta_{S_3}$ is optimally controlled when V_{S_7}, V_{S_2}, and V_{S_3} are added with $\xi_{S_7} = \xi_{S_2} = \xi_{S_3} = 1$.

Now we deal with $\eta_{S_{16}} = \eta_{S_\alpha} + \eta_{S_\beta}$. It is easy to verify that $M_0(S^R_{\alpha,\beta}) = M_0(p_{21}) = 2 > 1$, indicating that $\eta_{S_{16}} = \eta_{S_\alpha} + \eta_{S_\beta}$ is optimally controlled if monitor $V_{S_{12}}$ is added with $\xi_{S_{12}} = 1$. The optimal controllability of the T-vectors of other

Table 6.2 Strict minimal siphons found in the plant net model (N, M_0)

S	Places	S	Places
S_1	$p_{10}, p_{18}, p_{22}, p_{26}$	S_2	$p_4, p_9, p_{12}, p_{17}, p_{21}, p_{24}$
S_3	$p_2, p_4, p_8, p_{13}, p_{17}, p_{21}, p_{26}$	S_4	$p_2, p_4, p_8, p_{12}, p_{16}, p_{21}, p_{25}$
S_5^*	$p_2, p_4, p_8, p_{10}, p_{17}, p_{21}, p_{22}, p_{26}$	S_6^*	$p_4, p_9, p_{12}, p_{16}, p_{21}, p_{24}, p_{25}$
S_7^*	$p_4, p_9, p_{13}, p_{17}, p_{21}, p_{24}, p_{26}$	S_8^*	$p_2, p_4, p_8, p_{13}, p_{16}, p_{21}, p_{25}, p_{26}$
S_9	$p_4, p_{10}, p_{17}, p_{21}, p_{22}, p_{24}, p_{26}$	S_{10}^*	$p_4, p_9, p_{13}, p_{16}, p_{21}, p_{24}-p_{26}$
S_{11}^*	$p_2, p_4, p_8, p_{10}, p_{16}, p_{21}, p_{22}, p_{25}, p_{26}$	S_{12}	$p_2, p_4, p_8, p_{12}, p_{15}, p_{20}, p_{21}, p_{23}, p_{25}$
S_{13}^*	$p_4, p_{10}, p_{16}, p_{21}, p_{22}, p_{24}, p_{25}, p_{26}$	S_{14}^*	$p_4, p_9, p_{12}, p_{15}, p_{20}, p_{21}, p_{23}-p_{25}$
S_{15}^*	$p_2, p_4, p_8, p_{13}, p_{15}, p_{20}, p_{21}, p_{23}, p_{25}, p_{26}$	S_{16}^*	$p_4, p_9, p_{13}, p_{15}, p_{20}, p_{21}, p_{23}-p_{26}$
S_{17}^*	$p_2, p_4, p_8, p_{10}, p_{15}, p_{20}-p_{23}, p_{25}, p_{26}$	S_{18}^*	$p_4, p_{10}, p_{15}, p_{20} - p_{26}$

Table 6.3 T-vectors of the derived dependent and elementary siphons

S^*	η relationship	S^*	η relationship
S_5	$\eta_{S_5} = \eta_{S_1} + \eta_{S_3}$	S_6	$\eta_{S_6} = \eta_{S_2} + \eta_{S_4}$
S_7	$\eta_{S_7} = \eta_{S_2} + \eta_{S_3}$	S_8	$\eta_{S_8} = \eta_{S_3} + \eta_{S_4}$
S_{10}	$\eta_{S_{10}} = \eta_{S_2} + \eta_{S_3} + \eta_{S_4}$	S_{11}	$\eta_{S_{11}} = \eta_{S_1} + \eta_{S_3} + \eta_{S_4}$
S_{13}	$\eta_{S_{13}} = \eta_{S_9} + \eta_{S_4}$	S_{14}	$\eta_{S_{14}} = \eta_{S_2} + \eta_{S_{12}}$
S_{15}	$\eta_{S_{15}} = \eta_{S_3} + \eta_{S_{12}}$	S_{16}	$\eta_{S_{16}} = \eta_{S_2} + \eta_{S_3} + \eta_{S_{12}}$
S_{17}	$\eta_{S_{17}} = \eta_{S_1} + \eta_{S_3} + \eta_{S_{12}}$	S_{18}	$\eta_{S_{18}} = \eta_{S_9} + \eta_{S_{12}}$

dependent siphons can be accordingly verified, implying that there exists an $\mathcal{M}^*$ for the net shown in Fig. 6.10. The monitors are shown in Table 6.4.

Table 6.4 Monitors leading to an $\mathcal{M}^*$ for the net in Fig. 6.10 where $V_{S_{10}}$ and $V_{S_{16}}$ are redundant and can be removed via Algorithm 6.2

V_S	$M_0^\alpha(\cdot)$	Preset	Postset	V_S	$M_0^\alpha(\cdot)$	Preset	Postset
V_{S_1}	2	t_{10}, t_{16}	t_9, t_{15}	V_{S_2}	3	t_4, t_{13}	t_3, t_{11}
V_{S_3}	3	t_9, t_{17}	t_8, t_{16}	V_{S_4}	3	t_8, t_{18}	t_7, t_{17}
V_{S_5}	4	t_{10}, t_{17}	t_8, t_{15}	V_{S_6}	5	t_4, t_8, t_{13}, t_{18}	t_3, t_7, t_{11}, t_{17}
V_{S_7}	5	t_4, t_9, t_{13}, t_{17}	t_3, t_8, t_{11}, t_{16}	V_{S_8}	5	t_9, t_{18}	t_7, t_{16}
V_{S_9}	6	$t_5, t_{10}, t_{13}, t_{17}$	t_3, t_8, t_{11}, t_{15}	$V_{S_{10}}$	7	t_4, t_9, t_{13}, t_{18}	t_3, t_7, t_{11}, t_{16}
$V_{S_{11}}$	6	t_{10}, t_{18}	t_7, t_{15}	$V_{S_{12}}$	6	t_3, t_8, t_{19}	t_1, t_{17}
$V_{S_{13}}$	8	$t_5, t_{10}, t_{13}, t_{18}$	t_3, t_7, t_{11}, t_{15}	$V_{S_{14}}$	8	t_4, t_8, t_{13}, t_{19}	t_1, t_{11}, t_{17}
$V_{S_{15}}$	8	t_3, t_9, t_{19}	t_1, t_{16}	$V_{S_{16}}$	10	t_4, t_9, t_{13}, t_{19}	t_1, t_{11}, t_{16}
$V_{S_{17}}$	9	t_3, t_{10}, t_{19}	t_1, t_{15}	$V_{S_{18}}$	11	$t_5, t_{10}, t_{13}, t_{19}$	t_1, t_{11}, t_{15}

By Algorithm 6.2, $V_{S_{10}}$ and $V_{S_{16}}$ can be removed sequentially. As a result, for the net shown in Fig. 6.10, an $\mathcal{M}^*$ exists and consists of 16 monitors. Specifically, INA [27] indicates that the plant model in Fig. 6.10 has 108,105 reachable states, in which 11,696 are bad and deadlock states and 96,409 are good and dangerous states. Addition of the 16 or 18 monitors shown in Table 6.4 leads to an optimal controlled system with the 96,409 reachable states.

It is easy to find that there exists an $\mathcal{M}^*$ for the S^3PR net structure in Fig. 6.10 if $M_0(p_{21}) \geq 2$ and $M_0(p_{26}) \geq 2$.

Note that, in this FMS, if each of R1−R3 can hold one part only and M1−M4 can process two parts at a time, Algorithm 6.1 outputs "Undecided", indicating that we cannot synthesize an $\mathscr{M}^*$ for such a system by the proposed methods in Sect. 6.4. The outcome "Undecided" does not imply that there does not definitely exist an $\mathscr{M}^*$. Take the net in Fig. 6.5(a) as an example. When $M_0(p_{13}) = 1$, Algorithm 6.1 cannot sufficiently decide whether there exists an $\mathscr{M}^*$ for it. However, an $\mathscr{M}^*$ can be synthesized by the theory of regions.

6.6 Bibliographical Remarks

The theory of regions is first used to synthesize an optimal monitor-based liveness-enforcing supervisor for FMS modeled with Petri nets by Uzam in [28]. Briefly, Ghaffari et al. interpret it in a Petri net formalism by using plain and popular linear algebraic notions to design an optimal supervisor under liveness requirements. Recognizing its drawback, a hybrid approach that combines it with siphon control is proposed in [18] to reduce the number of marking/transition separation instances. Recent work has been done by Reveliotis et al. in which the theory of regions is used to design reversibility-enforcing supervisors for bounded Petri nets [23]. Similar work is performed by Xing et al. in [31]. The results in this chapter, however, are more general than that in [31]. Moreover, redundant monitors can be identified and removed.

Problems

6.1. Use the theory of regions to design an optimal liveness-enforcing supervisor for the nets shown in Fig. 2.7(a), and Fig. 6.1 with $M_0(p_{10}) = 2$.

6.2. Explore the initial marking condition under which there exists an optimal liveness-enforcing supervisor for the net structure shown in Fig. 5.16.

6.3. There are two S^3PR (N_1, M_{01}) and (N_2, M_{02}). Their INA files are shown below. Find their strict minimal siphons. Distinguish the elementary and dependent siphons. Discuss the existence of optimal liveness-enforcing supervisors for them. Note that in each of the two nets, p_1, p_7, and p_{13} are idle places, $p_{19}-p_{24}$ are resource places, and others are operation places.

INA description file of (N_1, M_{01}):

```
P M PRE , POST
1 5 6   , 1
2 0 1   , 2
3 0 2   , 3
4 0 3   , 4
5 0 4   , 5
```

```
6  0 5          , 6
7  5 12         , 7
8  0 7          , 8
9  0 8          , 9
10 0 9          , 10
11 0 10         , 11
12 0 11         , 12
13 5 18         , 13
14 0 13         , 14
15 0 14         , 15
16 0 15         , 16
17 0 16         , 17
18 0 17         , 18
19 1 11 17      , 10 16
20 1 10         , 9
21 1 6 9 16     , 5 8 15
22 1 2 4 12 14 , 1 3 11 13
23 1 15         , 14
24 1 3 5 8 18   , 2 4 7 17
```

INA description file of (N_2, M_{02}):

```
P  M PRE , POST
1  5  6    , 1
2  0  1    , 2
3  0  2    , 3
4  0  3    , 4
5  0  4    , 5
6  0  5    , 6
7  5  12   , 7
8  0  7    , 8
9  0  8    , 9
10 0  9    , 10
11 0  10   , 11
12 0  11   , 12
13 5  18   , 13
14 0  13   , 14
15 0  14   , 15
16 0  15   , 16
17 0  16   , 17
18 0  17   , 18
19 1 6            , 5
20 1 5 10 12 18 , 4 9 11 17
21 1 14 17        , 13 16
22 1 4 9          , 3 8
23 1 3 8 11 16    , 2 7 10 15
24 1 2 15         , 1 14
```

6.4. A correct computation of siphons in the nets in Problem 6.3 and Fig. 6.9 indicates that they contain no strongly dependent siphons. A careful inspection of these nets shows that for any weakly dependent siphon S with $\eta_S = \sum_{i=1}^{n} \eta_{S_i} - \sum_{j=n+1}^{m} \eta_{S_j}$, one has $\forall i, j \in \mathbb{N}_m$, $|S_i \cap S_j| \geq 2$. Consequently, each of the three nets has an optimal liveness-enforcing supervisor. Motivated by these facts, a natural conjecture is: If an S^{3}PR has no strongly dependent siphons, there exists an $\mathscr{M}^*$ for it. Can it be proved? If not, please find a counter-example.

6.5. There are a number of subclasses of Petri nets such as ES^{3}PR and S^{4}R that are more general than S^{3}PR. Investigate the existence of an optimal liveness-enforcing supervisor for them.

For example, the Petri net shown in Fig. 6.11 is an S^{4}R in which p_1 and p_4 are idle places, p_7 and p_8 are resources, and the others are operation places. $S = \{p_3, p_6, p_7, p_8\}$ is a minimal siphon with its complementary set $p_2 + p_5$. Note that $M_1 = 2p_1 + p2 + 2p_4 + p_5 + p_7 + p_8$ is a dead marking but $M_2 = p_1 + 2p_2 + 3p_4 + 2p_8$ and $M_3 = 3p_1 + p_4 + 2p_5 + 2p_7$ are not. From either M_2 or M_3, there exists a firable transition sequence leading the system to the initial marking. Try to compute a set of monitors whose addition eliminates the reachability of M_1 but keeps M_2 and M_3 to be reachable in the resulting supervisor.

Suppose that a forbidden marking can be transformed into a set of GMEC. The existence of an optimal supervisor is equivalent to a problem of finding a set of monitors that exactly enforce a set of GMEC. The solvability of this problem implies the existence of an optimal supervisor. Readers are recommended to tackle this problem from the viewpoint of enforcing a set of GMEC.

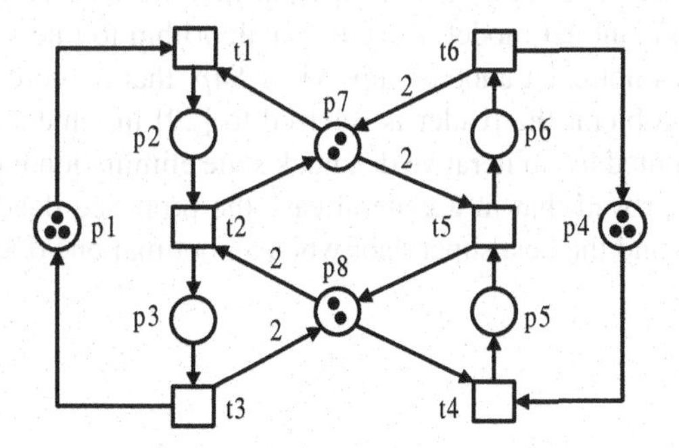

Fig. 6.11 An S^{4}R (N, M_0)

6.6. The Petri net shown in Fig. 6.12 is an S^{4}R model of an FMS, taken from [20]. There are two part types, namely J1 and J2, to be processed. The idle places are p_{10} and p_{20} and the resource places are p_{31}, p_{32}, and p_{33}. The set of operation places $P_A = \{p_{11} - p_{15}, p_{21} - p_{24}\}$. It is easy to verify by using INA that the plant net model has 363 reachable states among which 40 states are bad. In this sense, an optimal controlled system should have 323 reachable states. Try to use the theory of regions to check the existence of an optimal liveness-enforcing supervisor for the plant net model. Note that a finite-state automaton cannot always find a label-free Petri net implementation.

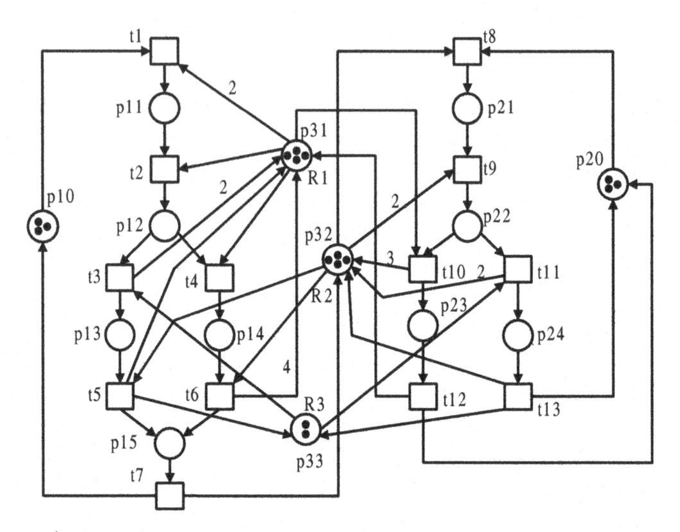

Fig. 6.12 An S^{4}R (N, M_0)

Suppose that there does not exist an optimal monitor-based liveness-enforcing supervisor given a plant net model. Develop an algorithm to find such a supervisor *Sup* that there does not exist another supervisor *Sup*$'$ that is more permissive than *Sup*. About this problem, the reader is referred to [29] in which a nearly optimal supervisor is computed by an iterative deadlock state elimination method. However, it is not formally proved that in a general case the proposed deadlock prevention policy can always find the best supervisor when an optimal one does not exist.

References

1. Abdallah, I.B., ElMaraghy, H.A. (1998) Deadlock prevention and avoidance in FMS: A Petri net based approach. *International Journal of Advanced Manufacturing Technology*, vol.14, no.10, pp.704–715.
2. Badouel, E., Darondeau, P. (1998) Theory of regions. *Lectures on Petri Nets I: Basic Models, Lecture Notes in Computer Science*, vol.1491, W. Reisig and G. Rozenberg (Eds.), pp.529–586.
3. Barkaoui, K., Lemaire, B. (1989) An effective characterization of minimal deadlocks and traps in Petri nets based on graph theory. In *Proc. 10th Int. Conf. on Application and Theory of Petri Nets*, pp.1–21.

4. Barkaoui, K., Abdallah, I.B. (1994) An efficient deadlock avoidance control policy in FMS using structural analysis of Petri nets. In *Proc. IEEE Int. Conf. on Systems, Man, and Cybernetics*, vol.1, pp.525–530.
5. Barkaoui, K., Abdallah, I.B. (1995) A deadlock prevention method for a class of FMS. In *Proc. IEEE Int. Conf. on Systems, Man, and Cybernetics*, vol.5, pp.4119–4124.
6. Barkaoui, K., Abdallah, I.B. (1995) Deadlock avoidance in FMS based on structural theory of Petri nets. In *Proc. INRIA/IEEE Symposium on Emerging Technologies and Factory Automation*, vol.2, pp.499–510.
7. Chu, F., Xie, X.L. (1997) Deadlock analysis of Petri nets using siphons and mathematical programming. *IEEE Transactions on Robotics and Automation*, vol.13, no.6, pp.793–804.
8. Cormen, T.H., Leiserson, C.E., Rivest, R.L. (1992) *Introduction to Algorithms*. The MIT Press/McGraw-Hill.
9. Darondeau, P. (2000) Region based synthesis of P/T-nets and its potential applications. In *Pro. 20th Int. Conf. on Applications and Theory of Petri Nets, Lecture Notes in Computer Science*, vol.1825, pp.16–23.
10. Ezpeleta, J, Colom, J.M., Martinez, J. (1995) A Petri net based deadlock prevention policy for flexible manufacturing systems. *IEEE Transactions on Robotics and Automation*, vol.11, no.2, pp.173–184.
11. Ghaffari, A., Rezg, N., Xie, X.L. (2003) Design of a live and maximally permissive Petri net controller using the theory of regions. *IEEE Transactions on Robotics and Automation*, vol.19, no.1, pp.137–142.
12. Giua, A., DiCesare, F., Silva, M. (1992) Generalized mutual exclusion constraints on nets with uncontrollable transitions. In *Proc. IEEE Int. Conf. on Systems, Man, and Cybernetics*, vol.2, pp.974–979.
13. Huang, Y.S., Jeng, M.D., Xie, X.L., Chung, S.L. (2001) Deadlock prevention policy based on Petri nets and siphons. *International Journal of Production Research*, vol.39, no.2, pp.283–305.
14. Lawley, M.A., Reveliotis, S.A. (2001) Deadlock avoidance for sequential resource allocation systems: Hard and easy cases. *International Journal of Flexible Manufacturing Systems*, vol.13, no.4, pp.385–404.
15. Li, Z.W., Zhou, M.C. (2004) Elementary siphons of Petri nets and their application to deadlock prevention in flexible manufacturing systems. *IEEE Transactions on Systems, Man, and Cybernetics, Part A*, vol.34, no.1, pp.38–51.
16. Li, Z.W., Zhou, M.C. (2006) Two-stage method for synthesizing liveness-enforcing supervisors for flexible manufacturing systems using Petri nets. *IEEE Transactions on Industrial Informatics*, vol.2, no.4, pp.313–325.
17. Li, Z.W., Liu, D. (2007) A correct minimal siphons extraction algorithm from a maximal unmarked siphon of a Petri net. *International Journal of Production Research*, vol.45, no.9, pp.2163–2167.
18. Li, Z.W., Zhou, M.C., Jeng, M.D. (2008) A maximally permissive deadlock prevention policy for FMS based on Petri net siphon control and the theory of regions. *IEEE Transactions on Automation Science and Engineering*, vol.5, no.1, pp.182–188.
19. Lindo, Premier Optimization Modeling Tools, http://www.lindo.com/.
20. Park, J., Reveliotis, S.A. (2001) Deadlock avoidance in sequential resource allocation systems with multiple resource acquisitions and flexible routings. *IEEE Transactions on Automatic Control*, vol.46, no.10, pp.1572–1583.
21. Paton, K. (1969) An algorithm for finding a fundamental set of cycles of a graph. *Communications of the ACM*, vol.12, no.9, pp.514–518.
22. Ramadge, P., Wonham, W.M. (1989) The control of discrete event systems. *Proceedings of the IEEE*, vol.77, no.1, pp.81–89.
23. Reveliotis, S.A., Choi, J.Y. (2006) Designing reversibility-enforcing supervisors of polynomial complexity for bounded Petri nets through the theory of regions. In *Proc. 27th Int. Conf. on Applications and Theory of Petri Nets and Other Models of Concurrency, Lecture Notes in Computer Science*, vol.4024, S. Donatelli and P. S. Thiagarajan (Eds.), pp.322–341.

24. Sreenivas, R.S. (1997) On Commoner's liveness theorem and supervisory policies that enforce liveness in free-choice Petri nets. *Systems Control Letters*, vol.31, no.1, pp.41–48.
25. Sreenivas, R.S. (1997) On the existence of supervisory control policies that enforce liveness in discrete-event dynamic systems modeled by controlled Petri nets. *IEEE Transactions on Automatic Control*, vol.42, no.7, pp.928–945.
26. Sreenivas, R.S. (1999) On supervisory policies that enforce liveness in completely controlled Petri nets with directed cut-places and cut-transitions. *IEEE Transactions on Automatic Control*, vol.44, no.6, pp.1221–1225.
27. Starke, P.H. (2003) *INA: Integrated Net Analyzer*. http://www2.informatik.hu-berlin.de/~starke/ina.html.
28. Uzam, M. (2002) An optimal deadlock prevention policy for flexible manufacturing systems using Petri net models with resources and the theory of regions. *International Journal of Advanced Manufacturing Technology*, vol.19, no.3, pp.192–208.
29. Uzam, M., Zhou, M.C. (2006) An improved iterative synthesis method for liveness enforcing supervisors of flexible manufacturing systems. *International Journal of Production Research*, vol.44, no.10, pp.1987–2030.
30. Uzam, M., Li, Z.W., Zhou, M.C. (2007) Identification and elimination of redundant control places in Petri net based liveness enforcing supervisors of FMS. *International Journal of Advanced Manufacturing Technology*, vol.35, no.1–2, pp.150–168.
31. Xing, K.Y., Hu, B.S. (2005) Optimal liveness Petri net controllers with minimal structures for automated manufacturing systems. In *Proc. IEEE Int. Conf. on Systems, Man and Cybernetics*, pp.282–287.

Chapter 7
Comparison of Deadlock Prevention Policies

Abstract This chapter, through a typical case study, reviews and compares a variety of Petri net-based deadlock prevention policies reported in the literature. Their comparison is done in terms of a resultant supervisor's structural complexity, behavior permissiveness, and computational complexity. This comparison study should help engineers to choose a suitable method for their industrial application cases. It is concluded that when behavior permissiveness is not a major concern, a policy with polynomial complexity should be the best choice for industrial-size automated systems. If the online computational cost is not a problem, an optimal deadlock prevention policy should be the best choice of industrial engineers. The deadlock prevention policies that use partial siphon enumeration achieve a good balance between the behavior permissiveness and computational complexity.

7.1 Introduction

Over the last two decades, a great deal of research has been focused on solving deadlock problems in resource allocation systems such as computer communication systems, workflow systems, and flexible manufacturing systems, resulting in a wide variety of approaches. Deadlock prevention is considered to be a well-defined problem in discrete-event systems literature, which is usually achieved by using an off-line computational mechanism to control the request for resources to ensure that deadlocks never occur. The goal of a deadlock prevention approach is to impose constraints on a system to prevent it from reaching deadlock states. In this case, the computation is carried out off-line in a static way and once the control policy is established, the system can no longer reach undesirable deadlock states. A major advantage of deadlock prevention algorithms is that they require no run-time cost since problems are solved in system design and planning stages. Due to its advantage over deadlock detection and recovery, and deadlock avoidance, deadlock prevention based on a Petri net formalism has received an enormous amount of attention in the literature. This chapter intends, through a typical case study, to review and compare a variety of Petri net-based deadlock prevention policies reported in the literature. Their comparison is done in terms of structural complexity, behavior permissiveness, and computational complexity. This comparison study should help engineers to choose a suitable method for their industrial application cases.

7.2 Applications of Deadlock Prevention Methods to a Case Study

The deadlock prevention policies that are investigated in this chapter are shown in Table 7.1. This section presents the supervisors for the FMS example in Sect. 5.4 by these deadlock prevention policies.

Table 7.1 Deadlock prevention policies

Researchers	Policy notation	Applicable nets
Abdallah and ElMaraghy [1]	AE-policy	S^4R
Barkaoui and Abdallah [2]	B1-policy	G-system
Barkaoui et al. [5]	B2-policy	S^3PR
Piroddi et al. [77]	P-policy	Petri nets
Ezpeleta et al. [19]	E-policy	S^3PR
Huang et al. [36]	H1-policy	S^3PR
Huang [38]	H2-policy	S^3PR
Li and Zhou [57]	L1-policy	S^3PR
Li et al. [63]	L2-policy	S^3PR
Li and Zhou [61]	L3-policy	S^3PR
Park and Reveliotis [73]	PR-policy	S^4R
Tricas et al. [89]	T-policy	S^4R
Uzam [92]	U1-policy	Petri nets
Uzam and Zhou [95]	U2-policy	Petri nets
Xing and Hu [101]	X-policy	S^3PR

The Petri net model of an FMS depicted in Fig. 5.8 is used to conduct the case study. It contains 18 strict minimal siphons shown in Table 7.2 again to facilitate the reader's understanding of these policies where a dependent one is marked with a star. Note that in the following tables in this chapter, monitor V_{S_i} or V_i does not necessarily correspond to siphon S_i since some policies do not need to explicitly control the siphons in Table 7.2.

7.2.1 Combination of Deadlock Prevention and Avoidance

Abdallah and ElMaraghy [1] propose a deadlock control method, AE-policy for short, for S^4R nets that are more general than S^3PR. This policy combines deadlock

Table 7.2 Strict minimal siphons in the case study ("*" means a dependent siphons after S_1, S_4, S_{10}, and $S_{16}-S_{18}$ are selected as elementary ones)

S	Places	S	Places
S_1	$p_{10}, p_{18}, p_{22}, p_{26}$	S_2^*	$p_4, p_{10}, p_{15}, p_{20}-p_{26}$
S_3^*	$p_4, p_{10}, p_{16}, p_{21}, p_{22}, p_{24}, p_{25}, p_{26}$	S_4	$p_4, p_{10}, p_{17}, p_{21}, p_{22}, p_{24}, p_{26}$
S_5^*	$p_4, p_9, p_{13}, p_{15}, p_{20}, p_{21}, p_{23}-p_{26}$	S_6^*	$p_4, p_9, p_{13}, p_{16}, p_{21}, p_{24}, p_{25}, p_{26}$
S_7^*	$p_4, p_9, p_{13}, p_{17}, p_{21}, p_{24}, p_{26}$	S_8^*	$p_4, p_9, p_{12}, p_{15}, p_{20}, p_{21}, p_{23}, p_{24}, p_{25}$
S_9^*	$p_4, p_9, p_{12}, p_{16}, p_{21}, p_{24}, p_{25}$	S_{10}	$p_4, p_9, p_{12}, p_{17}, p_{21}, p_{24}$
S_{11}^*	$p_2, p_4, p_8, p_{10}, p_{15}, p_{20}-p_{23}, p_{25}, p_{26}$	S_{12}^*	$p_2, p_4, p_8, p_{13}, p_{15}, p_{20}, p_{21}, p_{23}, p_{25}, p_{26}$
S_{13}^*	$p_2, p_4, p_8, p_{10}, p_{16}, p_{21}, p_{22}, p_{25}, p_{26}$	S_{14}^*	$p_2, p_4, p_8, p_{13}, p_{16}, p_{21}, p_{25}, p_{26}$
S_{15}^*	$p_2, p_4, p_8, p_{10}, p_{17}, p_{21}, p_{22}, p_{26}$	S_{16}	$p_2, p_4, p_8, p_{13}, p_{17}, p_{21}, p_{26}$
S_{17}	$p_2, p_4, p_8, p_{12}, p_{15}, p_{20}, p_{21}, p_{23}, p_{25}$	S_{18}	$p_2, p_4, p_8, p_{12}, p_{16}, p_{21}, p_{25}$

prevention and avoidance strategies. It consists of two stages. First, a monitor V_S is added for each strict minimal siphon S in a plant Petri net model (N, M_0) by the enforcement that $[S] \cup \{V_S\}$ is the support of a P-semiflow of the augmented net system (N^α, M_0^α), where $M_0^\alpha(V_S) = M_0(S) - 1$. In some cases, the first stage of AE-policy cannot eliminate deadlock states completely. When the occurrences of deadlocks remain possible in (N^α, M_0^α), an on-line controller that uses a dynamic resource allocation policy is employed such that the evolution of (N^α, M_0^α) can never enter an unsafe state that inevitably leads the system to deadlock.

For the net system in Fig. 5.8, 18 monitors are added due to AE-policy and their initial marking, presets and postsets are shown in Table 7.3.

Table 7.3 Monitors added using AE-policy

V_S	$M_0^\alpha(\cdot)$	Preset	Postset	V_S	$M_0^\alpha(\cdot)$	Preset	Postset
V_{S_1}	2	t_{10}, t_{16}	t_9, t_{15}	V_{S_2}	10	$t_5, t_{10}, t_{13}, t_{19}$	t_1, t_{11}, t_{15}
V_{S_3}	7	$t_5, t_{10}, t_{13}, t_{18}$	t_3, t_7, t_{11}, t_{15}	V_{S_4}	5	$t_5, t_{10}, t_{13}, t_{17}$	t_3, t_8, t_{11}, t_{15}
V_{S_5}	9	t_4, t_9, t_{13}, t_{19}	t_1, t_{11}, t_{16}	V_{S_6}	6	t_4, t_9, t_{13}, t_{18}	t_3, t_7, t_{11}, t_{16}
V_{S_7}	4	t_4, t_9, t_{13}, t_{17}	t_3, t_8, t_{11}, t_{16}	V_{S_8}	7	t_4, t_8, t_{13}, t_{19}	t_1, t_{11}, t_{17}
V_{S_9}	4	t_4, t_8, t_{13}, t_{18}	t_3, t_7, t_{11}, t_{17}	$V_{S_{10}}$	2	t_4, t_{13}	t_3, t_{11}
$V_{S_{11}}$	8	t_3, t_{10}, t_{19}	t_1, t_{15}	$V_{S_{12}}$	7	t_3, t_9, t_{19}	t_1, t_{16}
$V_{S_{13}}$	5	t_{10}, t_{18}	t_7, t_{15}	$V_{S_{14}}$	4	t_9, t_{18}	t_7, t_{16}
$V_{S_{15}}$	3	t_{10}, t_{17}	t_8, t_{15}	$V_{S_{16}}$	2	t_9, t_{17}	t_8, t_{16}
$V_{S_{17}}$	5	t_3, t_8, t_{19}	t_1, t_{17}	$V_{S_{18}}$	2	t_8, t_{18}	t_7, t_{17}

After the monitors are added, the augmented net system (N^α, M_0^α) contains deadlocks. As a result, the second stage is necessary for this example. Similar to AE-policy, a hybrid (prevention and avoidance) deadlock control approach is proposed by Barkaoui et al. [3].

Remark 7.1. The number of the monitors added in the first stage is equal to that of the strict minimal siphons in a plant model. Unfortunately, the number of such siphons in an S^3PR is theoretically exponential with respect to its size. This implies that in the first stage of AE-policy the number of the additional monitors is exponential with respect to the size of a plant model.

Remark 7.2. In AE-policy, the existence of deadlocks in $(N^{\alpha}, M_0^{\alpha})$ is confirmed by verifying the presence of strict minimal siphons in it. This means that one has to find them in $(N^{\alpha}, M_0^{\alpha})$, which becomes more time-consuming if in the first stage the concept of elementary siphons is not adopted. However, whether the second stage of AE-policy is necessary can be decided by pure structural analysis as stated in Chap. 6.

Remark 7.3. Chapter 6 also presents some conditions under which the first stage of this policy leads to an optimal liveness-enforcing supervisor. Furthermore, when the second stage is necessary, the concept of elementary siphons can be applied in the first stage, leading to a structurally simple intermediate Petri net. This would facilitate the siphon computation in the second stage. For the example investigated in this chapter, the first stage needs six monitors that are added for the elementary siphons with each siphon control variable being one. In this case, the dependent siphons in the plant model are implicitly controlled.

7.2.2 Modification of Initial Markings of Monitors

Barkaoui and Abdallah [2] develop a deadlock prevention method, called B1-policy for short, for S^3PR nets, which consists of two stages. The strict minimal siphons in a plant S^3PR are distinguished by basic strict minimal siphons and dependent strict minimal siphons. Its first stage is identical to that of AE-policy. When the deadlock-freedom or liveness in the augmented net system is not guaranteed, the second stage tries to modify the initial markings of the additional monitors added for dependent strict minimal siphons according to the proposed rules. Table 7.4 shows the monitors due to this policy for the example.

Table 7.4 Monitors added using B1-policy

V_S	$M_0^{\alpha}(\cdot)$	Preset	Postset	V_S	$M_0^{\alpha}(\cdot)$	Preset	Postset
V_{S_1}	2	t_{10}, t_{16}	t_9, t_{15}	V_{S_2}	9	$t_5, t_{10}, t_{13}, t_{19}$	t_1, t_{11}, t_{15}
V_{S_3}	6	$t_5, t_{10}, t_{13}, t_{18}$	t_3, t_7, t_{11}, t_{15}	V_{S_4}	5	$t_5, t_{10}, t_{13}, t_{17}$	t_3, t_8, t_{11}, t_{15}
V_{S_5}	8	t_4, t_9, t_{13}, t_{19}	t_1, t_{11}, t_{16}	V_{S_6}	5	t_4, t_9, t_{13}, t_{18}	t_3, t_7, t_{11}, t_{16}
V_{S_7}	3	t_4, t_9, t_{13}, t_{17}	t_3, t_8, t_{11}, t_{16}	V_{S_8}	6	t_4, t_8, t_{13}, t_{19}	t_1, t_{11}, t_{17}
V_{S_9}	3	t_4, t_8, t_{13}, t_{18}	t_3, t_7, t_{11}, t_{17}	$V_{S_{10}}$	2	t_4, t_{13}	t_3, t_{11}
$V_{S_{11}}$	8	t_3, t_{10}, t_{19}	t_1, t_{15}	$V_{S_{12}}$	6	t_3, t_9, t_{19}	t_1, t_{16}
$V_{S_{13}}$	5	t_{10}, t_{18}	t_7, t_{15}	$V_{S_{14}}$	3	t_9, t_{18}	t_7, t_{16}
$V_{S_{15}}$	3	t_{10}, t_{17}	t_8, t_{15}	$V_{S_{16}}$	2	t_9, t_{17}	t_8, t_{16}
$V_{S_{17}}$	5	t_3, t_8, t_{19}	t_1, t_{17}	$V_{S_{18}}$	2	t_8, t_{18}	t_7, t_{17}

Remark 7.4. Note that B1-policy may be problematic since it cannot lead to a liveness-enforcing supervisor for this example. This policy needs further investigation. Similar to AE-policy, the first stage of B1-policy can lead to an optimal liveness-enforcing supervisor under some conditions as presented in Chap. 6.

7.2.3 Deadlock Prevention via Proper Configuration of Initial Markings

The places in a manufacturing-oriented Petri net model are usually distinguished by idle, operation, and resource places [15], as done in augmented marked graphs [13], S^3PR [19], L-S^3PR [20], S^4R [1], S^4PR [89], WS^3PSR [87], PNR [44], RCN-merged nets [43], ERCN-merged nets [100], ERCN*-merged nets [45], S^3PGR^2 [73], G-task [5], and G-system [106]. Given a system with a fixed resource capacity, contrary to the monitor-based deadlock control methods, the Petri net model can be built by properly configuring the initial markings of its idle places such that the model itself is live. This idea can be traced back to the seminal works of Zhou, DiCesare, and Jeng in 1990s [40, 41, 103–105]. In the last decade, a fair amount of work in this direction has been done by Jeng and Xie [13, 43–45] in which the liveness of a plant model is tied to the non-existence of emptiable siphons. Chao [11] develops a maximal class of Petri nets, called non-virtual (NV) nets. An NV net is live iff there is no siphon that can become empty.

Motivated by the work mentioned above, Barkaoui et al. [5, 106] propose a deadlock prevention method, called B2-policy for short. It can derive a set of expressions about the initial markings of the idle and resource places in a G-system, under which all siphons are max-controlled and the net system itself is live or non-blocking. Since S^3PR nets are a subclass of G-systems, B2-policy can be applied to an S^3PR.

For an S^3PR (N, M_0), B2-policy first computes Π, the set of strict minimal siphons in N. $\forall S \in \Pi$, its controllability is ensured by constructing a P-invariant z_S such that $z_S^T M_0 > 0$ and $||z_S||^+ \subseteq S$. z_S is computed as follows:

$$z_S = g_S - \theta_S h_S,$$
$$g_S = \sum_{r \in S^R} I_r,$$
$$h_S = \sum_{p \in [S]} I_p,$$
$$\theta_S = max_{p \in [S] \cap ||h_S||} g_S(p),$$

where I_r is the minimal P-semiflow associated with resource place r, $[S]$ is the complementary set of siphon S, and I_p is the minimal P-semiflow associated with operation place p. I_p's support consists of idle and operation places but does not contain resource places.

Since every operation needs only one resource to perform, it is easy to prove that in an S^3PR $\forall S \in \Pi$, $\theta_S = 1$. For example, $S_1 = \{p_{10}, p_{18}, p_{22}, p_{26}\}$ is a siphon in our case study net. We have

$$I_{p_{22}} = p_{22} + p_{10} + p_{19},$$
$$I_{p_{26}} = p_{26} + p_{13} + p_{18},$$
$$g_{S_1} = I_{p_{22}} + I_{p_{26}} = p_{22} + p_{10} + p_{19} + p_{26} + p_{13} + p_{18},$$
$$[S_1] = \{p_{13}, p_{19}\},$$
$$h_{S_1} = \sum_{i=5}^{13} p_i + \sum_{i=14}^{19} p_i,$$
$$z_{S_1} = p_{22} + p_{26} - \sum_{i=5}^{9} p_i - \sum_{i=11}^{12} p_i - \sum_{i=14}^{17} p_i.$$

Considering that $\forall p \in P_A$, $M_0(p) = 0$, we have that S_1 is controlled if $z_{S_1}^T M_0 > 0$, i.e., $M_0(p_{22}) + M_0(p_{26}) - M_0(p_5) - M_0(p_{14}) > 0$. In summary, the controllability of the 18 strict minimal siphons is shown by (7.1)–(7.18), where for economy of space, we use μ_i to denote $M_0(p_i)$ where there is no confusion.

$$S_1: \mu_{22}+\mu_{26}-\mu_5-\mu_{14}>0, \tag{7.1}$$
$$S_2: \mu_{20}+\mu_{21}+\mu_{22}+\mu_{23}+\mu_{24}+\mu_{25}+\mu_{26}-\mu_1-\mu_5-\mu_{14}>0, \tag{7.2}$$
$$S_3: \mu_{21}+\mu_{22}+\mu_{24}+\mu_{25}+\mu_{26}-\mu_1-\mu_5-\mu_{14}>0, \tag{7.3}$$
$$S_4: \mu_{21}+\mu_{22}+\mu_{24}+\mu_{26}-\mu_1-\mu_5-\mu_{14}>0, \tag{7.4}$$
$$S_5: \mu_{20}+\mu_{21}+\mu_{23}+\mu_{24}+\mu_{25}+\mu_{26}-\mu_1-\mu_5-\mu_{14}>0, \tag{7.5}$$
$$S_6: \mu_{21}+\mu_{24}+\mu_{25}+\mu_{26}-\mu_1-\mu_5-\mu_{14}>0, \tag{7.6}$$
$$S_7: \mu_{21}+\mu_{24}+\mu_{26}-\mu_1-\mu_5-\mu_{14}>0, \tag{7.7}$$
$$S_8: \mu_{20}+\mu_{21}+\mu_{23}+\mu_{24}+\mu_{25}-\mu_1-\mu_5-\mu_{14}>0, \tag{7.8}$$
$$S_9: \mu_{21}+\mu_{24}+\mu_{25}-\mu_1-\mu_5-\mu_{14}>0, \tag{7.9}$$
$$S_{10}: \mu_{21}+\mu_{24}-\mu_1-\mu_5>0, \tag{7.10}$$
$$S_{11}: \mu_{20}+\mu_{21}+\mu_{22}+\mu_{23}+\mu_{25}+\mu_{26}-\mu_5-\mu_{14}>0, \tag{7.11}$$
$$S_{12}: \mu_{20}+\mu_{21}+\mu_{23}+\mu_{25}+\mu_{26}-\mu_5-\mu_{14}>0, \tag{7.12}$$
$$S_{13}: \mu_{21}+\mu_{22}+\mu_{25}+\mu_{26}-\mu_5-\mu_{14}>0, \tag{7.13}$$
$$S_{14}: \mu_{21}+\mu_{25}+\mu_{26}-\mu_5-\mu_{14}>0, \tag{7.14}$$
$$S_{15}: \mu_{21}+\mu_{22}+\mu_{26}-\mu_5-\mu_{14}>0, \tag{7.15}$$
$$S_{16}: \mu_{21}+\mu_{26}-\mu_5-\mu_{14}>0, \tag{7.16}$$
$$S_{17}: \mu_{20}+\mu_{21}+\mu_{23}+\mu_{25}-\mu_5-\mu_{14}>0, \tag{7.17}$$
$$S_{18}: \mu_{21}+\mu_{25}-\mu_5-\mu_{14}>0. \tag{7.18}$$

When each of the places modeling robots contains one token and each of the places representing machine tools has two tokens, the initial marking of idle places in this example, which satisfies (7.1)–(7.18), is $M_0(p_1)=M_0(p_5)=M_0(p_{14})=1$. As a result, the live initial marking derived by B2-policy is $M_0=p_1+p_5+p_{14}+p_{20}+p_{21}+p_{22}+2p_{23}+2p_{24}+2p_{25}+2p_{26}$. The number of reachable states under this marking is 166.

Similar to B2-policy, the live initial marking configuration approaches proposed by Chu, Xie, and Jeng [13, 43, 44, 100] fall under this class of deadlock prevention policies.

Remark 7.5. Although B2-policy does not need to add monitors to the original plant net model, the permissive behavior in the controlled system is considerably restricted. Specifically, the maximally permissive supervisor of this example has 21,581 reachable states as seen in the next subsection. B2-policy, however, only leads to 166 reachable states. That is to say, only 0.77% of maximally permissive behavior is preserved. We can conclude that this policy is rather conservative compared with the others.

Remark 7.6. The number of liveness requirement constraints derived from B2-policy is equal to that of the strict minimal siphons in an S^3PR. As a result, this number grows rapidly and in the worst case grows exponentially with respect to the size of a net model. However, there may exist redundant liveness requirement constraints whose removal does not influence the liveness condition. For example, the truth of (7.16) immediately leads to that of (7.15) due to $\mu_{22} > 0$ $(M_0(p_{22}) > 0)$. Specifically, (7.1), (7.7)–(7.10), (7.16), and (7.18) are essential for the example and then the others become redundant with respect to them.

7.2.4 A Selective Siphon Control Policy

The deadlock prevention policy, called P-policy for short, proposed by Piroddi et al. in [77] is performed in an iterative way. At each step, it needs the complete siphon enumeration. Siphons are divided into essential and dominated ones. The controllability of an essential siphon implies that of its dominated siphons. An essential siphon is distinguished by solving a set-covering problem that is known to be NP-hard. It is shown that in general P-policy cannot lead to an optimal liveness-enforcing supervisor. For the Petri net example, however, it gives an optimal supervisor whose corresponding optimal controlled system has 21,581 reachable states. The monitors are shown in Table 7.5.

Table 7.5 Monitors added using P-policy

V_S	$M_0^{\alpha}(\cdot)$	Preset	Postset
V_{S_1}	2	t_{10}, t_{16}	t_9, t_{15}
V_{S_2}	5	$t_5, t_{10}, t_{13}, t_{17}$	t_3, t_8, t_{11}, t_{15}
V_{S_3}	2	t_4, t_{13}	t_3, t_{11}
V_{S_4}	2	t_9, t_{17}	t_8, t_{16}
V_{S_5}	2	t_8, t_{18}	t_7, t_{17}
V_{S_6}	5	t_3, t_8, t_{19}	t_1, t_{17}
V_{S_7}	3	t_{10}, t_{17}	t_8, t_{15}
V_{S_8}	5	$2t_8, t_{18}$	$2t_7, t_{16}$
V_{S_9}	17	$t_3, t_5, t_8, 2t_{10}, t_{17}, 2t_{19}$	$2t_1, t_9, 2t_{15}, t_{18}$
$V_{S_{10}}$	12	$3t_8, t_{10}, 2t_{18}$	$4t_7, 2t_{15}$
$V_{S_{11}}$	27	$t_3, 2t_5, t_8, 3t_{10}, t_{13}, t_{18}, 2t_{19}$	$3t_1, t_9, t_{11}, 3t_{15}$
$V_{S_{12}}$	27	$t_3, 2t_5, t_9, 2t_{10}, t_{13}, t_{17}, 3t_{19}$	$3t_1, t_{11}, 3t_{15}, t_{18}$
$V_{S_{13}}$	8	$t_5, 2t_8, t_{13}, t_{18}$	$t_3, 2t_7, t_{11}, t_{15}$

Remark 7.7. At each iterative step, P-policy requires:

1. The complete enumeration of all minimal siphons.
2. The generation of all dominated markings.
3. The solution of a set-covering problem.

In theory, the computational complexity of each task mentioned above is exponential with respect to the size of a Petri net. However, it is claimed in [77] that:

1. An efficient siphon computation is proposed in [16] that can deal with a Petri net with more than 2×10^7 strict minimal siphons in a reasonable time.
2. The dominated markings can be found by solving an integer LPP whose size is usually very small or zero if only few essential siphons are found.
3. In practice, the set-covering problem is extremely small, and can be efficiently solved by using a mixed-integer-programming (MIP) solver.

P-policy assumes that the original Petri nets are ordinary. In some iterative step, the intermediate nets may become generalized. In this case, an intermediate net is transformed into an ordinary [47] or PT-ordinary one [39]. Then siphon control is performed. Finally, an inverse transformation is necessary.

7.2.5 Deadlock Prevention by Complete Siphon Enumeration

The work by Ezpeleta et al. [19] is usually considered to be the first one that designs monitor-based liveness-enforcing supervisors for FMS via the structural analysis of Petri nets. Their deadlock prevention approach, E-policy for short, first computes Π, the set of strict minimal siphons in an S^{3}PR (N, M_0). Then, $\forall S \in \Pi$, a monitor V_S is added such that any output arc of V_S points to the source transitions of the plant net model, which ensures the non-existence of emptiable strict minimal siphons in the augmented net system (N^α, M_0^α) and leads to a liveness-enforcing supervisor. Table 7.6 shows the monitors due to this policy for the example. The resultant controlled system has 6,287 reachable states with 18 monitors and 106 arcs.

Table 7.6 Monitors added using E-policy

V_S	$M_0^\alpha(\cdot)$	Preset	Postset	V_S	$M_0^\alpha(\cdot)$	Preset	Postset
V_{S_1}	2	t_2,t_{10},t_{16}	t_1,t_{15}	V_{S_2}	2	t_2,t_8,t_{18}	t_1,t_{15}
V_{S_3}	5	t_3,t_8,t_{19}	t_1,t_{15}	V_{S_4}	2	t_4,t_7,t_{13}	t_1,t_{11}
V_{S_5}	4	t_4,t_8,t_{13},t_{18}	t_1,t_{11},t_{15}	V_{S_6}	7	t_4,t_8,t_{13},t_{19}	t_1,t_{11},t_{15}
V_{S_7}	2	t_2,t_9,t_{17}	t_1,t_{15}	V_{S_8}	4	t_2,t_9,t_{18}	t_1,t_{15}
V_{S_9}	7	t_3,t_9,t_{19}	t_1,t_{15}	$V_{S_{10}}$	4	t_4,t_9,t_{13},t_{17}	t_1,t_{11},t_{15}
$V_{S_{11}}$	6	t_4,t_9,t_{13},t_{18}	t_1,t_{11},t_{15}	$V_{S_{12}}$	9	t_4,t_9,t_{13},t_{19}	t_1,t_{11},t_{15}
$V_{S_{13}}$	3	t_2,t_{10},t_{17}	t_1,t_{15}	$V_{S_{14}}$	5	t_2,t_{10},t_{18}	t_1,t_{15}
$V_{S_{15}}$	8	t_3,t_{10},t_{19}	t_1,t_{15}	$V_{S_{16}}$	5	t_5,t_{10},t_{13},t_{17}	t_1,t_{11},t_{15}
$V_{S_{17}}$	7	t_5,t_{10},t_{13},t_{18}	t_1,t_{11},t_{15}	$V_{S_{18}}$	10	t_5,t_{10},t_{13},t_{19}	t_1,t_{11},t_{15}

Remark 7.8. As is gradually recognized, E-policy suffers from a number of problems: structural complexity, behavior permissiveness, and computational complexity. First of all, the number of additional monitors is equal to that of strict minimal siphons. That is to say, the structural complexity of the supervisor is exponential with respect to the size of a plant net model. Second, this policy is rather

conservative. The supervisor resulting from E-policy for the example leads only to 6,287/21,581 = 29.13% of maximally permissive behavior. Finally, it needs complete siphon enumeration whose computation is time-consuming or impossible when the size of a plant is large. The subsequent deadlock control policies in the literature aim to address these problems.

Remark 7.9. It can be verified that some additional monitors in the controlled system obtained by E-policy can be removed, while the liveness of the resultant net system is preserved. Before the development of elementary siphons, fortunately, there is an established tool inside Petri net theory, which can be used to remove redundant places from a Petri net. These redundant places are called implicit places [32,78,85], which have the property that their addition to or removal from a net system does not change its behavior, i.e., an implicit place is redundant from the viewpoint of system behavior. However, the condition to decide the implicitness of a monitor in a net supervisor seems more difficult to meet than the one based on the elementary siphon theory (a formal proof has not been found yet). That is to say, an additional monitor can be removed by the elementary siphon-based methodology, but it does not satisfy the condition of an implicit place developed in [78] and [85].

7.2.6 Two-Stage Deadlock Control

As is well known, the number of siphons in a Petri net grows rapidly and in the worst case grows exponentially with respect to its size. In theory, the number of strict minimal siphons in an S^3PR is also exponential with respect to its size. Hence, their computation is expensive. Motivated by this well-recognized fact, Huang et al. [36] propose a two-stage deadlock prevention policy, called H1-policy for short, in which the complete siphon enumeration of a plant S^3PR model is avoided.

H1-policy proceeds in an iterative way. At each step, the MIP-based deadlock detection method that is first proposed in [13] is used to find a maximal emptiable siphon in an S^3PR (N, M_0), from which a strict minimal siphon, denoted by S, is derived. A monitor V_S is added for S by the enforcement that $[S] \cup \{V_S\}$ is the support of a P-semiflow of the augmented net system, as done in AE-policy. The siphon identification and monitor addition are performed iteratively until no siphons in (N, M_0) can be emptied. Then, the first stage terminates. This termination leads to an augmented net system (N^α, M_0^α).

The existence of deadlocks in (N^α, M_0^α) is verified by computing its strict minimal siphons, as done in AE-policy and B1-policy. If (N^α, M_0^α) has deadlock states, the second stage is initiated. Similarly, the MIP-based deadlock detection method is used to find a strict minimal siphon. Then a monitor is added to make it controlled. However, the siphon control method in the second stage of H1-policy is different from the one in the first. Any output arc of the monitors added in the second stage points to the source transitions of the plant model, as done in E-policy. Since the monitors added in the second stage cannot lead to problematic siphons any more, the second stage can converge rapidly with respect to the elimination of deadlock states. Note that a siphon is said to be problematic if its insufficient markedness (being empty in the case of ordinary Petri nets) is tied to a deadlock.

For the example, as shown in Table 7.7, H1-policy adds 15 monitors in the first stage and only one monitor in the second stage, which result in a liveness-enforcing supervisor, leading to 12,656 reachable states in the controlled system [58]. Note that V_{S^*} is added for siphon $S^* = \{p_{10}, p_{16}, p_{22}, p_{25}, p_{26}, V_{S_{16}}, V_{S_{18}}\}$ in the second stage.

Table 7.7 Monitors added using H1-policy

V_S	$M_0^\alpha(\cdot)$	Preset	Postset	V_S	$M_0^\alpha(\cdot)$	Preset	Postset
V_{S_1}	2	t_{10}, t_{16}	t_9, t_{15}	V_{S_4}	5	$t_5, t_{10}, t_{13}, t_{17}$	t_3, t_8, t_{11}, t_{15}
V_{S_6}	6	t_4, t_9, t_{13}, t_{18}	t_3, t_7, t_{11}, t_{16}	V_{S_7}	4	t_4, t_9, t_{13}, t_{17}	t_3, t_8, t_{11}, t_{16}
V_{S_8}	7	t_4, t_8, t_{13}, t_{19}	t_1, t_{11}, t_{17}	V_{S_9}	4	t_4, t_8, t_{13}, t_{18}	t_3, t_7, t_{11}, t_{17}
$V_{S_{10}}$	2	t_4, t_{13}	t_3, t_{11}	$V_{S_{11}}$	8	t_3, t_{10}, t_{19}	t_1, t_{15}
$V_{S_{12}}$	7	t_3, t_9, t_{19}	t_1, t_{16}	$V_{S_{13}}$	5	t_{10}, t_{18}	t_7, t_{15}
$V_{S_{14}}$	4	t_9, t_{18}	t_7, t_{16}	$V_{S_{15}}$	3	t_{10}, t_{17}	t_8, t_{15}
$V_{S_{16}}$	2	t_9, t_{17}	t_8, t_{16}	$V_{S_{17}}$	5	t_3, t_8, t_{19}	t_1, t_{17}
$V_{S_{18}}$	2	t_8, t_{18}	t_7, t_{17}	V_{S^*}	7	$t_2, t_5, 2t_{10}, 2t_{17}$	$2t_1, 2t_{15}$

Remark 7.10. H1-policy is the first one that does not need the complete siphon enumeration, which can lead to more permissive behavior than E-policy. It is worthy to note that the permissive behavior of the resultant supervisor depends on the siphons controlled in the second stage. This is to say, selecting different siphons in the second stage to control may lead to the supervisors with different permissive behavior, as shown in the next subsection.

Remark 7.11. For an S^3PR, the second stage of H1-policy in general leads to a generalized Petri net supervisor, whose liveness is ensured by the max-controllability of its siphons. Furthermore, it is clear that under some conditions the first stage of H1-policy can lead to an optimal liveness-enforcing supervisor. These conditions, for example, include the absence of dependent siphons in an S^3PR plant model.

7.2.7 Two-Stage Deadlock Control with Elementary Siphons

Recognizing the possible existence of dependent siphons in a Petri net model and their role in deadlock prevention, Huang [38] proposes a deadlock prevention policy, called H2-policy for short, for S^3PR nets, which is an improved version of H1-policy. The major difference of the two policies lies in their first stages. The first stage of H2-policy finds the set of elementary siphons in an S^3PR net model provided that all strict minimal siphons are known. Then, a monitor is added for each elementary siphon. Compared with H1-policy, this implies that fewer monitors are needed in the first stage of H2-policy. Table 7.8 shows the monitors due to H2-policy for the example.

In Table 7.8, V_{S_1}, V_{S_4}, $V_{S_{10}}$, and $V_{S_{16}}-V_{S_{18}}$ are the monitors added in the first stage. This is not surprising since there are only six elementary siphons in the example. As in H1-policy, the second stage of H2-policy adds one monitor only. The supervisor due to the additional monitors in Table 7.8 leads to 16,425 reachable states. The permissive behavior of the two supervisors resulting from H1- and H2-policies is different. This is caused by the fact that the siphon controlled in the second stage of H1-policy is different from the one of H2-policy.

Table 7.8 Monitors added using H2-policy

V_S	$M_0^\alpha(\cdot)$	Preset	Postset	V_S	$M_0^\alpha(\cdot)$	Preset	Postset
V_{S_1}	2	t_{10},t_{16}	t_9,t_{15}	V_{S_4}	5	t_5,t_{10},t_{13},t_{17}	t_3,t_8,t_{11},t_{15}
$V_{S_{10}}$	2	t_4,t_{13}	t_3,t_{11}	$V_{S_{16}}$	2	t_9,t_{17}	t_8,t_{16}
$V_{S_{17}}$	5	t_3,t_8,t_{19}	t_1,t_{17}	$V_{S_{18}}$	2	t_8,t_{18}	t_7,t_{17}
V_{S^*}	8	$2t_2,2t_{10},3t_{18}$	$2t_1,3t_{15}$				

Remark 7.12. Although H2-policy in general can lead to a structurally simple supervisor by considering the existence of elementary siphons in a plant model, the complete siphon enumeration becomes necessary again, which increases the computational complexity of this policy. However, the first stage of H2-policy can be improved, avoiding complete siphon enumeration, by the following steps:

1. Find a maximal unmarked siphon by the MIP-based deadlock detection approach, from which a strict minimal siphon S is derived.
2. If S is elementary, add V_S for it; otherwise, check the controllability of S. If it is not controlled, add a monitor for it.
3. Repeat the above two steps until all siphons in (N,M_0) are controlled.

Such an improved H2-policy has the same computational complexity with H1-policy, but in general it leads to a supervisor with less additional monitors.

Remark 7.13. In theory, it cannot be claimed that H2-policy is more permissive than H1-policy although it is true for the example. The permissive behavior of the supervisor due to H1(H2)-policy depends on the siphons controlled in its second stage. That is to say, if the siphons controlled in the second stage are identical, the resultant supervisors lead to the same permissive behavior.

7.2.8 A Policy Based on Elementary Siphons

According to E-policy, the number of additional monitors in the supervisor of an S^{3}PR is, in theory, exponential with respect to its size. That is to say, E-policy needs too many monitors when the size of a plant is large. To address this issue, Li and Zhou propose the concept of elementary and redundant siphons [57]. Redundant siphons are later renamed as dependent siphons [62]. In an S^{3}PR, a dependent siphon can be implicitly controlled by properly supervising its related elementary siphons.

For the example in this case study, the work in [57] shows that dependent siphons can never be emptied when we add monitors for elementary siphons only. As a result, the elementary siphon-based supervisor of the example has six monitors only that are shown in Table 7.9. The corresponding controlled system has 6,287 reachable states.

Table 7.9 Monitors added using L1-policy

V_S	$M_0^\alpha(\cdot)$	Preset	Postset	V_S	$M_0^\alpha(\cdot)$	Preset	Postset
V_{S_1}	2	t_2, t_{10}, t_{16}	t_1, t_{15}	V_{S_2}	2	t_4, t_7, t_{13}	t_1, t_{11}
V_{S_3}	2	t_2, t_9, t_{17}	t_1, t_{15}	V_{S_4}	2	t_2, t_8, t_{18}	t_1, t_{15}
V_{S_5}	5	$t_5, t_{10}, t_{13}, t_{17}$	t_1, t_{11}, t_{15}	V_{S_6}	5	t_3, t_8, t_{19}	t_1, t_{15}

Remark 7.14. In general cases, L1-policy can lead to a structurally simple liveness-enforcing supervisor. However, it suffers from the computational complexity and behavior permissiveness issues as in E-policy. Both need the complete siphon enumeration and have the same permissive behavior.

7.2.9 An Iterative Policy Based on Elementary Siphons

By using a small number of additional monitors, L1-policy can lead to a liveness-enforcing supervisor for an S^{3}PR, which leads to the same permissive behavior as E-policy. L1-policy suffers from two problems: computational complexity and restricted behavior. The development of L2-policy in [63] aims to address these two problems.

First, the MIP-based deadlock detection method is applied to a plant S^{3}PR net model, leading to a maximal unmarked siphon on condition that the plant model itself is not live. From the maximal unmarked siphon, a strict minimal siphon S is derived. If S is an elementary siphon, then a monitor is added such that S is controlled by the siphon control approach in E-policy. If S is dependent with respect to the elementary siphons that are already found, its controllability is ensured by properly setting the control depth variables of its elementary siphons. That is to say, we do not need to explicitly add monitors for dependent siphons in L2-policy.

The process of siphon identification and control proceeds iteratively until, at some step, the MIP-based deadlock detection method indicates that all siphons in the augmented net system (N^α, M_0^α) are controlled. The absence of emptiable siphons in (N^α, M_0^α) implies its liveness. Table 7.10 shows the monitors due to L2-policy for the example. The controlled system has 6,331 reachable states.

Remark 7.15. L2-policy improves the computational complexity of L1-policy by avoiding the complete siphon enumeration. In summary, this policy leads to a supervisor with a small number of monitors via partial siphon enumeration.

Table 7.10 Monitors added using L2-policy

V_S	$M_0^\alpha(\cdot)$	Preset	Postset	V_S	$M_0^\alpha(\cdot)$	Preset	Postset
V_{S_1}	2	t_2,t_{10},t_{16}	t_1,t_{15}	V_{S_2}	2	t_4,t_7,t_{13}	t_1,t_{11}
V_{S_3}	2	t_2,t_9,t_{17}	t_1,t_{15}	V_{S_4}	2	t_2,t_8,t_{18}	t_1,t_{15}
V_{S_5}	5	t_5,t_{10},t_{13},t_{17}	t_1,t_{11},t_{15}				

7.2.10 A More Permissive Policy Based on Elementary Siphons

Although L2-policy does not need complete siphon enumeration, the behavior of the resultant supervisor is quite restricted since the output arcs of the additional monitors point to the source transitions that represent the entry of raw parts of the system modeled by an S^3PR.

The deadlock prevention approach proposed in [61], called L3-policy for short, is in fact an improved version of L2-policy by the observation that, in many cases, a liveness-enforcing supervisor can be obtained even if the output arcs of the monitors point to non-source transitions. L3-policy consists of two stages: (1) siphon identification and control, and (2) rearranging the output arcs of the monitors.

By the MIP-based deadlock detection method, L3-policy first finds an elementary siphon without complete siphon enumeration. Then, a monitor is added such that it is controlled by the siphon control approach in E-policy.

This process is performed iteratively until the MIP-based deadlock detection method shows that there are no emptiable siphons in the augmented net, denoted by (N^β, M_0^β), i.e., the net is live.

The second stage of L3-policy is to rearrange the output arcs of the monitors in (N^β, M_0^β). This is done by making the arcs point to the transitions that are as far away from the source transitions as possible provided that this arrangement does not lead to emptiable siphons. Table 7.11 shows the monitors due to L3-policy for the FMS example. The supervisor leads to 15,999 reachable states.

Table 7.11 Monitors added using L3-policy

V_S	$M_0^\alpha(\cdot)$	Preset	Postset	V_S	$M_0^\alpha(\cdot)$	Preset	Postset
V_{S_1}	2	t_{10},t_{16}	t_9,t_{15}	V_{S_2}	2	t_4,t_{13}	t_3,t_{11}
V_{S_3}	2	t_9,t_{17}	t_8,t_{15}	V_{S_4}	2	t_8,t_{18}	t_7,t_{15}
V_{S_5}	5	t_5,t_{10},t_{13},t_{17}	t_3,t_8,t_{11},t_{15}	V_{S_6}	5	t_3,t_8,t_{19}	t_1,t_{15}

Remark 7.16. L3-policy aims to improve E-policy from the viewpoint of computational complexity, behavior permissiveness, and structural complexity. From both theoretical and practical standpoints, this policy seems to simultaneously address these issues in a reasonable way.

7.2.11 A Policy of Polynomial Complexity

Ideally, due to performance considerations, a deadlock prevention policy should be developed such that the behavior of the controlled system is as permissive as possible. However, it has been formally established that, in general, the implementation of an optimal (maximally permissive) deadlock control policy is an NP-hard problem. This is so for the class of Petri nets considered in this case study. As stated previously, researchers have used the MIP-based deadlock detection method to derive a number of deadlock prevention policies without the complete siphon enumeration. However, the deadlock control methods that are based on either complete or partial siphon enumeration are computationally expensive in theory. Although the case study in [13] and [61] shows that the MIP-based method is encouraging when it is applied to large systems, solving an MIP problem is, after all, NP-hard in theory.

Accordingly, it is significant, from both practical and theoretical standpoints, to develop a suboptimal but computationally efficient deadlock prevention policy for a system. The work by Park and Reveliotis [73] falls under this class. A key advantage and distinguished characteristic of the deadlock prevention policy in [73], PR-policy for short, with respect to the similar attempts existing in the literature is that PR-policy is of polynomial complexity.

PR-policy considers a system that is formally defined by a set of resources $R = \{r_i | i \in \mathbb{N}_m\}$ and a set of jobs $J = \{J_j | j \in \mathbb{N}_n\}$. Each resource type r_i has a capacity $C_i \in \mathbb{N}^+$. Each job type J_j is defined by a set of operations $\{p_{jk} | k \in \mathbb{N}_{\lambda_j}, \lambda_j \in \mathbb{N}^+\}$, which is partially ordered through a set of precedence constraints. Each job operation p_{jk} is associated with a conjunctive resource allocation requirement, formally expressed by an m-dimensional vector $a_{p_{jk}}$, with $a_{p_{jk}}[i]$, $i \in \mathbb{N}_m$, indicating how many units of resource r_i are required to support the operation execution. Such a system can be modeled by a class of Petri nets, S^{4}R that is more general than S^{3}PR. Let (N, M_0) denote an S^{4}R, where $N = (P^0 \cup P_A \cup P_R, T, F, W)$, and P^0 (P_A; P_R) is the set of idle (operation; resource) places. In order to fully understand its idea, the policy is reviewed as follows although it has been presented in Chap. 4.

Let $o_i \equiv O(r_i)$, $O : R \rightarrow \mathbb{N}_m$ be any partial order imposed on the resource set R. Given $p \in P_A$, let $\rho_p^{max} = max\{o_i | a_p[i] > 0, i \in \mathbb{N}_m\}$ and $\rho_p^{min} = min\{o_i | a_p[i] > 0, i \in \mathbb{N}_m\}$. Also, let $L_p = \{q | q \in p^{\bullet\bullet} \cap P_A \wedge \rho_q^{max} = min_{v \in p^{\bullet\bullet} \cap P_A} \rho_v^{max}\}$. By convention, $L_p = \emptyset$ if $p^{\bullet\bullet} \cap P^0 \neq \emptyset$. Then:

1. The neighborhood set N_p of $p \in P_A$ is defined by $N_p = \{p\} \cup \{q | q \in \cup_{v \in L_p} N_v \wedge \rho_p^{min} \leq \rho_q^{max}\}$.
2. The adjusted resource allocation requirement $\hat{a}_p$ for $p \in P_A$ under partial order $O()$ (resource ordering) is given by $\hat{a}_p[i] = max\{a_q[i] | q \in N_p\}$ if $o_i \geq \rho_p^{min}$; otherwise $\hat{a}_p[i] = 0$, $\forall i \in \mathbb{N}_m$.
3. The policy-imposed constraint on the system operation is expressed by the requirement that no resource is over-allocated with respect to the adjusted operation requirements specified by $\hat{a}_p[i]$.

The liveness requirements of a system can be represented by the inequality constraints taking the form of

$$\hat{A}_p \cdot M_P \leq f_p$$

where the column vector in $\hat{A}_p$ corresponding to an operation place p is $\hat{a}_p$, vector M_P is the restriction of marking M to the operation places, and f_p is the capacity vector of the resources, i.e., $f_p(i) = C_i$, $i \in \mathbb{N}_{|R|}$.

The behavior of the supervisor depends on the selected resource ordering. A method to find an optimal resource ordering is proposed such that the resultant supervisor is as permissive as possible. For the example investigated in this chapter, the resource ordering that we select is $o_1 = 1, o_2 = 1, o_3 = 1, o_4 = 2, o_5 = 2, o_6 = 2$, and $o_7 = 2$.

The conjunctive resource requirements of the operations are as follows:

$$\begin{aligned}
a_{p_2} &= (0,1,0,0,0,0,0)^T,\ a_{p_3} = (0,0,0,0,1,0,0)^T,\ a_{p_4} = (0,1,0,0,0,0,0)^T,\\
a_{p_6} &= (1,0,0,0,0,0,0)^T,\ a_{p_7} = (0,0,0,1,0,0,0)^T,\ a_{p_8} = (0,1,0,0,0,0,0)^T,\\
a_{p_9} &= (0,0,0,0,1,0,0)^T,\ a_{p_{10}} = (0,0,1,0,0,0,0)^T,\ a_{p_{11}} = (0,0,0,0,0,1,0)^T,\\
a_{p_{12}} &= (0,1,0,0,0,0,0)^T,\ a_{p_{13}} = (0,0,0,0,0,0,1)^T,\ a_{p_{15}} = (1,0,0,0,0,0,0)^T,\\
a_{p_{16}} &= (0,0,0,0,0,1,0)^T,\ a_{p_{17}} = (0,1,0,0,0,0,0)^T,\ a_{p_{18}} = (0,0,0,0,0,0,1)^T,\\
a_{p_{19}} &= (0,0,1,0,0,0,0)^T.
\end{aligned}$$

In this net, three job types are distinguished: $J_1 = \{p_2, p_3, p_4\}$, $J_2 = \{p_6, p_7, p_8, p_9, p_{10}, p_{11}, p_{12}, p_{13}\}$, and $J_3 = \{p_{15}, p_{16}, p_{17}, p_{18}, p_{19}\}$. Under the resource ordering $o_1 = 1, o_2 = 1, o_3 = 1, o_4 = 2, o_5 = 2, o_6 = 2$,and $o_7 = 2$, first, we obtain

$$\begin{aligned}
&\rho_{p_2}^{max} = \rho_{p_2}^{min} = 1,\ \rho_{p_3}^{max} = \rho_{p_3}^{min} = 2,\ \rho_{p_4}^{max} = \rho_{p_4}^{min} = 1,\ \rho_{p_6}^{max} = \rho_{p_6}^{min} = 1,\\
&\rho_{p_7}^{max} = \rho_{p_7}^{min} = 2,\ \rho_{p_8}^{max} = \rho_{p_8}^{min} = 1,\ \rho_{p_9}^{max} = \rho_{p_9}^{min} = 2,\ \rho_{p_{10}}^{max} = \rho_{p_{10}}^{min} = 1,\\
&\rho_{p_{11}}^{max} = \rho_{p_{11}}^{min} = 2,\ \rho_{p_{12}}^{max} = \rho_{p_{12}}^{min} = 1,\ \rho_{p_{13}}^{max} = \rho_{p_{13}}^{min} = 2,\ \rho_{p_{15}}^{max} = \rho_{p_{15}}^{min} = 1,\\
&\rho_{p_{16}}^{max} = \rho_{p_{16}}^{min} = 2,\ \rho_{p_{17}}^{max} = \rho_{p_{17}}^{min} = 1,\ \rho_{p_{18}}^{max} = \rho_{p_{18}}^{min} = 2,\ \rho_{p_{19}}^{max} = \rho_{p_{19}}^{min} = 1.
\end{aligned}$$

The L_p sets on the stages of job type J_1 are: $L_{p_4} = \emptyset$, $L_{p_3} = \{p_4\}$, and $L_{p_2} = \{p_3\}$.

The L_p sets on the stages of job type J_2 are: $L_{p_{10}} = \emptyset$, $L_{p_{13}} = \{p_{10}\}$, $L_{p_{12}} = \{p_{13}\}$, $L_{p_{11}} = \{p_{12}\}$, $L_{p_9} = \{p_{10}\}$, $L_{p_8} = \{p_9\}$, $L_{p_7} = \{p_8\}$, and $L_{p_6} = \{p_7, p_{11}\}$.

The L_p sets on the stages of job type J_3 are: $L_{p_{15}} = \emptyset$, $L_{p_{16}} = \{p_{15}\}$, $L_{p_{17}} = \{p_{16}\}$, $L_{p_{18}} = \{p_{17}\}$, and $L_{p_{19}} = \{p_{18}\}$.

The neighborhood sets on the stages of job type J_1 are: $N_{p_4} = \{p_4\}$, $N_{p_3} = \{p_3\}$, and $N_{p_2} = \{p_2, p_3\}$.

The neighborhood sets on the stages of job type J_2 are: $N_{p_{10}} = \{p_{10}\}$, $N_{p_{13}} = \{p_{13}\}$, $N_{p_{12}} = \{p_{12}, p_{13}\}$, $N_{p_{11}} = \{p_{11}, p_{13}\}$, $N_{p_9} = \{p_9\}$, $N_{p_8} = \{p_8, p_9\}$, $N_{p_7} = \{p_7, p_9\}$, and $N_{p_6} = \{p_6, p_7, p_9, p_{11}, p_{13}\}$.

The neighborhood sets on the stages of job type J_3 are: $N_{p_{15}} = \{p_{15}\}$, $N_{p_{16}} = \{p_{16}\}$, $N_{p_{17}} = \{p_{16}, p_{17}\}$, $N_{p_{18}} = \{p_{16}, p_{18}\}$, and $N_{p_{19}} = \{p_{16}, p_{18}, p_{19}\}$.

Once the stage neighborhood sets are computed, the stage adjusted resource allocation requirements are found directly as follows:

$\hat{a}_{p_2} = (0,1,0,0,1,0,0)^T$, $\hat{a}_{p_3} = (0,0,0,0,1,0,0)^T$, $\hat{a}_{p_4} = (0,1,0,0,0,0,0)^T$,
$\hat{a}_{p_6} = (1,0,0,1,1,1,1)^T$, $\hat{a}_{p_7} = (0,0,0,1,1,0,0)^T$, $\hat{a}_{p_8} = (0,1,0,0,1,0,0)^T$,
$\hat{a}_{p_9} = (0,0,0,0,1,0,0)^T$, $\hat{a}_{p_{10}} = (0,0,1,0,0,0,0)^T$, $\hat{a}_{p_{11}} = (0,0,0,0,0,1,1)^T$,
$\hat{a}_{p_{12}} = (0,1,0,0,0,0,1)^T$, $\hat{a}_{p_{13}} = (0,0,0,0,0,0,1)^T$, $\hat{a}_{p_{15}} = (1,0,0,0,0,0,0)^T$,
$\hat{a}_{p_{16}} = (0,0,0,0,0,1,0)^T$, $\hat{a}_{p_{17}} = (0,1,0,0,0,1,0)^T$, $\hat{a}_{p_{18}} = (0,0,0,0,0,1,1)^T$,
$\hat{a}_{p_{19}} = (0,0,1,0,0,1,1)^T$.

Thus the modified set of linear inequality constraints expressed by state vector can be shown below:

$$\hat{A}_p = \begin{bmatrix} 0\,0\,0\,1\,0\,0\,0\,0\,0\,0\,0\,0\,1\,0\,0\,0\,0 \\ 1\,0\,1\,0\,0\,1\,0\,0\,0\,1\,0\,0\,0\,1\,0\,0 \\ 0\,0\,0\,0\,0\,0\,0\,1\,0\,0\,0\,0\,0\,0\,0\,1 \\ 0\,0\,0\,1\,1\,0\,0\,0\,0\,0\,0\,0\,0\,0\,0\,0 \\ 1\,1\,0\,1\,1\,1\,1\,0\,0\,0\,0\,0\,0\,0\,0\,0 \\ 0\,0\,0\,1\,0\,0\,0\,0\,1\,0\,0\,0\,1\,1\,1\,1 \\ 0\,0\,0\,1\,0\,0\,0\,0\,1\,1\,1\,0\,0\,0\,1\,1 \end{bmatrix}. \tag{7.19}$$

Table 7.12 shows the monitors for the example due to PR-policy. The corresponding controlled system has 2,480 reachable states.

Table 7.12 Monitors added using PR-policy

V_S	$M_0^\alpha(\cdot)$	Preset	Postset	V_S	$M_0^\alpha(\cdot)$	Preset	Postset
V_1	1	t_2,t_7,t_{20}	t_1,t_{19}	V_2	1	$t_4,t_9,t_{12},t_{14},t_{18}$	$t_3,t_8,t_{11},t_{13},t_{17}$
V_3	1	t_6,t_{16}	t_5,t_{10},t_{15}	V_4	2	t_3,t_7	t_1
V_5	2	t_5,t_7,t_{13}	t_1,t_{11}	V_6	2	t_2,t_8,t_{19}	t_1,t_{15}
V_7	2	t_2,t_{10},t_{17}	t_1,t_{15}				

Remark 7.17. Although this policy is not optimal, as far as the authors of this book know, it is the first polynomial algorithm to design a monitor-based liveness-enforcing supervisor for S^{3}PR. The supervisor computed by PR-policy is not necessarily optimal in structure. That is to say, there may exist redundant monitors whose removal keeps the liveness of the resultant net system, as stated in Chap. 4. For example, monitor V_3 is certainly redundant since it has the same incidence vector and initial marking with resource place p_{22} in the original plant net model.

7.2.12 An Iterative Deadlock Prevention Policy

Deadlock control in an iterative way is an old and intuitive idea. The latest iterative deadlock prevention policy, called T-policy for short, is reported by Tricas et al. [90,91]. Similar approaches can be found in the literature [5,7,36,39,89,106].

T-policy deals with S^{4}R nets that are more general than S^{3}PR. The policy distinguishes siphons in an S^{4}R by good and bad ones. The insufficient markedness of the latter is closely related to potential deadlocks. At each iteration, a bad siphon S is obtained by solving an integer LPP, and then a monitor V_S is accordingly added such that S is max-controlled [4, 8, 107]. The iterative process proceeds until no bad siphons exist in the resultant net system.

For the example, this policy detects and controls S_1, S_4, S_{10}, S_{16}, S_{17}, S_{18}, $S^* = \{p_2, p_4, p_8, p_{10}, p_{17}, p_{21}, p_{22}, V_{S_1}, V_{S_{16}}\}$, and $S^{**} = \{p_2, p_3, p_8, p_9, p_{12}, p_{13}, p_{16}, p_{17}, V_{S_4}, V_{S_{17}}\}$ sequentially. After the eight siphons are controlled, the resultant net system is live.

Table 7.13 shows the additional monitors due to T-policy for the example. The process terminates after eight iterations. The resultant controlled system has 14,850 reachable states.

Table 7.13 Monitors added using T-policy

V_S	$M_0^\alpha(\cdot)$	Preset	Postset	V_S	$M_0^\alpha(\cdot)$	Preset	Postset
V_{S_1}	2	t_{10}, t_{16}	t_9, t_{15}	V_{S_4}	5	$t_5, t_{10}, t_{13}, t_{17}$	t_3, t_8, t_{11}, t_{15}
$V_{S_{10}}$	2	t_4, t_{13}	t_3, t_{11}	$V_{S_{16}}$	2	t_9, t_{17}	t_8, t_{16}
$V_{S_{17}}$	5	t_3, t_8, t_{19}	t_1, t_{17}	$V_{S_{18}}$	2	t_8, t_{18}	t_7, t_{17}
V_S^*	5	$t_9, t_{10}, t_{16}, t_{17}$	$2t_8, 2t_{15}$	V_S^{**}	3	t_3, t_8, t_{17}	t_1, t_{15}

Remark 7.18. In theory, T-policy is still NP-complete since the siphon controlled in each iteration is found by solving an integer LPP. The number of additional monitors and the permissive behavior of the resultant supervisor of a plant net model are unknown in advance, which depend on the siphons identified and controlled at each iteration.

7.2.13 An Optimal Deadlock Prevention Policy Based on Theory of Regions

The permissive behavior is one of the most important criteria in evaluating the performance of a liveness-enforcing or non-blocking supervisor. In terms of the theory of regions, Uzam [92] develops an optimal liveness-enforcing supervisor synthesis method, called U1-policy for short. Details of this policy can be found in Chap. 6.

The plant model of the FMS example has 26,750 reachable states, 21,581 of which are legal, i.e., either good or dangerous states. There are 5,299 elements in Ω, the set of marking/transition separation instances. This implies that we have to solve 5,299 LPPs. This policy can in theory lead to an optimal liveness-enforcing supervisor that results in 21,581 reachable states in the controlled system since such a supervisor exists from P-policy [77].

Remark 7.19. U1-policy (as well as the one proposed in [34]) can be considered the only strategy to design an optimal monitor-based supervisor although, due to the inherent complexity, it is computationally too expensive. In theory, $|\Omega|$ is also exponential with respect to the size of a plant model. In practice, we have to solve $|\Omega|$ LPPs, which is clearly infeasible for either a sizable net or a small net with a sizable initial marking.

7.2.14 A Suboptimal Deadlock Prevention Policy

The application of the theory of regions to deadlock prevention is widely extended after the work in [92] and [34]. The deadlock prevention methods proposed in [93] and [96] are claimed to require less computational cost to obtain the monitors. They proceed in an iterative way. At each iteration, a marking called first-met bad marking (FBM) is identified from the reachability graph of a Petri net model. An FBM is such a marking in $\mathcal{M}_F$ that there exists a father node of it in $\mathcal{M}_L$. Then, a monitor is added to prevent the marking from being reached via a well-established invariant-based control approach [102]. Uzam and Zhou's method, called U2-policy for short, in [95] can be considered as an improved version of the deadlock control approaches in [93] and [96]. The improvement is made since (1) Petri net reduction techniques are used to simplify a net model so as to find the reachability graph with less computational overhead; and (2) a simplification for the invariant-based control approach is proposed. For convenience of discussion, U2-policy is briefly presented as follows:

Input: a plant model G

Output: controlled system G'

Step 1. $i := 1$.

Step 2. Obtain a reduced net model G_i from G by using reduction techniques.

Step 3. Compute the reachability graph RG_i for G_i.

Step 4. If G_i is live then $G' := G_i$; go to Step 8.

Step 5. Find a first-met bad marking FBM_i in G_i.

Step 6. Add monitor V_i by defining a P-invariant such that FBM_i can never be reached; denote the resultant net system by G.

Step 7. $i := i+1$; go to Step 2.

Step 8. Output controlled system G'.

For the example in the case study, the monitors computed by U2-policy are shown in Table 7.14 after 19 iterations. The corresponding supervisor leads to 21,562 reachable states in the controlled system, and only 0.088% (21,581 − 21,562 = 19 out of 21,581) of the maximally permissive behavior is lost.

Remark 7.20. The significance of U2-policy deserves careful considerations from the standpoint of computational complexity. It is well known that the computation of a reachability graph is of exponential complexity. As indicated in [34], given the reachability graph of a plant net model, the computation of a monitor is polynomial. That is to say, U1-policy needs to compute the reachability graph only once and solve $|\Omega|$ LPP, while, in U2-policy, at each iteration, one has to compute a reachability graph until all FBM are excluded in the resultant net system. Let k denote the number of FBM in a net system. We conclude that in theory the number of times of

Table 7.14 Monitors added using U2-policy

V_S	$M_0^{\alpha}(\cdot)$	Preset	Postset	V_S	$M_0^{\alpha}(\cdot)$	Preset	Postset
V_1	2	t_{13}	t_{11}	V_2	2	t_4,t_{13}	t_3,t_{12}
V_3	2	t_8,t_{18}	t_7,t_{17}	V_4	2	t_9,t_{17}	t_8,t_{16}
V_5	3	t_8,t_{17}	t_7,t_{16}	V_6	2	t_{10},t_{16}	t_9,t_{15}
V_7	4	t_9,t_{17}	t_7,t_{15}	V_8	3	t_{10},t_{17}	t_8,t_{15}
V_9	4	t_8,t_{10},t_{17}	t_7,t_9,t_{15}	V_{10}	5	t_5,t_{13},t_{17}	t_4,t_{11},t_{15}
V_{11}	5	t_3,t_8,t_{19}	t_1,t_{17}	V_{12}	5	t_5,t_{13},t_{17}	t_3,t_{12},t_{15}
V_{13}	6	t_5,t_8,t_{13},t_{17}	t_4,t_7,t_{11},t_{15}	V_{14}	5	t_5,t_{10},t_{13},t_{17}	t_4,t_9,t_{11},t_{15}
V_{15}	6	t_5,t_8,t_{13},t_{17}	t_3,t_7,t_{12},t_{15}	V_{16}	5	t_5,t_{10},t_{13},t_{17}	t_3,t_9,t_{12},t_{15}
V_{17}	9	$t_5,t_8,t_{10},t_{17},t_{19}$	t_1,t_9,t_{15},t_{18}	V_{18}	9	$t_3,t_5,t_8,t_{10},t_{18}$	t_1,t_4,t_9,t_{15}
V_{19}	9	$t_3,t_5,t_9,t_{17},t_{19}$	t_1,t_4,t_{15},t_{18}				

computing reachability graphs in U2-policy is equal to $k+1$, where k is in theory exponential with respect to the size of a net model. To summarize, the complexity of U2-policy is in theory much worse than that of U1-policy. In practice, different FBM may have common monitor solutions under which they can never be reached. As a result, the number of times to compute reachability graphs in U2-policy is (much) smaller than $k+1$. In addition, we do notice that some improvements have been made in U2-policy. However, they cannot significantly reduce its computational complexity in theory. To summarize, we are led to infer the following points concerning the two policies:

1. In theory, U1-policy needs to compute the reachability graph only once and to solve $|\Omega|$ LPPs. U2-policy needs to compute the reachability graphs $k+1$ times, where k is the number of FBM in $\mathcal{M}_F$. $|\Omega|$ and k have the same order of magnitude, which, in theory, are exponential with respect to the size of a net.
2. In practice, both of them first compute the reachability graph of a plant. Then, U1-policy has to solve $|\Omega|$ LPPs in order to decide whether different marking/transition instances have the common monitor solutions. While, the number of times to compute reachability graphs in U2-policy may be small due to the fact that different FBMs often have common monitor solutions. The practical examples even suggest that it is exponentially smaller than k as each FBM often represents the first one that leads to a large portion of bad markings. Its control means this portion's control.

In theory U1-policy as well as the one proposed in [34], in which the reachability graph of a plant net model needs to be computed only once, remains to be an elegant strategy in designing a liveness-enforcing supervisor with maximally permissive behavior (provided that such a supervisor exists). However, in practice, the superiority of the one over the other remains open. Compared with U1-policy, U2-policy seems promising when we deal with a net in which $|\Omega|$ is large but the state space is small. However, it is taken for granted that a small state space of a plant net model usually has a small $|\Omega|$. In this scenario, intuitively, U2-policy is computationally better than U1-policy. More work is needed to compare them by the study of examples whose state spaces range from being small to large.

7.2.15 An Optimal Policy Based on Complete Siphon Enumeration

For S^3PR nets, the deadlock prevention policy, called X-policy for short, proposed by Xing and Hu in [101] aims, by structural analysis, to develop a liveness-enforcing supervisor with maximally permissive behavior and a minimized number of additional monitors.

The concepts of perfect resource transition circuits (PRT-circuits) and their saturated states are proposed. A saturated PRT-circuit implies the existence of circular wait that is tied to deadlock states. The liveness of an S^3PR net is characterized by the fact that no PRT-circuit can reach a saturated state at any reachable marking of the system.

Starting from PRT-circuits, elementary maximal PRT-circuits and center resources are accordingly defined. A center resource has a capacity of one and is contained in the resource sets of at least two elementary maximal PRT-circuits. The following facts are claimed [101]:

1. For an S^3PR without center resources, an optimal liveness-enforcing supervisor can be obtained by computing a set of monitors.
2. An S^3PR net model with center resources can be reduced to a simple one SPN(r) by removing center resources according to a set of rules. An optimal liveness-enforcing supervisor PC(r) can be found for SPN(r). The composition of PC(r) and SPN(r) can derive a supervisory policy ρ under which the system is live.
3. Each elementary maximal PRT-circuit can be used to derive a strict minimal siphon and vice versa.
4. X-policy needs the complete siphon enumeration and hence is of exponential complexity.

For an elementary maximal PRT-circuit θ with its resource place set $R[\theta]$, a monitor p_θ is added by the siphon control approach of AE-policy with $M_0^\alpha(p_\theta) = M_0(R[\theta]) - 1$.

The example has a center resource $R_2(p_{21})$. The reduced Petri net model resulting from removing p_{21} is shown in Fig.7.1. Table 7.15 shows the monitors of PC(r) for the reduced net model SPN(r) of the example. A supervisory policy derived from the synchronous synthesis of PC(r) and SPN(r) can lead to the controlled system that has 16,276 reachable states.

Table 7.15 Monitors added using X-policy in PC(r)

V_S	$M_0^\alpha(\cdot)$	Preset	Postset	V_S	$M_0^\alpha(\cdot)$	Preset	Postset
V_1	2	t_{10}, t_{16}	$t_{8,9}, t_{15}$	V_2	3	$t_{8,9}, t_{17,18}$	t_7, t_{16}
V_3	4	$t_{10}, t_{17,18}$	t_7, t_{15}	V_4	9	t_5, t_{10}, t_{19}	t_1, t_{15}

Remark 7.21. The rationality of X-policy is worthy of being further investigated when a plant net model contains center resources although it is theoretically correct that the removal of the center resources can lead to an optimal supervisor.

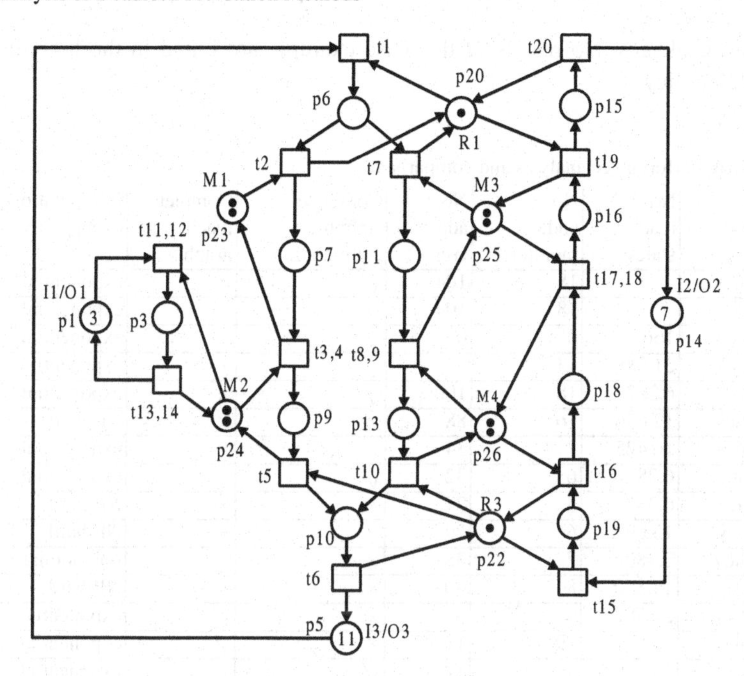

Fig. 7.1 The reduced Petri net model SPN(r) of the example

Remark 7.22. While X-policy guarantees the optimality of deadlock control for S^3PR without center resources, some S^3PR with center resources may also be optimally controlled. For example, the net depicted in Fig. 6.9 has center resources but an optimal liveness-enforcing supervisor can be found by computing four monitors for the strict minimal siphons.

7.3 Analysis of Deadlock Prevention Methods

Table 7.16 summarizes the performance of the supervisors resulting from the policies recounted in the last section. The first column lists the policies and the second denotes the number of reachable states of the corresponding supervisor. The third and fourth columns show the number of the additional monitors and arcs, respectively. The fifth (sixth) indicates whether the complete siphon (state) enumeration is necessary in a policy ("√" means "yes" and "×" means "no"). The seventh column exhibits the computational complexity. We do not present here the specific computation time of these policies for this example since the necessary computation required by each of them can be finished within a few seconds or, at most, a few minutes (except for U1-policy) in a personal computer. Note that for the example in this case study the monitors and arcs due to U1-policy can be computed with commercial software packages. No results of this case example are found in the literature by using U1-policy.

Table 7.16 Performance analysis and comparison

Policy in the literature	No. reach. states	No. add. monitors	No. add. arcs	Complete siphon enumeration	Complete reach. graph	Computational Complexity
AE-policy	×	18	104	√	×	exponential
B1-policy	×	18	104	√	×	exponential
B2-policy	166	0	0	√	×	exponential
P-policy	21,581	13	82	√	×	exponential
E-policy	6,287	18	106	√	×	exponential
H1-policy	12,656	16	88	×	×	NP-hard
H2-policy	16,425	7	34	√	×	exponential
L1-policy	6,287	6	32	√	×	exponential
L2-policy	6,331	5	27	×	×	NP-hard
L3-policy	15,999	6	29	×	×	NP-hard
PR-policy	2,480	7	38	×	×	polynomial
T-policy	14,850	8	40	×	×	NP-hard
U1-policy	21,581	×	×	×	√	exponential
U2-policy	21,562	19	112	×	√	exponential
X-policy	15,098	4	17	√	×	exponential

The performance analysis of the policies is carried out by considering the following three criteria: behavior permissiveness, computational complexity, and structural complexity.

7.3.1 Reachability-Graph-Based Policies

U1- and U2-policies fall under this class of deadlock prevention methods. The supervisors derived from them can usually have maximally permissive behavior if such supervisors exist. Since the computation of the reachability graph for a plant net model is necessary, these policies are of exponential complexity. The number of additional monitors is, in theory, equal to that of marking/transition separation instances or that of FBMs. Although, as indicated by the case study, different marking/transition separation instances or FBMs can have common monitor solutions, as far as the authors know, there is no formal proof indicating that in practice the number of monitors is polynomial with respect to the size of the plant. These policies cannot deal with a large-sized net with a large initial marking due to the state explosion problem [14, 17]. By INA [86], we have tried to find the reachability graph for an S^3PR net model with 68 places and 54 transitions. The computation was carried out on a Toshiba notebook with 1.7 GHz processor speed and 512 M of memory under the Windows XP operating system. After having generated 1,764,263 states in more than six days, the computation aborted due to memory overflow.

It is worthy of note that symbolic techniques based on binary decision diagrams [9, 10] have emerged as an efficient strategy for the reachability analysis and computation of Petri nets [71, 74, 75]. These methods are expected to be used in the resolution of deadlocks in FMSs in the future.

7.3.2 Complete-Siphon-Enumeration-Based Policies

As a structural object, siphons are extensively used in the analysis and control of deadlocks in resource allocation systems [15, 42, 79–82], leading to a large number of siphon-based deadlock control strategies [59,64–66,88]. In this chapter, AE-, B1-, B2-, P-, E-, H2-, L1-, and X-policies fall under the class of deadlock prevention methods that need the complete siphon enumeration of a plant net model.

In general, a siphon-based deadlock prevention policy cannot lead to a maximally permissive liveness-enforcing supervisor except for the nets with particular structures or initial markings. Siphons in a Petri net are a purely structural object whose computation is independent of markings. However, it is well known that their number grows fast and in the worst case grows exponentially with respect to the size of a net [18,46]. It seems that siphon-based methods can deal well with the deadlock problems in a small net structure with any initial marking. In this sense, a siphon-based method, compared with the ones based on reachability graphs, behaves much better from the standpoint of computational complexity.

For the example investigated in this chapter, it is easy to see that the structure of a supervisor can be simplified if the elementary siphons are taken into account when one designs a deadlock prevention policy. An elegant result concerning elementary siphon theory is that a dependent siphon can be controlled by properly supervising its elementary siphons. This implies that, in theory, monitors can be added for elementary siphons only. Since the number of the elementary siphons is bounded by the smaller of place and transition counts, it can be concluded that an elementary siphon-based deadlock prevention policy can usually lead to a structurally simple liveness-enforcing supervisor.

7.3.3 Partial-Siphon-Enumeration-Based Policies

In theory, behavior permissiveness of a supervisor does not depend on whether or not a policy needs complete siphon enumeration. From the case study, a deadlock prevention policy that needs partial siphon enumeration may lead to a supervisor with more permissive behavior and simplified net structure. H1-, L2-, L3-, PR-, and T-policies belong to this class of methods. From the standpoint of performance of the supervisors, these policies make a favorable balance between computational complexity and behavior permissiveness. For example, as shown in [61], it takes more than 6 hours to compute all 169 strict minimal siphons in an S^3PR with 68 places and 54 transitions by using INA. However, L3-policy just needs to compute 24 strict minimal siphons and to control 13 of them, taking 178 seconds only, where Lindo [68] is used to solve the MIP problems. Except the PR-policy, which is of polynomial complexity, most deadlock prevention policies requiring no complete siphon enumeration need to solve MIP problems, which is NP-hard. However, the experimental study of many cases shows that PR-policy is over-conservative.

7.3.4 Exponential Complexity and NP-Hardness

Except PR-policy [73], which is of polynomial-time, other existing deadlock prevention policies need either the (partial or complete) siphon enumeration or reachability graph. A policy is classified into the class with exponential complexity if it needs the complete siphon enumeration or reachability graph. As is well-known, the number of siphons in a Petri net grows in the worst case exponentially with respect to its size and the size of reachability graph grows exponentially with the size of the net as well as initial markings. A policy is of NP-hard complexity if it involves the solution of an MIP problem.

Recognizing the inefficiency and intractability of computing the complete siphon enumeration and reachability graph for a Petri net, the MIP-based method [13] is a fast deadlock detection approach for structurally bounded nets whose deadlocks are tied to unmarked siphons. Although an MIP problem is NP-hard and is difficult to solve [33], the case study conducted by Chu and Xie [13] shows that its computational efficiency is relatively insensitive to the initial marking and it seems to be much more efficient than the classical complete state or siphon enumeration methods.

H1-, L2-, L3-, and T-policies use the MIP-based deadlock detection method to avoid the complete siphon enumeration. In each of these policies, the number of times of solving MIP problems is in theory exponential with respect to the size of a net [88]. In practice, the number of the iteration steps is in general much smaller than that of deadlock states since different deadlock states may share a common monitor solution, which leads to the relatively high efficiency of these deadlock prevention policies.

Remark 7.23. When behavior permissiveness is not a major concern, PR-policy should be the best choice for industrial-size automated systems as it is the only deadlock prevention approach in the literature with polynomial complexity. If the off-line computational cost is not a problem and the computation is feasible, U1- or U2-policies should be the first choice of industrial engineers. The deadlock prevention policies that use partial siphon enumeration achieve a good balance between the behavior permissiveness and computational complexity [84].

Remark 7.24. A natural extension to the current work is to improve the computational efficiency of deadlock prevention methods that are based on the theory of regions [83]. Future work can also focus on the development of elementary siphon-based deadlock control methods. As a conjecture, if a polynomial algorithm is developed to find a set of elementary siphons for a particular class of Petri nets [97], it is possible that we can develop a number of polynomial deadlock control methods for the net class, which are supposed to be more permissive than PR-policy.

7.4 Bibliographical Remarks

Most of the material in this chapter is from the recent work [67]. It is hard to compare, from a theoretical point of view or by a formal proof, the behavior permissiveness of two deadlock prevention strategies if both of them are not optimal. As a result, experimental study seems to be an effective way to find an answer to this problem. It is a time-consuming and tedious task since we need a fair number of Petri nets generated randomly.

The paper by Fanti and Zhou [29] compares the deadlock control approaches for automated manufacturing systems under different formalisms such as Petri nets, graph theory [12, 22–30, 98, 99], automata [69], and the combination of these formalisms [48–56, 70, 76, 94]. Not much work except [6, 21, 31, 35, 37, 60, 72] is found in the literature to compare different deadlock control strategies for FMS.

Problems

7.1. Design liveness-enforcing supervisors for S^3PR shown in Fig. 5.3 by the deadlock prevention policies in Sect. 7.2. Compare the resultant supervisors in terms of structural complexity, computational cost, and permissive behavior. Note that an optimal liveness-enforcing supervisor of the Petri net model should lead to a controlled system that has 60 reachable states.

7.2. Design liveness-enforcing supervisors for S^3PR shown in Fig. 5.16 by the deadlock prevention policies in Sect. 7.2. Compare the resultant supervisors in terms of structural complexity, computational cost, and permissive behavior. Note that an optimal controlled system should have 205 reachable states.

7.3. Design liveness-enforcing supervisors for S^3PR shown in Fig. 6.7 by the deadlock prevention policies in Sect. 7.2. Compare the resultant supervisors in terms of structural complexity, computational cost, and permissive behavior. Note that an optimal controlled system of the Petri net model should have 48 reachable states. A fact can be easily observed that some of the policies lead to optimal liveness-enforcing supervisors.

7.4. Design a liveness-enforcing supervisor for S^3PR shown in Fig. 6.7 by P-policy proposed in [77].

7.5. As seen in Subsect. 7.2.5, E-policy adds 18 monitors for the FMS example in this chapter. Find the implicit places in the controlled system.

References

1. Abdallah, I.B., ElMaraghy, H.A. (1998) Deadlock prevention and avoidance in FMS: A Petri net based approach. *International Journal of Advanced Manufacturing Technology*, vol.14, no.10, pp.704–715.
2. Barkaoui, K., Abdallah, I.B. (1995) A deadlock prevention method for a class of FMS. In *Proc. IEEE Int. Conf. on Systems, Man, and Cybernetics*, pp.4119–4124.
3. Barkaoui, K., Abdallah, I.B. (1995) Deadlock avoidance in FMS based on structural theory of Petri nets. In *Proc. INRIA/IEEE Symposium on Emerging Technologies and Factory Automation*, pp.499–510.
4. Barkaoui, K., Pradat-Peyre, J.F. (1996) On liveness and controlled siphons in Petri nets. In *Proc. 17th Int. Conf. on Applications and Theory of Petri Nets Lecture Notes in Computer Science*, vol.1091, pp.57–72.
5. Barkaoui, K., Chaoui, A., Zouari, B. (1997) Supervisory control of discrete event systems based on structure theory of Petri nets. In *Proc. IEEE Int. Conf. on Systems, Man, and Cybernetics*, pp.3750–3755.
6. Barkaoui, K., Chaoui, A., Benamara, R. (1997) The performance of alternative strategies for dealing with deadlocks in FMS. In *Proc. 6th Int. Conf. on Emerging Technologies and Factory Automation*, pp.281–286.
7. Barkaoui, K., Petrucci, L. (1998) Structural analysis of workflow nets with shared resources. In *Proc. Workshop on Workflow Management: Net-based Concepts, Models, Techniques and Tools*, pp.82–95.
8. Barkaoui, K., Couvreur, J.M., Klai, K. (2005) On the equivalence between liveness and deadlock-freeness in Petri nets. In *Proc. 26th Int. Conf. on Applications and Theory of Petri Nets and Other Models of Concurrency, Lecture Notes in Computer Science*, vol.3536, pp.90–107.
9. Bryant, R.E. (1986) Graph-based algorithms for Boolean function manipulation. *IEEE Transactions on Computers*, vol.35, no.8, pp.677–691.
10. Bryant, R.E. (1992) Symbolic boolean manipulation with ordered binary-decision diagrams. *ACM Computing Surveys*, vol.24, no.3, pp.293–318.
11. Chao, D.Y. (2006) Maximal class of weakly live nets without emptiable siphons. *IEEE Transactions on Systems, Man, and Cybernetics, Part B*, vol.36, no.6, pp.1332–1341.
12. Cho, H., Kumaran, T.K., Wysk, R.A. (1995) Graph-theoretic deadlock detection and resolution for flexible manufacturing systems. *IEEE Transactions on Robotics and Automation*, vol.11, no.3, pp.413–421.
13. Chu, F., Xie, X.L. (1997) Deadlock analysis of Petri nets using siphons and mathematical programming. *IEEE Transactions on Robotics and Automation*, vol.13, no.6, pp.793–804.
14. Ciardo, G. (2004) Reachability set generation for Petri nets: can brute force be smart? In *Proc. 25th Int. Conf. on Applications and Theory of Petri Nets, Lecture Notes in Computer Science*, vol.3099, pp.17–34.
15. Colom, J.M. (2003) The resource allocation problem in flexible manufacturing systems. In *Proc. Int. Conf. on Applications and Theory of Petri Nets*, W. van der Aalst and E. Best (Eds.), *Lecture Notes in Computer Science*, vol. 2679, pp.23–35.
16. Cordone, R., Ferrarini, L., Piroddi, L. (2005) Enumeration algorithms for minimal siphons in Petri nets based on place constraints. *IEEE Transactions on Systems, Man and Cybernetics, Part A*, vol.35, no.6, pp.844–854.
17. Esparza, J. (1998) Decidability and complexity of Petri net problems – an introduction. In *Lectures on Petri Nets I: Basic Models, Lecture Notes in Computer Science*, vol.1491, G. Rozenberg and W. Reisig (Eds.), pp.374–428.

18. Ezpeleta, J., Couvreur, J.M., Silva, M. (1993) A new technique for finding a generating family of siphons, traps, and st-components: Application to colored Petri nets. In *Advances in Petri Nets, Lecture Notes in Computer Science*, vol.674, G. Rozenberg (Eds.), pp.126–147.
19. Ezpeleta, J, Colom, J.M., Martinez, J. (1995) A Petri net based deadlock prevention policy for flexible manufacturing systems. *IEEE Transactions on Robotics and Automation*, vol.11, no.2, pp.173–184.
20. Ezpeleta, J., García-Vallés, F., Colom, J.M. (1998) A class of well structured Petri nets for flexible manufacturing systems. In *Proc. 19th Int. Conf. on Applications and Theory of Petri Nets, Lecture Notes in Computer Science*, vol.1420, J. Desel and M. Silva (Eds.), pp.64–83.
21. Fanti, M.P., Maione, B., Mascolo, S., Turchiano, B. (1996) Performance of deadlock avoidance algorithms in flexible manufacturing systems. *Journal of Manufacturing Systems*, vol.15, no.3, pp.164–178.
22. Fanti, M.P., Maione, B., Mascolo, S., Turchiano, B. (1997) Event-based feedback control for deadlock avoidance in flexible production systems. *IEEE Transactions on Robotics and Automation*, vol.13, no.3, pp.347–363.
23. Fanti, M.P., Maione, B., Turchiano, B. (2000) Comparing digraph and Petri net approaches to deadlock avoidance in FMS. *IEEE Transactions on Systems, Man, and Cybernetics, Part B*, vol.30, no.5, pp.783–798.
24. Fanti, M.P., Maione, B., Turchiano, B. (2001) Distributed event-control for deadlock avoidance in automated manufacturing systems. *International Journal of Production Research*, vol.39, no.9, pp.1993–2021.
25. Fanti, M.P., Turchiano, B. (2001) Deadlock avoidance in automated guided vehicle systems. In *Proc. IEEE/ASME International Conference on Advanced Intelligent Mechatronics*, vol.2, pp.1017–1022.
26. Fanti, M.P. (2002) Event-based controller to avoid deadlock and collisions in zone-control AGVS. *International Journal of Production Research*, vol.40, no.6, pp.1453–1478.
27. Fanti, M.P., Maione, B., Turchiano, B. (2002) Design of supervisors to avoid deadlock in flexible assembly systems. *International Journal of Flexible Manufacturing Systems*, vol.14, no.2, pp.157–175.
28. Fanti, M.P. (2004) Deadlock resolution strategy for automated manufacturing systems including conjunctive resource service. *IEEE Transactions on Systems, Man, and Cybernetics, Part A*, vol.34, no.1, pp.80–92.
29. Fanti, M.P., Zhou, M.C. (2004) Deadlock control methods in automated manufacturing systems. *IEEE Transactions on Systems, Man, and Cybernetics, Part A*, vol.34, no.1, pp.5–22.
30. Fanti, M.P., Zhou, M.C. (2005) Deadlock control methods in automated manufacturing systems. In *Deadlock Resolution in Computer-Integrated Systems*, New York: Marcel Dekker, pp.1–22.
31. Ferrarini, L., Piroddi, L. (2005) The effect of modeling and control techniques on the management of deadlocks in FMS. In *Deadlock Resolution in Computer-Integrated System*, M. C. Zhou and M. P. Fanti, (Eds)., New York: Marcel Dekker, pp.407–444.
32. García-Vallés, F., Colom, J.M. (1999) Implicit places in net systems. In *Proc. 8th Int. Workshop on Petri Nets and Performance Models*, pp.104–113.
33. Garey, M.R., Johnson, D.S. (1979) *Computers and Intractability: A Guide to the Theory of NP-Completeness*. New York: W. H. Freeman.
34. Ghaffari, A., Rezg, N., Xie, X.L. (2003) Design of a live and maximally permissive Petri net controller using the theory of regions. *IEEE Transactions on Robotics and Automation*, vol.19, no.1, pp.137–142.
35. Hosack, B., Mahmoodi, F., Mosier, C.T. (2003) A comparison of deadlock avoidance policies in flexible manufacturing systems. *International Journal of Production Research*, vol.41, no. 13, pp.2991–3006.
36. Huang, Y.S., Jeng, M.D., Xie, X.L., Chung, S. L. (2001) Deadlock prevention policy based on Petri nets and siphons. *International Journal of Production Research*, vol.39, no.2, pp.283–305.

37. Huang, Y.S., Lin, J.H., Hsu, C.N., (2004) Comparison of deadlock prevention policies in FMS based on Petri nets siphons. In *Proc. IEEE Int. Conf. on Systems, Man, and Cybernetics*, pp.4867–4872.
38. Huang, Y.S. (2007) Design of deadlock prevention supervisors for FMS using Petri nets. *International Journal of Advanced Manufacturing Technology*, vol.35, no.3–4, pp.349–362.
39. Iordache, M.V., Moody, J.O., Antsaklis, P.J. (2002) Synthesis of deadlock prevention supervisors using Petri nets. *IEEE Transactions on Robotics and Automation*, vol.18, no.1, pp.59–68.
40. Jeng, M.D., DiCesare, F. (1995) Synthesis using resource control nets for modeling shared-resource systems. *IEEE Transactions on Robotics and Automation*, vol.11, no.3, pp.317–327.
41. Jeng, M.D. (1997) A Petri net synthesis theory for modeling flexible manufacturing systems. *IEEE Transactions on Systems, Man and Cybernetics, Part B*, vol.27, no.2, pp.169–183.
42. Jeng, M.D., Peng, M.Y., Huang, Y.S. (1999) An algorithm for calculating minimal siphons and traps in Petri nets. *International Journal of Intelligent Control and Systems*, vol.3, no.3, pp.263–275.
43. Jeng, M.D., Xie, X.L. (1999) Analysis of modularly composed nets by siphons. *IEEE Transactions on Systems, Man, and Cybernetics, Part A*, vol.29, no.4, pp.399–406.
44. Jeng, M.D., Xie, X.L., Peng, M.Y. (2002) Process nets with resources for manufacturing modeling and their analysis. *IEEE Transactions on Robotics and Automation*, vol.18, no.6, pp.875–889.
45. Jeng, M.D., Xie, X.L., Chung, S.L. (2004) ERCN* merged nets for modeling degraded behavior and parallel processes in semiconductor manufacturing systems. *IEEE Transactions on Systems, Man, and Cybernetics, Part A*, vol.34, no.1, pp.102–112.
46. Lautenbach, K. (1987) Linear algebraic calculation of deadlocks and traps. In *Concurrency and Nets*, K. Voss, H. J. Genrich and G. Rozenberg (Eds.), pp.315–336.
47. Lautenbach, K., Ridder, H. (1996) The linear algebra of deadlock avoidance – a Petri net approach. No.25-1996, Technical Report, Institute of Software Technology, University of Koblenz-Landau, Koblenz, Germany.
48. Lawley, M.A., Reveliotis, S.A., Ferreira, P.M. (1997) Design guidelines for deadlock-handling strategies in flexible manufacturing systems. *International Journal of Flexible Manufacturing Systems*, vol.9, no.1, pp.5–30.
49. Lawley, M.A., Reveliotis, S.A., Ferreira, P.M. (1998) Flexible manufacturing system structural control and the neighborhood policy, part 2. Generalization, optimization, and efficiency. *IIE Transactions*, vol.29, no.10, pp.889–899.
50. Lawley, M.A., Reveliotis, S.A., Ferreira, P.M. (1998) A correct and scalable deadlock avoidance policy for flexible manufacturing systems. *IEEE Transactions on Robotics and Automation*, vol.14, no.5, pp.796–809.
51. Lawley, M.A., Reveliotis, S.A., Ferreira, P.M. (1998) Flexible manufacturing system structural control and the neighborhood policy, part 1. Correctness and Scalability. *IIE Transactions*, vol.29, no.10, pp.877–887.
52. Lawley, M.A., Reveliotis, S.A., Ferreira, P.M. (1998) The application and evaluation of banker's algorithm for deadlock-free buffer space allocation in flexible manufacturing systems. *International Journal of Flexible Manufacturing Systems*, vol.10, no.1, pp.73–100.
53. Lawley, M.A. (1999) Deadlock avoidance for production systems with flexible routing. *IEEE Transactions on Robotics and Automation*, vol.15, no.3, pp.497–509.
54. Lawley, M.A. (2000) Integrating flexible routing and algebraic deadlock avoidance policies in automated manufacturing systems. *International Journal of Production Research*, vol.38, no.13, pp.2931–2950.
55. Lawley, M.A., Reveliotis, S.A. (2001) Deadlock avoidance for sequential resource allocation systems: Hard and easy cases. *International Journal of Flexible Manufacturing Systems*, vol.13, no.4, pp.385–404.
56. Lawley, M.A., Sulistyono, W. (2002) Robust supervisory control policies for manufacturing systems with unreliable resources. *IEEE Transactions on Robotics and Automation*, vol.18, no.3, pp.346–359.

57. Li, Z.W., Zhou, M.C. (2004) Elementary siphons of Petri nets and their application to deadlock prevention in flexible manufacturing systems. *IEEE Transactions on Systems, Man, and Cybernetics, Part A*, vol.34, no.1, pp.38–51.
58. Li, Z.W., Uzam, M., Zhou, M.C. (2004) Comments on "Deadlock prevention policy based on Petri nets and siphons". *International Journal of Production Research*, vol.42, no.24, pp.5253–5254.
59. Li, Z.W., Zhou, M.C. (2005) Elementary siphon of Petri nets for effective deadlock control in flexible manufacturing systems. In *Deadlock Resolution in Computer-Integrated Systems*, M. C. Zhou and M. P. Fanti (Eds.), New York: Marcel Dekker, pp. 309–348.
60. Li, Z.W., Zhou, M.C. (2005) Comparison of two deadlock prevention methods for different-size flexible manufacturing systems. *International Journal of Intelligent Control and Systems*, vol.10, no.3, pp.235–243.
61. Li, Z.W., Zhou, M.C. (2006) Two-stage method for synthesizing liveness-enforcing supervisors for flexible manufacturing systems using Petri nets. *IEEE Transactions on Industrial Informatics*, vol.2, no.4, pp.313–325.
62. Li, Z.W., Zhou, M.C. (2006) Clarifications on the definitions of elementary siphons of Petri nets. *IEEE Transactions on Systems, Man, and Cybernetics, Part A*, vol.36, no.6, pp.1227–1229.
63. Li, Z.W., Hu, H.S., Wang, A.R. (2007) Design of liveness-enforcing supervisors for flexible manufacturing systems using Petri nets. *IEEE Transactions on Systems, Man, and Cybernetics, Part C*, vol.37, no.4, pp.517–526.
64. Li, Z.W., Zhou, M.C., Uzam, M. (2007) Deadlock control policy for a class of Petri nets without complete siphon enumeration. *IET Control Theory and Applications*, vol.1, no.6, pp.1594–1605.
65. Li, Z.W., Zhang, J, Zhao, M. (2007) Liveness-enforcing supervisor design for a class of generalized Petri net models of flexible manufacturing systems. *IET Control Theory and Applications*, vol.1, no.4, pp.955–967.
66. Li, Z.W., Sheptalni, M. (2009) A smart deadlock prevention policy for flexible manufacturing systems using Petri nets. To appear in *IET Control Theory and Applications*.
67. Li, Z.W., Zhou, M.C. (2008) A survey and comparison of Petri net-based deadlock prevention policies for flexible manufacturing systems. *IEEE Transactions on Systems, Man, and Cybernetics, Part C*, vol.38, no.2, pp.172–188.
68. Lindo, Premier Optimization Modeling Tools, http://www.lindo.com/.
69. Ma, C., Wonham, W.M. (2005) *Nonblocking Supervisory Control of State Tree Structures*. Berlin: Springer.
70. Maione, G., DiCesare F. (2005) Hybrid Petri net and digraph approach for deadlock prevention in automated manufacturing systems. *International Journal of Production Research*, vol.43, no.24, pp.5131–5159.
71. Miner, A.S., Ciardo, G. (1999) Efficient reachability set generation and storage using decision diagrams. In *Proc. Int. Conf. on Application and Theory of Petri Nets, Lecture Notes in Computer Science*, vol.1639, pp.6–25.
72. Mohan, S., Yalcin, A., Khator, S. (2004) Controller design and performance evaluation for deadlock avoidance in automated flexible manufacturing cells. *Robotics and Computer-Integrated Manufacturing* vol.20, no.6, pp.541–551.
73. Park, J., Reveliotis, S.A. (2001) Deadlock avoidance in sequential resource allocation systems with multiple resource acquisitions and flexible routings. *IEEE Transactions on Automatic Control*, vol.46, no.10, pp.1572–1583.
74. Pastor, E., Cortadella, J., Pena, M.A. (1999) Structural methods to improve the symbolic analysis of Petri nets. In *Proc. Int. Conf. on Application and Theory of Petri Nets, Lecture Notes in Computer Science*, vol.1639, pp.26–45.
75. Pastor, E., Cortadella, J., Roig, O. (2001) Symbolic analysis of bounded Petri nets. *IEEE Transactions on Computers*, vol.50, no.5, pp.432–448.
76. Pinzon, L.E., Hanisch, H.M., Jafari, M.A., Boucher, T. (1999) A comparative study of synthesis methods for discrete event controllers. *Formal Methods in System Design*, vol.15, no.2, pp.123–167.

77. Piroddi, L., Cordone, R., Fumagalli, I. (2008) Selective siphon control for deadlock prevention in Petri nets. *IEEE Transactions on Systems, Man, and Cybernetics, Part A*, vol. 38, no. 6, pp.1337–1348.
78. Recalde, L., Teruel, E., Silva, M., (1997) Improving the decision power of rank theorems. In *Proc. IEEE Int. Conf. on Systems, Man, and Cybernetics*, pp.3768–3773.
79. Reveliotis, S.A. (2002) Liveness enforcing supervision for sequential resource allocation systems: state of the art and open issues. In *Synthesis and Control of Discrete Event Systems*, B. Caillaud, X. L. Xie, P. Darondeau, and L. Lavagno (Eds.), Boston, MA: Kluwer, pp.203–212.
80. Reveliotis, S.A. (2003) On the siphon-based characterization of liveness in sequential resource allocation systems. In *Proc. Int. Conf. on Applications and Theory of Petri Nets, Lecture Notes in Computer Science*, vol.2679, W. M. P. van der Aalst and E. Best (Eds.), pp.241–255.
81. Reveliotis, S.A. (2005) On the siphon-based characterization of liveness and liveness-enforcing supervision for sequential resource allocation systems. In *Deadlock Resolution in Computer-Integrated Systems*, pp.283–307, New York: Marcel Dekker.
82. Reveliotis, S.A. (2005) *Real-time Management of Resource Allocation Systems: A Discrete Event Systems Approach*, New York: Springer.
83. Reveliotis, S.A., Choi, J.Y. (2006) Designing reversibility-enforcing supervisors of polynomial complexity for bounded Petri nets through the theory of regions. In *Proc. 27th Int. Conf. on Applications and Theory of Petri Nets and Other Models of Concurrency, Lecture Notes in Computer Science*, vol.4024, S. Donatelli and P. S. Thiagarajan (Eds.), pp.322–341.
84. Reveliotis, S.A., Roszkowska, E., Choi, J.Y. (2007) Generalized algebraic deadlock avoidance policies for sequential resource allocation systems. *IEEE Transactions on Automatic Control*, vol.52, no.12, pp.2345–2350.
85. Silva, M., Teruel, E., Colom, J.M. (1998) Linear algebraic and linear programming techniques for the analysis of place/transition net systems. In *Lectures on Petri Nets I: Basic Models, Lectures Notes in Computer Science*, vol.1491, W. Reisig and G. Rozenberg (Eds.), pp.309–373.
86. Starke, P.H. (2003) *INA: Integrated Net Analyzer*. http://www2.informatik.hu-berlin.de/~starke/ina.html.
87. Tricas, F., Martinez, J. (1995) An extension of the liveness theory for concurrent sequential processes competing for shared resources. In *Proc. IEEE Int. Conf. on Systems, Man, and Cybernetics*, pp.3035–3040.
88. Tricas, F., Colom, J.M., Ezpeleta, J. (1997) A solution to the problem of deadlocks in concurrent systems using Petri nets and integer linear programming. Research Report, RR-GISI-06, Departamento de Informática e Ingeniería de Sistemas, Universidad de Zaragoza, Spain.
89. Tricas, F., García-Vallés, F., Colom, J.M., Ezpeleta, J. (2000) An iterative method for deadlock prevention in FMSs. In *Proc. 5th Workshop on Discrete Event Systems*, R. Boel and G. Stremersch (Eds.), pp.139–148.
90. Tricas, F., García-Vallés, F., Colom, J.M., Ezpeleta, J. (2005) Using linear programming and the Petri net structure for deadlock prevention in sequential resource allocation systems. *XIII Jornadas de Concurrencia y Sistemas Distribuidos*, pp.65–77.
91. Tricas, F., García-Vallés, F., Colom, J.M., Ezpeleta, J. (2005) A Petri net structure-based deadlock prevention solution for sequential resource allocation systems. In *Proc. IEEE Int. Conf. on Robotics and Automation*, pp.271–277.
92. Uzam, M. (2002) An optimal deadlock prevention policy for flexible manufacturing systems using Petri net models with resources and the theory of regions. *International Journal of Advanced Manufacturing Technology*, vol.19, no.3, pp.192–208.
93. Uzam, M., Zhou, M.C. (2004) Iterative synthesis of Petri net based deadlock prevention policy for flexible manufacturing systems. In *Proc. IEEE Int. Conf. on Systems, Man, and Cybernetics*, pp.4260–4265.
94. Uzam, M., Wonham, W.M. (2006) A hybrid approach to supervisory control of discrete event systems coupling RW supervisors to Petri nets. *International Journal of Advanced Manufacturing Technology*, vol.28, no.7–8, pp.747–760.

95. Uzam, M., Zhou, M.C. (2006) An improved iterative synthesis method for liveness enforcing supervisors of flexible manufacturing systems. *International Journal of Production Research*, vol.44, no.10, pp.1987–2030.
96. Uzam, M., Zhou, M.C. (2007) An iterative synthesis approach to Petri net based deadlock prevention policy for flexible manufacturing systems. *IEEE Transactions on Systems, Man, and Cybernetics, Part A*, vol.37, no.3, pp.362–371.
97. Wang, A.R., Li, Z.W., Jia, J.Y., Zhou, M.C. (2009) An effective algorithm to find elementary siphons in a class of Petri nets. To appear in *IEEE Transactions on Systems, Man, and Cybernetics, Part A*.
98. West, D. B. (2001) *Introduction to Graph Theory*. Pearson Education Inc.
99. Wysk, R.A., Yang, N.S., Joshi, S. (1991) Detection of deadlocks in flexible manufacturing cells. *IEEE Transactions on Robotics and Automation*, vol.7, no.6, pp.853–859.
100. Xie, X.L., Jeng, M.D. (1999) ERCN-merged nets and their analysis using siphons. *IEEE Transactions on Robotics and Automation*, vol.15, no.4, pp.692–703.
101. Xing, K.Y., Hu, B.S. (2005) Optimal liveness Petri net controllers with minimal structures for automated manufacturing systems. In *Proc. IEEE Int. Conf. on Systems, Man and Cybernetics*, pp.282–287.
102. Yamalidou, E., Moody, J.O., Antsaklis, P.J. (1996) Feedback control of Petri nets based on place invariants. *Automatica*, vol.32, no.1, pp.15–28.
103. Zhou, M.C., DiCesare, F. (1991) Parallel and sequential exclusions for Petri net modeling for manufacturing systems. *IEEE Transactions on Robotics and Automation*, vol.7, no.4, pp.515–527.
104. Zhou, M.C., DiCesare, F. (1992) A hybrid methodology for synthesis of Petri nets for manufacturing systems. *IEEE Transactions on Robotics and Automation*, vol.8, no.3, pp.350–361.
105. Zhou, M.C., DiCesare, F. (1993) *Petri Net Synthesis for Discrete Event Control of Manufacturing Systems*. Boston, MA: Kluwer.
106. Zouari, B., Barkaoui, K. (2003) Parameterized supervisor synthesis for a modular class of discrete event systems. In *Proc. IEEE Int. Conf. on Systems, Man, and Cybernetics*, pp.1874–1879.
107. Zouari, B. (2006) A structure causality relation for liveness characterisation in Petri nets. *Journal of Universal Computer Science*, vol.12, no.2, pp.214–232.

Chapter 8
Conclusions and Future Research

Abstract This chapter concludes the book by providing a number of interesting problems in the area of deadlock prevention for automated manufacturing systems, which is based on Petri nets. These problems include the development of deadlock prevention methods by considering the existence of uncontrollable and unobservable transitions in a plant model, polynomial algorithms to find a set of elementary siphons in a Petri net, and the analysis of elementary Petri net subclasses such as free-choice nets by using elementary siphon theory. System productivity comparison of different strategies by considering time factor in a Petri net model represents an important research area.

As a formal modeling tool, Petri nets are increasingly becoming a popular mathematical formalism to investigate the modeling, control, analysis, and performance evaluation of a discrete-event system [2,4–6,15–23,27,51,58]. Petri net theory has been one of the most interesting and hot topics in the area of computer science. They also find wide application in contemporary technical systems. The systems around us can be communication protocols, computer networks, traffic systems, distributed database, software, production systems [56], C^3I (command, control, communication, and intelligence), Internet web services, social services, and even logistics security systems and workflow [24–26,52,57,65–69]. Petri nets are employed by many academic researchers and engineers as a mathematical framework to investigate the deadlock control problems in a variety of resource allocation systems that are a theoretical abstract of real-world discrete-event systems. Particularly, Petri nets have become a popular and effective tool for the design and management of modern automated manufacturing systems [3,7–9,12,14,31–39,42,43,46–49,53–55,59–64, 71–73,76,80–84].

The work conducted in the last two decades shows that the siphon-based characterization dominates among the methodologies that deal with the deadlock analysis and control of resource allocation systems [50]. Traditionally, a deadlock prevention policy can be evaluated by a number of performance criteria: structural complexity, behavioral permissiveness, and computational complexity. That is to say, a perfect liveness-enforcing supervisor is the one that has a simple control structure, is optimal, and can be computed without prohibitive computational cost. Unfortunately, such a perfect supervisor does not exist in a general large-scale case.

The development of elementary siphons of Petri nets is motivated by the need to explore the avenue of simplifying the structure of the Petri net supervisors resulting from the existing deadlock control policies in the literature. In essence, by using a small space linear with system size, the concept of elementary siphons aims to characterize the whole net structure information related to the deadlock control purpose. The supervisory control problem of a discrete-event system in a Petri net formalism is to establish a mechanism that is implemented through properly controlling the firing of transitions such that the controlled system satisfies the control specifications. In this sense, the characteristic T-vector matrix gives a global characterization for a given control requirement. Due to the linear dependency of the T-vectors, a basis of the matrix, to some extent, carries the information of the control specifications. As a result, it is not surprising that the concept of elementary siphons can be easily extended to the monitor-based implementation of a set of generalized mutual exclusion constraints.

Although many important contributions have been made in recent years, deadlock prevention and avoidance are still an open research area. In the case of very large-scale systems, most existing approaches suffer one or more problems such as structural complexity, behavioral permissiveness, and computational complexity [44].

The motivation of elementary siphons is just to tackle the structural complexity problem of the supervisors. In theory, it cannot definitely improve the behavioral permissiveness and computational complexity. However, all existing deadlock control approaches based on siphons can be greatly improved if the concept of elementary siphons is fully considered during the deadlock analysis and siphon control process.

To end this chapter as well as this book, the following problems deserve further consideration.

- The theory of elementary siphons of Petri nets indicates that their number is bounded by the smaller of place and transition counts. Moreover, the controllability of dependent siphons can be implicitly ensured by explicitly supervising its elementary siphons. These facts naturally inspire one to explore the possibility of (1) the development of a polynomial algorithm to find a set of elementary siphons in a Petri net and (2) providing a formal proof that dependent siphons are implicitly controlled if the elementary siphons are controlled by explicitly adding monitors. This may lead to a deadlock-free or liveness-enforcing supervisor computed in polynomial time.
- At present, most work on handling deadlock problems arising in discrete-event systems by using Petri net techniques does not account for uncontrollable and unobservable transitions within the plant models. However, uncontrollable and unobservable events (or transitions in a Petri net formalism) are a standard feature in the supervisory control framework based on automata [45].
- The existence and synthesis of an optimal liveness-enforcing supervisor remain noteworthy although this problem can be decided by the theory of regions [13]. In

this direction, the known results through structural but not enumeration analysis are concerned with PPN [74] and S^3PR [75] only.

- For a number of Petri net subclasses that can model manufacturing systems, their liveness conditions are established through a set of inequalities with respect to the initial markings of its resource and idle places [1, 34, 83]. In theory, each inequality is associated with a strict minimal siphon, leading to the fact that the number of inequalities is exponential with respect to the size of the plant Petri net model. A potential extension to the concept of elementary siphons is to reduce such inequalities by considering the existence of redundant or dependent constraints.
- At present, the interests of elementary siphons have focused on the application-oriented Petri net subclasses. Interesting work includes their application to such subclasses as free-choice nets and asymmetric choice nets. Specifically, the characterization of their behavioral properties based on elementary siphons is an interesting problem.
- Deadlock prevention is a pessimistic approach in which the possibility of a deadlock is broken statically at the price of restricting the concurrency of the plant. However, once a deadlock prevention policy is established for a plant, the controlled system can never reach deadlock states without online decision time. On the contrary, a deadlock avoidance policy [11, 70] takes an online decision procedure that examines each resource allocation request to ensure that it cannot lead to a deadlock. Deadlock avoidance usually leads to a high throughout [77–79] but requires maintaining the global or local state enumeration. For a given system to be controlled, it is significant for the engineers to choose a deadlock control policy that results in high productivity rate. The throughput comparison between deadlock prevention and avoidance strategies is also an interesting area.
- Timed discrete-event systems and their supervisory control are an interesting research topic. The system productivity must be measured when time is introduced. They will inevitably involve various scheduling and planning methods. Integration of two sets of methodologies for industrial systems represents a very important research and development area.

Problems

8.1. A transition in a plant is said to be uncontrollable if its firing cannot be inhibited by an external action. It is called unobservable if its firing cannot be directly measured. A Petri net supervisor cannot have any connections to an unobservable transition, thus all unobservable transitions are implicitly uncontrollable [41]. In the deadlock prevention policy developed by Ezpeleta et al. [10], all output arcs of the additional monitors point to the source transitions that are included in the postset of idle places. As a result, the source transitions in general should be controllable and observable. This is true in practice as they often represent the entry of a job into a system. The deadlock prevention policy in [28] consists of two phases. The first is to make every siphon S in the plant controlled by the enforcement that $V_S + [S]$ is a P-semiflow of the resultant net through the addition of monitor V_S. The second phase aims to control the control-induced siphons by monitors, in which the output arcs of the monitors point to the source transitions so that the iterative control process converges as fast as possible. For an S^3PR under the deadlock prevention policy in [28], develop an algorithm to identify the transitions that must be controllable.

8.2. Discuss the existence of an algorithm finding the transitions that are controllable in an S^3PR (N, M_0), under which there exists an optimal monitor-based liveness-enforcing supervisor. For other subclasses of Petri nets, discuss the existence of such an algorithm.

8.3. Most of the deadlock prevention policies in the literature do not consider the existence of uncontrollable and unobservable transitions in a plant. Improve the deadlock control approaches in [28–30], and [40] by considering the existence of uncontrollable transitions in the plant models.

References

1. Barkaoui, K., Chaoui, A., Zouari, B. (1997) Supervisory control of discrete event systems based on structure theory of Petri nets. In *Proc. IEEE Int. Conf. on Systems, Man, and Cybernetics*, pp.3750–3755.
2. Ben-Naoum, L., Boel, R., Bongaerts, L., De Schutter, B., Peng, Y., Valckenaers, P., Vandewalle, J., Wertz, V. (1995) Methodologies for discrete event dynamic systems: A survey. *Journal A*, vol.36, no.4, pp.3–14.
3. Bogdan, S., Lewis, F.L., Kovacic, Z., Mireles, J. (2006) *Manufacturing Systems Control Design*. London: Springer.
4. David, R., Hassane, A. (1994) Petri nets for modeling of dynamic systems-A survey. *Automatica*, vol.30, no.2, pp.175–202.
5. David, R. (1995) Grafcet: A powerful tool for specification of logical controllers. *IEEE Transactions on Control Systems Technology*, vol.3, no.3, pp.253–268.
6. David, R., Hassane, H. (2004) *Discrete, Continuous, and Hybrid Petri Nets*. Berlin: Springer.
7. DiCesare, F., Harhalakis, G., Porth, J.M., Vernadat, F.B. (1993) *Practice of Petri Nets in Manufacturing*. Chapman and Hall.
8. Dotoli, M., Fanti, M.P. (2004) Coloured timed Petri net model for real-time control of automated guided vehicle systems. *International Journal of Production Research*, vol.42, no.9, pp.1787–1814.
9. Dotoli, M., Fanti, M.P. (2007) Deadlock detection and avoidance strategies for automated storage and retrieval systems. *IEEE Transactions on Systems, Man, and Cybernetics, Part C*, vol.37, no.4, pp.541–552.
10. Ezpeleta, J, Colom, J.M., Martinez, J. (1995) A Petri net based deadlock prevention policy for flexible manufacturing systems. *IEEE Transactions on Robotics and Automation*, vol.11, no.2, pp.173–184.
11. Ferrarini L., Maroni, M. (1998) Deadlock avoidance control for manufacturing systems with multiple capacity resources. *International Journal of advanced manufacturing Technology*, vol.14, no.4, pp.729–736.
12. Gebraeel, N.Z., Lawley, M.A. (2001) Deadlock detection, prevention, and avoidance for automated tool sharing systems. *IEEE Transactions on Robotics and Automation*, vol.17, no.3, pp.342–356.

13. Ghaffari, A., Rezg, N., Xie, X.L. (2003) Design of a live and maximally permissive Petri net controller using the theory of regions. *IEEE Transactions on Robotics and Automation*, vol.19, no.1, pp.137–142.
14. Girault, C., Valk, R. (Eds). (2003) *Petri Nets for Systems Engineering: A Guide to Modeling, Verification, and Applications*. Berlin: Springer.
15. Giua, A. (1992) Petri nets as discrete event models for supervisory control. Ph.D Dissertation, Rensselaer Polytechnic Institute, Troy, New York.
16. Giua, A., DiCesare, F., Silva, M. (1992) Generalized mutual exclusion constraints on nets with uncontrollable transitions. In *Proc. IEEE Int. Conf. on Systems, Man, and Cybernetics*, pp.974–979.
17. Giua, A., DiCesare, F. (1992) On the existence of Petri net supervisors. In *Proc. 31st IEEE Conf. on Decision and Control*, pp.3380–3385.
18. Giua, A., DiCesare, F. (1993) A class of Petri nets with a convex reachability set. In *Proc. IEEE Int. Conf. on Robotics and Automation*, pp.578–583.
19. Giua, A., DiCesare, F., Silva, M. (1993) Petri net supervisors for generalized mutual exclusion constraints. In *Proc. 12th IFAC World Congress*, pp.267–270.
20. Giua, A., DiCesare, F. (1994) Blocking and controllability of Petri nets in supervisory control. *IEEE Transactions on Automatic Control*, vol.39, no.4, pp.818–823.
21. Giua, A., DiCesare, F. (1994) Petri net structural analysis for supervisory control. *IEEE Transactions on Robotics and Automation*, vol.10, no.2, pp.185–195.
22. Giua, A. Seatzu, C. (2001) Supervisory control of railway networks with Petri nets. In *Proc. 40th IEEE Int. Conf. on Decision and Control*, pp.5004–5009.
23. Giua, A., Seatzu, C. (2007) A systems theory view of Petri nets. In *Advances in Control Theory and Applications, Lecture Notes in Control and Information Science*, vol.353, C. Bonivento et al. (Eds.), pp.99–127.
24. Hee, K.V., Sidorova, N., Voorhoeve, M. (2003) Soundness and separability of workflow nets in the stepwise refinement approach. In *Proc. 24th Int. Conf. on Applications and Theory of Petri Nets, Lectures Note in Computer Science*, vol.2679, W. M. P. van der Aalst and E. Best (Eds.), pp.337–356.
25. Hee, K.V., Sidorova, N., Voorhoeve, M. (2004) Generalised soundness of workflow nets is decidable. In *Proc. 25th Int. Conf. on Applications and Theory of Petri Nets, Lecture Notes in Computer Science*, vol.3099, J. Cortadella and W. Reisig (Eds.), pp.197–216.
26. Hee, K.V., Serebrenik, A., Sidorova, N., Voorhoeve, M. (2005) Soundness of resource-constrained workflow nets. In *Proc. 27th Int. Conf. on Applications and Theory of Petri Nets, Lecture Notes in Computer Science*, vol.3536, G. Ciardo and P. Darondeau (Eds.), pp.250–267.
27. Hruz, B., Zhou, M.C (2007) *Modeling and Control of Discrete-Event Dynamic Systems: With Petri Nets and Other Tools*. London: Springer.
28. Huang, Y.S., Jeng, M.D., Xie, X.L., Chung, S.L. (2001) Deadlock prevention policy based on Petri nets and siphons. *International Journal of Production Research*, vol.39, no.2, pp.283–305.
29. Huang, Y.S. (2007) Design of deadlock prevention supervisors for FMS using Petri nets. *International Journal of Advanced Manufacturing Technology*, vol.35, no.3–4, pp.349–362.
30. Huang, Y.S. (2007) Deadlock prevention for flexible manufacturing systems in sequence resource allocation systems. *Journal of Information Science and Engineering*, vol.23, no.1, pp.215–231.
31. Jeng, M.D., DiCesare, F. (1993) A review of synthesis techniques for Petri nets with applications to automated manufacturing systems. *IEEE Transactions on Systems, Man, and Cybernetics, Part A*, vol.23, no.1, pp.301–312.
32. Jeng, M.D., DiCesare, F. (1995) Synthesis using resource control nets for modeling shared-resource systems. *IEEE Transactions on Robotics and Automation*, vol.11, no.3, pp.317–327.
33. Jeng, M.D. (1997) A Petri net synthesis theory for modeling flexible manufacturing systems. *IEEE Transactions on Systems, Man and Cybernetics, Part B*, vol.27, no.2, pp.169–183.

34. Jeng, M.D., Peng, M.Y., Huang, Y.S. (1999) An algorithm for calculating minimal siphons and traps in Petri nets. *International Journal of Intelligent Control and Systems*, vol.3, no.3, pp.263–275.
35. Jeng, M.D., Huang, Y.S. (1999) Petri nets for modeling and analysis of manufacturing systems with local operation cycles. In *Proc. IEEE Int. Conf. on Systems, Man, and Cybernetics*, pp.793–797.
36. Jeng, M.D., Xie, X.L. (1999) Analysis of modularly composed nets by siphons. *IEEE Transactions on Systems, Man, and Cybernetics, Part A*, vol.29, no.4, pp.399–406.
37. Jeng, M.D., DiCesare, F., Xie, X.L. (2000) Corrections to "Synthesis using resource control nets for modeling shared-resource systems". *IEEE Transactions on Robotics and Automation*, vol.16, no.2, pp.202–203.
38. Jeng, M.D., Xie, X.L. (2001) Modeling and analysis of semiconductor manufacturing systems with degraded behavior using Petri nets and siphons. *IEEE Transactions on Robotics and Automation*, vol.17, no.5, pp.576–588.
39. Lewis, F., Gurel, A., Bogdan, S., Doganalp, A., Pastravanu, O. (1998) Analysis of deadlock and circular waits using a matrix model for flexible manufacturing systems. *Automatica*, vol.34, no.9, pp.1083–1100.
40. Li, Z.W., Zhou, M.C. (2006) Two-stage method for synthesizing liveness-enforcing supervisors for flexible manufacturing systems using Petri nets. *IEEE Transactions on Industrial Informatics*, vol.2, no.4, pp.313–325.
41. Moody, J.O., Antsaklis, P.J. (1998) *Supervisory Control of Discrete Event Systems Using Petri Nets*. Boston, MA: Kluwer.
42. Moore, K.E., Gupta, S.M. (1996) Petri net models of flexible manufacturing systems: A survey. *International Journal of Production Research*, vol.34, no.11, pp.3001–3035.
43. Narahari, Y., Viswanadham, N. (1985) A Petri net approach to the modelling and analysis of flexible manufacturing systems. *Annals of Operations Research*, vol.3, no.8, pp.449–472.
44. Pablo, J., Colom, J.M. (2006) Resource allocation systems: Some complexity results on the S^4PR class. In *Proc. IFIP International Federation for Information Processing, Lecture Notes in Computer Science*, vol.4229, E. Najm et al. (Eds.), pp.323–338.
45. Ramadge, P., Wonham, W.M. (1989) The control of discrete event systems. *Proceedings of the IEEE*, vol.77, no.1, pp.81–89.
46. Recalde, L., Silva, M., Ezpeleta, J., Teruel, E. (2004) Petri nets and manufacturing systems: An examples-driven tour. In *Lectures on Concurrency and Petri Nets: Advances in Petri Nets, Lecture Notes in Computer Science*, vol.3098, J. Desel, W. Reisig, and G. Rozenberg (Eds.), pp.742–788.
47. Reveliotis, S.A., Ferreira, P.M. (1997) Deadlock avoidance policies for automated manufacturing cells. *IEEE Transactions on Robotics and Automation*, vol.12, no.6, pp.845–857.
48. Reveliotis, S.A., Lawley, M.A., Ferreira, P.M. (1997) Polynomial-complexity deadlock avoidance policies for sequential resource allocation systems. *IEEE Transactions on Automatic Control*, vol.42, no.10, pp.1344–1357.
49. Reveliotis, S.A., Lawley, M.A., Ferreira, P.M. (2001) Structural control of large-scale flexibly automated manufacturing systems. In *The Design of Manufacturing Systems*, C. T. Leondes (Ed.), pp.4-1-4-34. CRC Press.
50. Reveliotis, S.A. (2003) On the siphon-based characterization of liveness in sequential resource allocation systems. In *Proc. Int. Conf. on Applications and Theory of Petri Nets, Lecture Notes in Computer Science*, vol.2679, W. M. P. van der Aalst and E. Best (Eds.), pp.241–255.
51. Roszkowska, E. (2004) Supervisory control for deadlock avoidance in compound processes, *IEEE Trans. on Syst., Man, Cybern., Part A*, vol.34, no.1, pp.52–64.
52. Salimifard, K., Wright, M. (2001) Petri net-based modelling of workflow systems: An overview. *European Journal of Operational Research*, vol.134, no.3, pp.664–676.
53. Silva, M., Valette, R. (1990) Petri nets and flexible manufacturing. In *Proc. Int. Conf. on Applications and Theory of Petri Nets, Lecture Notes in Computer Science*, vol.424, G. Rozenberg (Ed.), pp.374–417.

54. Silva, M. (1993) Introducing Petri nets. In *Practice of Petri Nets in Manufacturing*, pp.1–62, Chapman & Hall.
55. Silva, M., Teruel, E. (1997) Petri nets for the design and operation of manufacturing systems. *European Journal of Control*, vol.3, no.3, pp.182–199.
56. Silva, M., Teruel, E., Valette, R., Pingaud, H. (1998) Petri nets and production systems. In *Lectures in Petri Nets II: Applications, Lecture Notes in Computer Science*, vol.1492, G. Rozenberg and W. Reisig (Eds.), Springer, pp.85–124.
57. Tiplea, F.L., Marinescu, D.C. (2005) Structural soundness of workflow nets is decidable. *Information Processing Letters*, vol.96, no.2, pp.54–58.
58. Tsitsiklis, J.N. (1986) On the control of discrete-event dynamic systems. *International Journal of Control*, vol.42, no.2, pp.475–491.
59. Valette, R., Courvoisier, M., Mayeux, D. (1982) Control of flexible production systems and Petri nets. In *Proc. Application of Theory of Petri nets*, Informatik-Fachberichte, no.66, pp.264–277.
60. Valette, R., Courvoisier, M., Demmou, H., Bigou, J.M., Desclaux, C. (1985) Putting Petri nets to work for controlling flexible manufacturing systems. In *Proc. IEEE Int. Symp. on Circuits and Systems*, Kyoto, Japan, pp.929–932.
61. Valette, R. (1987) Nets in production systems. In *Petri Nets: Applications and Relationships to Other Models of Concurrency, Lecture Notes in Computer Science*, vol.255, Springer, pp.191–217.
62. Valette, R., Cardoso, J., Atabakhone, H., Courvoisier, M., Lemaire, T. (1988) Petri nets and production rules for decision levels in FMS control. In *Proc. 12th IMACS World Congress on Scientific Computation*, Paris, Juillet, pp.522–524.
63. Valette, R. (1989) Monitoring manufacturing systems by means of Petri nets with imprecise markings. In *Proc. IEEE Int. Symp. on Intelligent Control*, Albany, NY, pp.233–238.
64. Vattle, R. (1997) Some issues about Petri net application to manufacturing and process supervisory control. In *Lecture Notes in Computer Science*, vol.1248, P. Azéma and G. Balbo (Eds.), pp.23–41.
65. Van der Aalst, W.M.P. (1996) Structural characterization of sound workflow nets. Computer Science Report 96/23, Eindhoven University of Technology.
66. Van der Aalst, W.M.P. (1997) Verification of workflow nets. In *Lecture Notes in Computer Science*, vol.1248, P. Azema and G. Balbo (Eds.), pp.407–426.
67. Van der Aalst, W.M.P. (1998) The application of Petri nets to workflow management. *Journal of Circuits, Systems, and Computers*, vol.8, no.1, pp.21–66.
68. Van der Aalst, W.M.P. (2000) Workflow verification: Finding control-flow errors using Petri-net-based techniques. In *Lecture Notes in Computer Science*, vol.1806, W. M. P. van der Alst, et al. (Eds.), pp.162–183.
69. Van der Aalst, W.M.P., Van Dongen, B.F., Herbst, J., Maruster, L., Schimm, G., Weijters, A.J.M.M. (2003) Workflow mining: A survey of issues and approaches. *Data and Knowledge Engineering* vol.47, no.2, pp.237–267.
70. Viswanadham, N., Narahari, Y., Johnson, T. (1990) Deadlock prevention and deadlock avoidance in flexible manufacturing systems using Petri net models. *IEEE Transactions on Robotics and Automation*, vol.6, no.6, pp.713–723.
71. Viswanadham, N., Narahari, Y. (1992) *Performance Modelling of Automated Manufacturing Systems*. Englewood Cliffs, NJ: Prentice Hall.
72. Wu, N.Q., Zhou, M.C. (2007) Deadlock resolution in automated manufacturing systems with robots. *IEEE Transactions on Automation Science and Engineering*, vol.4, no.3, pp.474–480.
73. Wu, N.Q., Zhou, M.C. Zhou, Li, Z.W. (2008) Resource-oriented Petri net for deadlock avoidance in flexible assembly systems. *IEEE Transactions on System, Man, and Cybernetics, Part A*, vol.38, no.1, pp.56–69.
74. Xing, K.Y., Hu, B.S., Chen, H.X. (1996) Deadlock avoidance policy for Petri-net modelling of flexible manufacturing systems with shared resources. *IEEE Transactions on Automatic Control*, vol.41, no.2, pp.289–295.

75. Xing, K.Y., Hu, B.S. (2005) Optimal liveness Petri net controllers with minimal structures for automated manufacturing systems. In *Proc. IEEE Int. Conf. on Systems, Man and Cybernetics*, pp.282–287.
76. Xing, K.Y., Zhou, M.C., Liu, H.X., Tian, F. (2009) Optimal Petri net-based polynomial-complexity deadlock avoidance policies for automated manufacturing systems. To appear in *IEEE Transactions on Systems, Man, and Cybernetics, Part A*.
77. Zhang, W., Judd, R.P., Deering, P. (2004) Necessary and sufficient conditions for deadlocks in flexible manufacturing systems based on a digraph model. *Asian Journal of Control*, vol.6, no.2, pp.217–228.
78. Zhang, W., Judd, R.P. (2007) Deadlock avoidance for flexible manufacturing systems with choices based on digraph. *Asian Journal of Control*, vol.9, no.2, pp.111–120.
79. Zhang, W., Judd, R.P. (2008) Deadlock avoidance algorithm for flexible manufacturing systems by calculating effective free space of circuits. *International Journal of Production Research*, vol.46, no.13, pp.3441–3457.
80. Zhou, M.C., DiCesare, F., Rudolph, D. (1992) Design and implementation of a Petri net supervisor for a flexible manufacturing system. *Automatica*, vol.28, no.6, pp.1199–1208.
81. Zhou, M.C. (1998) Modeling, analysis, simulation, scheduling, and control of semiconductor manufacturing systems: A Petri net approach. *IEEE Transactions on Semiconductor Manufacturing*, vol.11, no.3, pp.333–357.
82. Zhou, M.C., Venkatesh, K. (1998) *Modelling, Simulation and Control of Flexible Manufacturing Systems: A Petri Net Approach*. Singapore: World Scientific.
83. Zouari, B., Barkaoui, K. (2003) Parameterized supervisor synthesis for a modular class of discrete event systems. In *Proc. IEEE Int. Conf. on Systems, Man, and Cybernetics*, pp.1874–1879.
84. Zurawski, R., Zhou, M.C. (1994) Petri nets and industrial applications: A tutorial. *IEEE Transactions on Industrial Electronics*, vol.41, no.6, pp.567–583.

Symbols

2^X	The power set of set X
b	A non-negative integer
B	The vector of the constraint constants of a set of GMECs
$EP(x_1, x_n)$	An elementary path from x_1 to x_n
F	Flow relation of a Petri net
$H(r)$	The set of holders using resource r
I	a P-vector or P-invariant
$\|\|I\|\| = \{p \in P \| I(p) \neq 0\}$	The support of P-vector I
$\|\|I\|\|^+ = \{p \in P \| I(p) > 0\}$	The positive support of P-vector I
$\|\|I\|\|^- = \{p \in P \| I(p) < 0\}$	The negative support of P-vector I
l	A P-vector
(l, b)	A generalized mutual exclusion constraint (GMEC)
(L, B)	A set of GMECs
$max_{p^\bullet}$	$max\{W(p,t) \| t \in p^\bullet\}$
M	A marking of a Petri net
$M(p)$	The number of tokens in place p under marking M
M_0	An initial marking of a Petri net
$M = M_0 + [N]\overrightarrow{\sigma}$	The state equation of Petri net (N, M_0)
$M_{min}(S)$	$min\{M(S) \| M \in R(N, M_0)\}$
$M_{max}(S)$	$max\{M(S) \| M \in R(N, M_0)\}$
$M^{min}(S)$	$min\{M(S) \| M = M_0 + [N]Y, M \geq 0, Y \geq 0\}$
$M^{max}(S)$	$max\{M(S) \| M = M_0 + [N]Y, M \geq 0, Y \geq 0\}$
$M[t\rangle$	t is enabled under marking M
$\mathcal{M}^*$	An optimal monitor-based liveness-enforcing supervisor
$\mathcal{M}_F$	The set of forbidden markings
$\mathcal{M}_L$	The set of legal markings
$\mathcal{M}(l, b)$	The set of legal markings defined by GMEC (l, b)
$\mathbb{N}^+ = \{1, 2, \cdots\}$	The set of positive integers
$\mathbb{N} = \{0, 1, 2, \cdots\}$	The set of non-negative integers
$\mathbb{N}^k$	Set of k-dimensional non-negative integer vectors
$\mathbb{N}_k$	$\{1, 2, \ldots, k\}$

N	a Petri net (structure)
(N, M_0)	a (marked) Petri net
$[N]$	the incidence matrix of Petri net N
$[N](p, \cdot)$	incidence vector of place p in Petri net N
$[N](\cdot, t)$	incidence vector of transition t in Petri net N
$N_i(N_j)$	a Petri net
$N_i \circ N_j$	composition of nets N_i and N_j via shared places
$N_i \otimes N_j$	synchronous synthesis of N_i and N_j via shared transitions
p	a place in a Petri net
p^0	an idle process place
p_i^0	the ith idle process place
P^0	the set of idle process places
P_A	the set of operation (activity) places
P_R	the set of resource places
P_V	the set of monitors
P	the set of places of a Petri net
(P, T, F, W)	a generalized Petri net
(P, T, F)	an ordinary Petri net
$\mathscr{P}_S$	the adjoint set of a siphon S
$R(N, M_0)$	the set of reachable markings of (N, M_0)
$R^S(N, M_0)$	the linearized reachability set of (N, M_0)
$rank(A)$	the rank of matrix A
S	a (minimal) siphon
S^R	the set of resource places in siphon S
S^A	the set of operation places in siphon S
$[S]$	complementary set of siphon S
$\langle S \rangle$	a set of equivalent siphons
t	a transition
T	the set of transitions of a Petri net
$Th(S)$	complementary set of S in the form of multisets
V_S	monitor for siphon S
$V(V_i)$	a monitor
W	$(P \times T) \cup (T \times P) \rightarrow \mathbb{N}$
$W(x, y)$	the weight of arc (x, y)
${}^\bullet x = \{y \in P \cup T \mid (y, x) \in F\}$	the preset of node $x \in P \cup T$
$x^\bullet = \{y \in P \cup T \mid (x, y) \in F\}$	the postset of node $x \in P \cup T$
${}^\bullet X = \cup_{x \in X} {}^\bullet x$	the preset of set $X \subseteq P \cup T$
$X^\bullet = \cup_{x \in X} x^\bullet$	the postset of set $X \subseteq P \cup T$
$X(Z)$	a set
$\lvert X \rvert$	the cardinality of a set X
$X \setminus Z$	$\{x \mid x \in X, x \notin Z\}$ (set difference)
$\mathbb{Z}$	the set of integers
$\mathbb{Z}^+$	the set of non-negative integers
Π	the set of minimal siphons in a net
Π_E	the set of elementary siphons in Π

Π_D	the set of dependent siphons in Π
$\pi(p)$	structural bound of place p
ξ_S	the control depth variable of siphon S
$\Delta^+(t)$	downstream siphons of a transition t
$\Delta^-(t)$	upstream siphons of a transition t
λ_S	the characteristic P-vector of S
$\eta_S = [N]^T \lambda_S$	the characteristic T-vector of S
$[\eta]$	the characteristic T-vector matrix
$[\lambda]$	the characteristic P-vector matrix
σ	a transition sequence
$\vec{\sigma}$	the Parikh vector of transition sequence σ
$\vec{\sigma}(t)$	the number of times that t appears in σ

Index

Other titles published in this series (continued):

Soft Sensors for Monitoring and Control of Industrial Processes
Luigi Fortuna, Salvatore Graziani, Alessandro Rizzo and Maria G. Xibilia

Adaptive Voltage Control in Power Systems
Giuseppe Fusco and Mario Russo

Advanced Control of Industrial Processes
Piotr Tatjewski

Process Control Performance Assessment
Andrzej W. Ordys, Damien Uduehi and Michael A. Johnson (Eds.)

Modelling and Analysis of Hybrid Supervisory Systems
Emilia Villani, Paulo E. Miyagi and Robert Valette

Process Control
Jie Bao and Peter L. Lee

Distributed Embedded Control Systems
Matjaž Colnarič, Domen Verber and Wolfgang A. Halang

Precision Motion Control (2nd Ed.)
Tan Kok Kiong, Lee Tong Heng and Huang Sunan

Optimal Control of Wind Energy Systems
Iulian Munteanu, Antoneta Iuliana Bratcu, Nicolaos-Antonio Cutululis and Emil Ceangă

Identification of Continuous-time Models from Sampled Data
Hugues Garnier and Liuping Wang (Eds.)

Model-based Process Supervision
Arun K. Samantaray and Belkacem Bouamama

Diagnosis of Process Nonlinearities and Valve Stiction
M.A.A. Shoukat Choudhury, Sirish L. Shah, and Nina F. Thornhill

Magnetic Control of Tokamak Plasmas
Marco Ariola and Alfredo Pironti

Real-time Iterative Learning Control
Jian-Xin Xu, Sanjib K. Panda and Tong H. Lee

Model Predictive Control Design and Implementation Using MATLAB®
Liuping Wang

Fault-tolerant Flight Control and Guidance Systems
Guillaume Ducard
Publication due May 2009

Design of Fault-tolerant Control Systems
Hassan Noura, Didier Theilliol, Jean-Christophe Ponsart and Abbas Chamseddine
Publication due June 2009

Predictive Functional Control
Jacques Richalet and Donal O'Donovan
Publication due June 2009

Advanced Control and Supervision of Mineral Processing Plants
Daniel Sbárbaro and René del Villar (Eds.)
Publication due July 2009

Stochastic Distribution Control System Design
Lei Guo and Hong Wang
Publication due August 2009

Detection and Diagnosis of Stiction in Control Loops
Mohieddine Jelali and Biao Huang (Eds.)
Publication due October 2009

Active Braking Control Design for Road Vehicles
Sergio M. Savaresi and Mara Tanelli
Publication due November 2009